ALONGSHORE

JOHN R. STILGOE

ALONG

Yale University Press 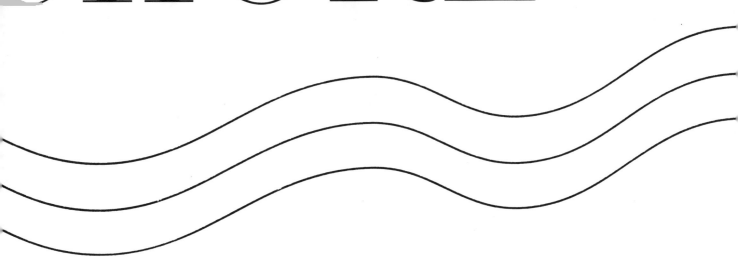 New Haven & London

SHORE

Published with assistance from the foundation established in memory of Philip Hamilton McMillan of the Class of 1894, Yale College.

Designed by Richard Hendel. Set in Bodoni Book and Gill types by Tseng Information Systems, Inc. Printed in the United States of America by Vail-Ballou Press Binghamton, New York.

A catalogue record for this book is available from the British Library.

The paper in this book meets the guidelines for permanence and durability of the Committee on Production Guidelines for Book Longevity of the Council on Library Resources.

10 9 8 7 6 5 4 3 2 1

Library of Congress Cataloging-in-Publication Data

Stilgoe, John R., 1949–
    Alongshore / John R. Stilgoe.
        p.   cm.
    Includes bibliographical references and index.
    ISBN 0-300-05909-4
    1. Cape Cod (Mass.)—Description and travel.   2. Coasts—Massachusetts—Cape Cod.   3. Seashore.   4. Landscape—Massachusetts—Cape Cod.   I. Title.
F72.C3S79   1994
917.44'92—dc20   93-21434
                                    CIP

for Debra, *with love and a smile*

CONTENTS

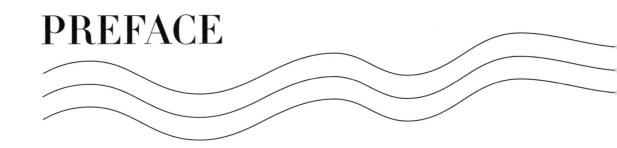

PREFACE

When I walk, I drift east, downhill to the salt marsh, the estuary, the beach. Westward lies almost all of North America, its wildernesses and small towns, cities and farms, railroads and suburbs. Eastward lies no man's land, no woman's either. Through the trees begins the salt marsh sliced by estuaries, then low dunes and cobble beaches, then sandbars and ledges exposed only at low tide, then ocean itself, no land at all. Through the trees begins this book.

However designated, the seashore, the coast, the marge, the coastal zone, the littoral, the limicole realm, the theater of this book deserves prolonged scrutiny. For all that tourists and tourist-trappers wax eloquent about the timelessness of the sea, the seacoast itself is deeply historic, the threshold first touched by the Europeans and Africans who settled North America. For that matter, prehistoric Asians crossed the threshold, too, wading the shallows that opened on frigid uplands. The seacoast is the threshold of American prehistory and history, of American culture, and like most well-passed thresholds, it is hollowed and worn.

And historians routinely ignore it. History begins after the landing at Jamestown, after the *Mayflower* drops anchor, after some nameless liner disembarks immigrants at Ellis Island. Once historians abandon it to ecologists, conchologists, and other scientists, or dismiss it as a summertime playground, a child's place, a realm to be visited during nonthinking vacations, the seacoast floats in the national imagination as timeless as the sea, but somehow quaint and weather-beaten, too. Visual images, chiefly postcards, travel posters, and advertising, especially clothing-catalogue layouts, compound the dereliction of historians. A half-dozen constituent elements, a fishing boat or two, a lighthouse, a bikinied girl, a tottering wharf, a dune, an old salt or two playing checkers outside a boatshop, a bronze-skinned man hoisting a sail, such are enough to signal beach or harbor or coast. How such elements fit into a larger concatenation intrigues few, often not even the inhabitants of the alongshore realm, who frequently accept their environment, natural and built, as unthinkingly as they take the tourists arrived to glimpse it. Certainly elements beyond the popular-image ones attract remarkably little notice, and prompt few summer-afternoon visits or even questions aimed at locals. World War II–era lookout towers, condominia crowding out boatyards, steel piers, fiberglass boats, rocky hazards just beyond the surf, all such things receive slight notice—almost never sustained historical inquiry. All alongshore lies one of the most visited, most noticed, most pictured, and least scrutinized places in North America.

Thresholds, especially very old, heavily passed thresholds, require some attention now and then. Over time, use and misuse wear them, hollow them, until some night the northeast gale roars through the gap between door-bottom and flooring. Suddenly the thresholds receive notice, then the complicated ministrations of carpenters mumbling about rehanging doors and ripping up adjacent joists, all to improve the fit of the restored or replaced thresholds so long neglected. At the threshold of the twenty-first century, the alongshore realm suddenly attracts attention, not for symbolic reasons perhaps, but simply because much of it has worn and frayed and altered until its fragility scares not only locals but even tourists. More than sandy beaches appear to erode, more than wooden wharves appear to totter. A whole visual image seems ready to fade, to vanish with the next recession, the next gale. Change occurs fast, and more change lies in the offing.

What follows, therefore, is a long backward look at alongshore things from salt marshes to wooden boats, from wharves to sand dunes, from harbors to coast-artillery emplacements, and a look, too, at alongshore behavior and perceptions over time, everything from swimsuit etiquette to estimating horizon distances. A backward look befits any subject so richly rewarding the slightest historical scrutiny. And it is the only proper look I can provide. For when I get through the trees to the salt marsh, I take off my shoes and row my boat, and in rowing my boat I look almost always at what lies astern.

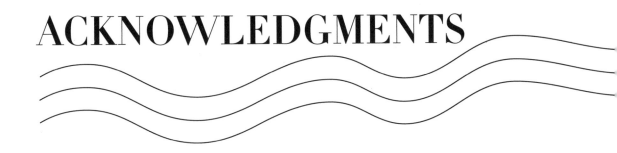

ACKNOWLEDGMENTS

I thank my wife, Debra, and our twin ten-year-old sons, Nathaniel and Adam, at the beginning, because they have been with me nearly every step, splash, slog, and stroke of the way, winter and summer, fair weather and foul. Debra noticed innumerable things I missed, and for her comments I am especially grateful.

In my solitary explorations of libraries, I have had much help. Especially I thank Harvard librarians David Cobb, Mary Daniels, Judy Genthner, Marion Schoon, and Hinda Sklar for their precise efforts. At the Museum of Modern Art, Mary Corliss provided the most thoughtful assistance, as did Philip L. Budlong and Peggy Tate Smith at Mystic Seaport Museum and Samuel W. Shogren at Penobscot Marine Museum. Douglas Cogger did wonders with many salt-stained negatives and photographs.

My colleagues and friends helped, too, often by suggesting that I deal with matters about which they had wondered. William Bachman, Michael Binford, TenBroeck Davidson, Gwen Lexow, Elizabeth Myers, and Zachary Schrag offered useful observations indeed, and along the coast from Harvard, my friends and fellow townsfolk offered more. Two locals, Thomas Armstrong and Robert Belyea, several times kept me a cable's tow from shore so that in traveling east I might find a better perspective on my subject, and for their winter efforts I am especially grateful.

Bits and pieces of two chapters appeared in *Geographical Review* and *Orion: People and Nature*. I thank their editors, Douglas McManis and Aina Niemela, for the help they gave me in earlier days. And I thank especially my editor at Yale University Press, Judy Metro, and her associates, Laura Jones Dooley and Alex Blanton, for wise and timely counsel.

INTRODUCTION

To anyone remotely familiar with shingle beaches, this image is immediately recognizable. The ocean must lie just beyond the man.

But does its implied presence make this image a seascape? (Gardner Photograph Collection, Harvard University)

Land ends at low-tide mark. Where the sheen of wet mud or sand deepens to salt water, to water moved by waves or wind or current, land becomes bottom only, something seen through the intervening prism, something felt with toes or down-thrust oar. Twice each day the tide reaches landward, covering the exposed ground, momentarily creating a new edge to land. But that edge holds always the promise of the ebb, the revelation of damp ground stretching to the water, the regular possibility of walking dry-shod somewhat further toward ocean.

Landscape ends sooner. Topography shrinks from the sea. Dunes shift and slink along barrier beaches, harbors silt up, cliffs and marsh-edges tumble, even high ground sometimes gives way. And no built form endures long in the wilderness of the intertidal zone, or even on the dunes and rocks just inland from the farthest reach of high tide. Piers, breakwaters, mooring bollards, summer cottages, lighthouses, even fortifications pass away, sometimes over centuries, sometimes overnight. Natural forces, sea-born powers, govern the natural and built forms routinely touched by tide and wind. Permanent landscape begins, fitfully, just inland, just past the reach of the highest spring tides, just beyond reach of massive winter spray and late-summer hurricanes. A ribbon of discrete landscape, a mix of wind-twisted trees and motels, drawbridges and summer cottages, boatyards and salt marshes, parallels the ocean and its adjacent wilderness, a ribbon part of something loosely designated *seashore*.

Contemporary Americans miss the awesome importance of the term. The sea rules the limicole zone and intermittently reaches far into the landscape ribbon that parallels it. Indeed the reach of the sea perplexes, and sometimes stuns, any diligent inquirer. Sometimes the reach is a mere mile or so, the distance ebb and flow make an estuary of a river; sometimes it is the five miles smothered in ocean fog; sometimes—rarely—it is as long as the grasp of the hurricane that in 1938 frosted Vermont windows with salt spray. But more explains *seashore* than the reach of ocean into land, on calm days or in storm.

The sea dominates the visual environment of the shore, making everything adjacent bound it. "The sea was meant to be looked at from shore," mused James Russell Lowell in the 1850s, "as mountains from the plain."[1] Hear the landsman argue. Note that the land is not made to be looked at from the sea. The land, any land, say mountains, is to be looked at from land, perhaps land different in topography, but land. And the sea is different. Any seashore, sandy beach or rocky coast, does equally well as viewing ground, for the sea is all the same, seamless in its homogeneity.

Lowell is wrong.

Language fails to assess the visual domination. Consider the confusion implicit in *seascape*. A landscape painting may include rivers or ponds or lakes and remain a landscape, but how much ocean—or how little—can intrude before the landscape painting becomes a seascape? A fragment of exposed ledge, even ledge smothered in breakers, perhaps a hint of lighthouse or fog-bound cliff, is such enough to make *landscape* the correct designation? Is a *seascape* any painting in which the sea appears, any painting in which the sea dominates the view, or any painting of ocean only? In answering, even specialized lexicons fail, or rely on other, more specific terms. As early as 1891, *Adeline's Art Dictionary* defined *seascape* as "pictures

An almost stereotypical seascape, Francis Augustus Silva's **Schooner Passing Castle Island, Boston Harbor** *(1874), emphasizes the sea, a schooner, and a "local boat," not the fortifications. (The Bostonian Society)*

or drawings representing maritime scenes or views of the sea," a bold definition, considering the newness of the term in English—as late as 1911 the *Century Dictionary* labeled it "recent"—and its overwhelming of the more precise term *maritime*.[2] A *maritime* painting, after all, is one of sea and seashore, something different from a *marine*, a painting of ocean unencumbered with land but perhaps graced with ships, what older generations, especially seafaring folk, called a *sea-piece*. "He led me up to a painting, a sea-piece," commented one American fisherman-turned-author in 1868, "a schooner, riding at her anchor, at sunset, far out at sea, no land in sight, sails down, all but a little patch of storm-sail fluttering wildly in the gale, and heavily pitching on a grand, rolling sea." Only in 1864 did the English neologism *seascape* appear in the unabridged *Webster's American Dictionary*, defined brusquely as "a picture representing a scene at sea," and as British in origin.[3]

But by then the British grasped the word, albeit as "arty," somehow an affectation, as the English novelist William Makepeace Thackeray demonstrated in 1840 in *A Shabby Genteel Story*. When commonsensical, plain-speaking Brandon discovers the celebrated Andrea Fitch at Margate, "perched on the cliff, his fingers blue with cold," the eccentric artist is "employed in sketching a land or a sea-scape on a sheet of gray paper." Brandon gazes out across the windswept sea, notes the tossing steamship, and wonders aloud why the artist is not indoors " 'in such a freezing storm as this.' " Fitch retorts that " 'a true artist is never so happy as when he can have the advantage to gaze upon yonder tempestuous hocean in one of its

hangry moods,'" a stilted answer that puts Brandon in mind of the endangered steamship passengers.[4] In a few sentences, Thackeray ridicules the ostentation of *seascape*, reasserts the overwhelming importance of the sea, and swipes at the cliff-top artist secure from seasickness and drowning.

A seascape takes its identity from the vantage point of the artist, not from its subject. A painting only of ocean, of ocean bereft of ships and even navigation buoys, is perhaps properly a seascape only if it implies a land-bound painter, or viewer, distant from immediate contact, say the immediate contact enjoyed or endured by the steamship passengers. Implication may come from height—assuming few painters work from mastheads—or from depiction of waves, for example. A high view implies cliffs like those favored by the cold but comfortable Fitch, and breakers, not rollers, imply a shelving strand. Nowadays, of course, *seascape* designates any picture of the open sea, and the old term *marine* is forgotten, along with *maritime*, except in the designation of certain Canadian provinces. But Thackeray's original understanding remains a powerful one, for it grasps the wonderful difference between landsman and seaman, the one stable, looking outward at ocean, and the other forever in motion, in danger always, sometimes in danger of striking land. *Seascape* implies too much, and provides too little.

Seascape cannot properly designate the subject of this book, for all that its cousin *landscape* proves useful in designating land shaped by people for their own needs and in classifying certain sorts of pictures or views. The whole concept of seascape reeks of lubberly bias,

Although the scruffiness of early twentieth-century small harbors is immediately apparent in this image, the patch of ocean confined to the dock may nonetheless make this a seascape. (author's collection)

the artist-as-landsman or the viewer-as-landsman, and connotes, too, the bias implicit in the landsman distant from seaborne activity. *Seascape* always suggests a viewer with a continent of land behind, a whole inland territory from which to view the ocean beyond low-tide mark, a viewer with no business other than to observe. Lowell is biased indeed in his view of the sea, something that did not escape the editors of the September 1899 issue—the "Salt-Water Number"—of *Century Magazine*. "Lowell is usually so broad in his sympathies that one can pardon this surprising lack of faith in one who was essentially a landsman, and a landsman of the cultivated places of the earth, a lover of man and the park and the orchard rather than of the earth in its wilder aspects." At the turn of the century, the editors remark, under "the insistence of steam, and the Hydrographic Office," old concepts of sea, seafaring, and seascape must change.[5] But about the land next to the sea, the editors are strangely silent, and the saltwater number contains little prose and scant poetry focused on it, perhaps because the turn-of-the-century American seashore had suddenly become an infuriatingly complex place.

That complexity haunts it still, and keeps it from having a solid, seaworthy, satisfactory name. No word, not *littoral,* not *coast,* not *beach* even begins to designate it, and the jangling of names used and misused echoes in disciplines from law to marine ecology. Legal dictionaries define terms like *tidal flats* and *foreshore* with ever finer precision, explaining that tidal rivers, for example, need not be salt, but they include no definition of such equally important terms as *estuary.* Dictionaries of maritime terminology define *shore* as coastline or land adjacent to water, but not even topographical dictionaries define *guzzle,* the Plymouth Bay and

Not every alongshore image welcomes the viewer to partake of the sea. In this view, the landing stages stand as mazes and are explicitly off-limits to casual observation. (JRS)

Cape Cod Bay localism Thoreau uses in his *Cape Cod*. Without a solid vocabulary, without a vocabulary accepted by lexicographers and easily learned before arrival, newcomers in the place find themselves not only lost for words but ignoring items of importance. At the close of the nineteenth century, when Americans discovered the seashore as a place for summertime recreation, magazine editors discovered not only the awkwardness of Lowell's view but the awkwardness of discussing a deliciously intricate, infuriatingly modern, resiliently traditional place. Would new terms improve discourse? Would "the shore" do? Could "seashore" or "sea-coast" serve the new century?

A century later the seashore region remains ill defined, and essentially undesignated. Localisms like *guzzle* remain in use, but as lively localisms only, and older, less provincial terms for the larger seashore zone, "all alongshore," for example, now survive mostly as corruptions, as in "longshoreman" for "stevedore."[6] Summer visitors laugh or grimace at the occasional rare word, and locals struggle with new official terminology—say "coastal zone management area" or "coastlands." Churchgoers sense the subtle confusion, too. Jeremiah

31 in the King James version resonates in phrases like "the coasts of the earth" and "the isles afar off," while the Revised Standard version phrases, "the farthest parts of the earth" and "the coastlands far off," merely tinkle. To reach beyond the vocabulary problem, however, is to uncover a remarkably interesting, remarkably unstudied place indeed—the subject of this book.

That place is home. My earlier books focus on large expanses of territory, the proto-national landscape from 1580 to 1845, the metropolitan corridor shaped by the national railroad network between 1880 and 1940, the American suburb from 1820 to 1939, all subjects dear to my heart, but far broader than home. Only my slim *Shallow-Water Dictionary: A Grounding in Estuary English*, deals with the history of United States dictionary publishing and the language of the salt marshes in which my family and I spend so many of our days, year-round. But the sea and its environs have always been close to me, for I live as I always have, near the beaches, near the estuaries, near the salt marshes mentioned in this book. I knew the word *guzzle* before I read Thoreau. I see flood tide or ebb each morning before I see my university. I am close to my subject, too close for relaxation, for comfort.

Ostensibly that subject is a reach or stretch of Massachusetts coastline and its environs, seaward and landward, from Gurnet Light to Minot Ledge. From Gurnet to Minot is south to north, a contrary way of speaking in these post–Civil War times when places are listed north to south. But here in Massachusetts, at the edge of land and the twentieth century both, local speech still emphasizes upwind origins. Steam power changed cruising, as the *Century* editors guessed, but it did nothing to obliterate the language of sail. Maine is "down east" from here, downwind from Gurnet, downwind from Minot Ledge, easy sailing, most of the time.

But my other subject is not easy sailing, not ever. Marginality never is.

This book focuses on a zone between what seamen call "open ocean" and what landsmen might call "ordinary inland landscape." As a simile, the margins at the left and right of these words work well, but not perfectly. In printing, a margin is that odd space between the edge of the text and the edge of the page, a space that may shrink or expand slightly without troubling the reader, a space the author may fill with gloss, a space the reader may embellish with hand-scrawled notes. *Margin* in fact derives from *marge*, an old term for coast or shore, strong and vital in Spenser's time but obsolescent by the nineteenth century.[7] *Marginalia*

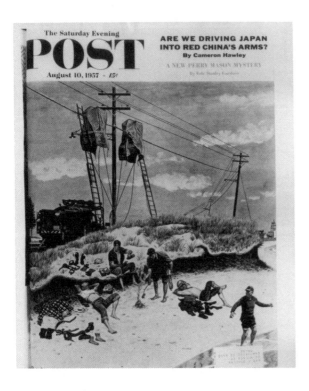

Sometimes activity peculiar to coasts, say the way sandy beaches entice adults to act momentarily as children, appears in images that deserve the designation seascape. (Harvard College Library)

now connotes things unimportant, extrinsic, nonessential.[8] That is unfortunate, for perhaps what happens in the marginal zone is exactly that which is important, intrinsic, essential, that which illuminates not only larger issues of landscape, of environmental presentiments, but whole components of American culture.

Wonderfully odd things happen in the margin. Consider a typical summer day. Total strangers sleep nearly naked next to one another, college-student lifeguards scan the water for sharks and flailing swimmers, blowing sand sticks to freshly sun-screen-slathered skin, children scream from jellyfish stings, teenage boys steer four-wheel-drive vehicles near the incoming tide, fiberglass motorboats run aground or collide with windsurfers, lobstermen haul pots just offshore, marine ecologists collect samples and warn hikers off the dunes, Coast Guard cutters search for smugglers and overdue sailboats, photographers set up tripods and curse kite-flying adolescents, dogs swim and fetch and swim again, surfers pose and scan the waves, joggers run barefoot mile after mile, overweight men suck in their beer bellies and stride toward the surf, elderly women collect plastic litter, old-timers beachcomb with metal detectors and shovels, young women smile in bikinis briefer than sunglass lenses, boatyard crews strain at a marine railway, a man hammers away at the yacht he has been building for three decades, and everywhere locals and summer people alike enjoy the momentary suspension of inland rules that legitimates marginal behavior. Where else do strangers sleep within arm's reach of each other? Where else do children play within grasp of voracious wild animals? Where else is a zone so marginal in every way?

And through it all moves the barefoot historian, wondering about the behavior, the zone itself, the shifting nature of propriety and sand, the traditions of buried treasure, natural-history guidebooks, and small-boat management, hearing always the rote of the sea, the ceaseless sound of waves hitting sand and rock, the *clamor nauticus*, trying to put a name to the place, right in the midst of it all, so unlike Thackeray's above-it-all artist.

Of course, the barefoot historian has a name, in fact several. *Beachcomber* smacks of romance, the South Seas romance Somerset Maugham details in his *Moon and Sixpence*. "The society of beach-combers always repays the small pains you need be at to enjoy it," writes Maugham. "They are easy of approach and affable in conversation. They seldom put on airs." But then again, beachcombers are not wholly to be trusted. "The extent of their experience is pleasantly balanced by the fertility of their imagination," he warns, and they have only "a tolerant respect for the law, when the law is supported by strength."[9] In early twentieth-century Tahiti, Maugham encountered the beachcombers Robert Louis Stevenson knew, the men Stevenson called by the older, more traditional name *wreckers*.[10] A wrecker picks apart wrecks, salvages the useful and discards the trash, but his harmless activities somehow belie the name. Just as *tow-truck* has replaced the older term *wrecker* and *pry-bar* has replaced *wrecking bar*, so *beachcomber* has replaced *wrecker*, but not for casual reasons. No one suspects the wrecker of causing the automobile collision or the wrecking bar of precipitating the train wreck, but somehow the beach-walking wrecker might, just possibly, somehow cause a shipwreck, somehow lure honest mariners ashore on a stormy night, perhaps by lighting a Judas lantern and so imitating a lighthouse. No one really trusts a wrecker, for all that Thoreau calls him "the true monarch of the beach."[11] How can landsmen trust someone so much at ease in the alongshore wilderness, someone looming out of the sleet on a winter workday morning, someone so idle on a summer afternoon, someone walking barefoot across masses of shattered shells?

And how much can a local, a native in faded blue shirt and threadbare khaki pants, know of the aliens called "summer people"? Can one who knows the coast year-round, the beach in winter, the salt marsh in autumn, know the summer beach as summer people know it? Can one who pulls so softly in his glistening wooden yacht-tender know much of jet-skiers pounding the waves or pounding their fists against technological failure? Can one who has always known the sea, smelled the salt air, know anything of tourists who walk seaward through the dunes and suddenly see ocean for the first time? Can one whose ocean is always green or slate-gray but never turquoise know anything of beaches in Malibu? No, of course not.

But he can play his hunches, spend winter days in library stacks, in dusty, ill-heated waterfront archives, in museumlike ship chandleries, digging out fragments of seacoast literature and seacoast illustration. And in summer, behind his sunglasses, he can pull about in his small boats, examining the ruins exposed at low tide, and can walk the sands, musing on the fragile interaction of place and people, on wharves and marine railways, on sand dunes and quicksand, on pirates and bikinis, local boats and factory-made speedboats, on quaint harbor scenery and World War II watchtowers, on salt haying and physical fitness, on alongshore human history.

Natural history has a place in this book, but only a very small one, and that fragile. So many writers follow the course set by Thoreau in *Cape Cod*, by Henry Beston in *The Outermost House*, by John and Mildred Teal in *Life and Death of the Saltmarsh*, by Charlton Ogburn in *The Winter Beach*, by dozens of other observers that what is built, what is people-made, strikes me as in danger of devaluation, of slipping away from sustained notice. Long before my wife taught me that seaside goldenrod is *Solidago sempervirens*, I knew it simply as the salt-marsh plant whose bright yellow flowerhead warns that school is about to begin, that easterly gales are due, maybe hurricanes, that hunters will soon be building duck blinds. I know something of cormorants and herring gulls, horseshoe crabs and striped bass, but not nearly enough, and even less of coastal oddities like the migrating English cuckoo that crashed into a local oak before falling into caring Yankee hands. Yet perhaps the gull perched atop the sun-bleached, half-toppled piling deserves attention—in some small degree—because it perches atop a piling, something driven by men long ago at the edge of the salt marsh, its one-time use now long forgotten. Perhaps what ecologists and ornithologists designate "limicole" needs another look, lest it endure merely as that half-land, half-water zone across which piping plovers skitter, and not as the theater for human activity, too.

After all, if the automobile and the truck so dramatically changed the United States landscape, what did the fiberglass speedboat do to the seashore? And how did Imperial Japanese Navy airplanes bomb the Oregon coast in April 1942? And why do old wharves collapse?

Who is the contemporary dark-visaged pirate? What kinds of risks are run by the bikinied women who run across the sandbar?

And what of the limicole activity reported in two local newspapers, one called the *Mariner?* Even if I never walked east, I would have the newspaper stories to remind me of the sea. Articles clipped day after day, year after year acquire an almost tidal rhythm suggested even by the headlines.

"Storm Pounds Coast"

"Boater Rescued in New Inlet Offers Advice to Others"

"Toplessness—Emancipation Vs. Imagination"

"Fishing Boat Sinks; Captain Missing"

"Yachtsman Presumed Dead; Coast Guard Calls Off Search"

"Marshfield Man Missing in Treacherous Seas"

"Freighter Breaks Up Off Cape"

"Drunken Boaters Facing Crackdown"

"Carver Fisherman Traps Aqua Lobster"

"Bikinis: Bombshells Hitting the Beaches"

"Police Rescue Dennis Man from Muddy Sinkhole"

"Weymouth Harbormaster Finds Adventure Along 17.5-Mile Coast"

"Fisherman Saved from Icy Waters"

"State Targets Unlicensed Dock Owners"

"Man Lost in Thick Fog Saved from Rising Tide"

"Storm Wrecks Cars, Floods Coast Roads"

"Storms, Neglect Crumble Sea Walls"

Winter gales, summer zaniness, drowning after drowning, escapades in salt marshes and escapes in mudflats, bare breasts and odd catches, all make the local dailies and weeklies, and over the years merge into a seamless seasonal story, a chronicle of human activity all alongshore. But even assembled over three decades, the stories remain a chronicle, nothing more.

What follows is no chronicle, but no "history of the American seashore" either. It is a book twisted by the east wind and winter surf of the North Atlantic Ocean, by the full force of autumn gales that shake the house in which it took form, by the summer lunacies of locals and tourists playing in beautiful, dangerous places. It is an exploration of marginality warped by firsthand observation of contemporary alongshore things, and angled to reach to coasts far from here, to the muddy sands of Apalachicola Bay, to the Hollywood surf of Redondo, to the echoing rocks of Teahwit Head, to the magic sands of Kahoolawe. Even beaches and inlets as ordinary, as commonplace as those noticed here can be putting-off places for more spectacular sites lined with hotels, parking lots, and surfers, the famous beaches examined in books like William M. Varrell's *Summer by the Sea: The Golden Era of Victorian Beach Resorts*, and Charles E. Funnell's *By the Beautiful Sea: The Rise and High Times of That Great American Resort, Atlantic City.* Any scrutiny of this or any other stretch of ordinary coastline may illuminate the long-ago shores analyzed in books like Arthur P. Middleton's *Tobacco Coast: A Maritime History of Chesapeake Bay in the Colonial Era*, and Roger F. Duncan's *Coastal Maine: A Maritime History*, for the ordinary seashore remains remarkably unchanged five centuries after European contact. So this book reaches backward, toward origins, wading the guzzle of time that lies everywhere and nowhere in the marge, the guzzle that now and then swallows the modern attitudes expressed in works ranging from *Coastal Zone Management Journal* to conference proceedings entitled *Visual Quality and the Coastal Zone* to Douglas Alvord's small-craft advisory *Beachcruising* to John Casey's wrenching novel *Spartina*. It deliberately ignores those few constituents of the marge that attract so much attention nowadays, say the lighthouses of Sarah C. Gleason's *Kindly Lights: A History of the Lighthouses of Southern New England*, and so many other books, as well as the artistic renderings scrutinized in Roger B. Stein's *Seascape and the American Imagination* and other art-history monographs. It is both a historical inquiry and a close look, something like the mix of land and water so poorly designated by *seashore* or *coast* or *seascape*, and perhaps best designated *coastal realm*, for the mix is ruled indeed, ruled by the whole concept of marge, of coast.

And always it is a personal book, based on local observation, grounded in the "local knowledge" federal chartmakers counsel long-distance mariners to obtain before closing the nearby inlet, before moving too near inshore. "The estuaries of rivers appeal strongly to an adventurous imagination," muses Joseph Conrad in *The Mirror of the Sea*. "From the offing the open estuary promises every possible fruition to adventurous hopes." But to the aesthete biased toward traditional beauty of landscape, the estuary which "resembles a breach in a sand rampart" seems utterly worthless, "a repulsive mask" forcing away the viewer. "There are estuaries of a particularly dispiriting ugliness," Conrad continues, "lowlands, mudflats, or perhaps barren sandhills without beauty of form or amenity of aspect, covered with a

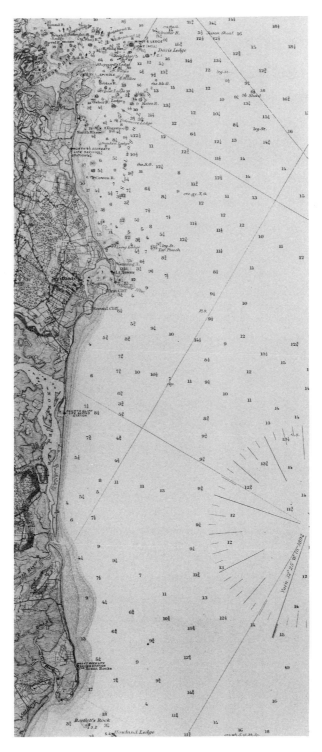

The explorer who needs charts might just as well study this stretch of coast using Chart 109, a federal government explication of Massachusetts Bay dating from 1898. Little has changed in this ordinary coastal realm. (Harvard College Library)

Old Light Tower Scituate

Minots Ledge Light, bearing S. by E. distant 2¾ miles.

Cohasset Harbor

Elevations published on the margins of Chart 109 reveal a place as undistinguished as any imagined by Conrad, a coastal realm of low sandy hills punctuated by exclamation-point-like lighthouses. (Harvard College Library)

shabby and scanty vegetation conveying the impression of poverty and uselessness." [12] Along such an estuary, among such barren sandhills and dunes, in the great marshland behind the barrier beach, occur the daily observations and trespassings and wonderings that shape this book, efforts made sometimes afloat, sometimes on foot, but always in ordinary, unremarkable space, the sort aesthetes dismiss at once.

Others dismiss it, too, exactly as Conrad might. "For most of this distance it is a mere strip of low, sandy beach," reports the *United States Coast Pilot* in 1903 of the coastal realm examined here, or it is "undulating land of moderate height, thickly settled, dotted here and there with woods, but for the most part grassy and bare of trees." The estuary "is of no importance, and is only used by vessels of extremely light draft whose masters are well acquainted with the locality." Moreover, the whole region is deadly dangerous, "rendered extremely hazardous to navigation by the large number of rocks and detached ledges lying off it" and so little visited by anyone but locals that the lighthouse stands unlit at night. Nothing special anywhere, nothing for the oceangoing mariner, nothing for the well-to-do vacationer, nothing, absolutely nothing for the aesthete, just one "low and flat headland with several higher patches of ground" after another. [13] As period photographs suggest, as firsthand observations convince, since 1903 scarcely anything has changed. Mariners might just as well trust the *Coast Pilot* and aesthetes might just as well travel elsewhere. Nothing seems likely to render this piece of coastal realm less dangerous or more beautiful. Adventurous imagination or not, even Conrad would find little to lure him inshore, and Thackeray's artist would find scant reason to set up easel and paint.

Yet something might be found here, something akin to pirate treasure, the treasure every beachcomber searches for every day, even when about on other business. Aboard the bilge-bottom rowboat *Essay*, built ostensibly as the "expedition research vessel" for this book, and the varnished, fast-rowing wooden yacht-tender *Orion*, the barefoot historian moves easily along the salt creeks, among the glitzy speedboats, across the shallow-water harbors and rocky channels, and takes the ground in a hundred places, jumping ashore to walk and look and dig. From those smallest of small boats, land becomes something seen from something

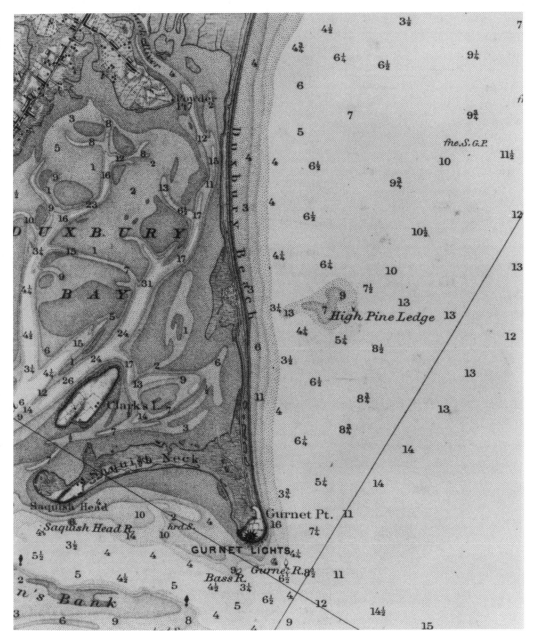

Chart 110, another federal government publication of 1898, marks some of the subjects of this book, a long sandy barrier beach, a great salt marsh laced with creeks, the square earthwork fortifications next to Gurnet Light. (Harvard College Library)

else, something bounded, something defined by the sea from which landings are made. "The sea does not reward those who are too anxious, too greedy, or too impatient," asserts Anne Morrow Lindbergh in her shell-like testament *Gift from the Sea*. "To dig for treasures shows not only impatience and greed, but lack of faith." But beachcombers dig, and now and again plunge into the surf to drag ashore some bleached timber, some skein of manila rigging, some gorgeous white-ash oar. Perhaps "one should lie empty, open, choiceless as a beach—waiting for a gift from the sea."[14] But perhaps one should walk and row, and probe the wrack, and all alongshore, everywhere in the coastal realm, watch—and dig. Perhaps one must dig, if one must know the edge of the sea, the edge of the land, the frontier where ocean and land mingle, the coastal realm.

GLIM

Only rarely, say when a replica of an eighteenth-century warship passes close to the beach, does the five-foot, three-inch-tall woman find an opportunity to test old guides to estimating distances at sea. (JRS)

L̲ate afternoon finds her standing at the very edge of the sea, waves just touching her toes, the rising onshore breeze lifting her hair, sunlight glowing against her skin and faded neon bikini. One of the locals, one of the women who bring no accessories to the edge of the world, stares seaward, watching something invisible to the summer people who walk behind her, between her back and the dunes. Now and then some inlander stops to follow her stare, focusing and refocusing on the immensity of waves beyond the surf, then gives up and strolls on, content to look a few yards ahead. Only other locals know that the woman watches vastness.

Vastness cheats watchers all alongshore, even locals. Proper vastness, dictionary vastness, lacks edges, stunning eye and numbing brain with boundary-less immensity, with infinite extension. Coastal vastness ends at the horizon, even on the clearest, most colorful, sunniest days that torment eye-shielding beachgoers scrutinizing what only seamen truly know, or feel. Sky and sea meet at the "horizon line," or so newcomers think as they remove sunglasses and squint, determined to see some limit, some line, marking the edge of infinity.

Ocean beaches front extraordinary vastness, opening on encompassed vistas that at first surprise, then unnerve, then bore and bore and bore. However many beachgoers watch swimmers or sailboats or even squint toward the lobsterboat or rare steamship far beyond, few watch the sea for long. To the uninitiated, a boatless sea is simply empty, a visual blankness that not only fails to reward sustained scrutiny but mocks the most experienced of landlubber observers.

How far away is the horizon, the "line" about which inlanders speak so certainly? Mariners and locals offer little immediate aid. At high tide on a clear day, the five-foot, three-inch-tall woman standing with her toes just touching the water can see the top of something—say the head of a swimmer—floating on the surface of a calm sea roughly two and a half nautical miles before her. But that same woman can see the masthead of a sixteen-foot-high sailboat cruising much further off, slightly more than seven miles, in fact, and she can discern the topsails of a sailing vessel with hundred-foot-high masts far further, some thirteen and a half miles away. In perfect weather she might see the tip of an object—the rim of a volcano—328 feet above sea level, almost twice as distant, twenty-three and a half miles beyond her toe-hold at the edge of land, and she can see the top of an eight-thousand-foot-high mountain a hundred miles away, the last something that makes the feats of Odysseus and other classical navigators far easier to comprehend. Mount Ida in Crete is eight thousand feet high, a wonderful landmark for mariners in small boats, almost as wonderful as the clouds that hover two miles above Polynesian atolls, the clouds understood by traditional Polynesian navigators as exclamation points in the *kapesani lemetau*, the speech of the sea remarked in canoes floating nearly at sea level.

But let the woman climb slightly higher than the ordinary reach of the flood tide, let her clamber to the top of a ten-foot-high dune, and her range of vision lengthens indeed. Now she sees the swimmer's head almost four and a half miles off, the same little sailboat nine miles from shore, and the tall ship more than fifteen and a half miles away. Such are the distances R. C. Carrington painstakingly tested, then published in his late-nineteenth-century *Marine Survey of India*, distances still used today.[1] Of course, she has not yet climbed very high at all, not to the top of the cliff, to the crowns of cliff-top trees, to the lantern room of the cliff-top lighthouse. Precision proves difficult in the coastal realm of shifting sand. Rarely is the ocean flat calm. Usually the long rollers toss all but the biggest ships up and down so much that observers intrigued with distance must calculate average vertical movement. Then, too, barefoot observers staring seaward often ignore the gentle undulations of the beach, and too easily assume "sea level" when in fact they have missed it altogether.

A fair approximation, the mid-nineteenth-century "deck of the vessel" rule, is that from dry sand an adult sees the top of a sixty-foot-tall object roughly ten miles away. The old rule reassures novice beachgoers willing to dismiss questions of farsightedness and average vertical movement, and unwilling to consider the rest of the rule: objects one mile in height are visible at deck level from about one hundred nautical miles away.[2] Such simple rules ram home the visual significance of Melville's crow's-nest passages in *Moby-Dick*, where Ishmael sways in great arcs, his eyes sharpened for blowing whales, or the circumference of scrutiny available to the lookout of the USS *Constitution,* "Old Ironsides," the frigate built in 1797 whose mainmast towers 220 feet above sea level.

Aboard that frigate, a sharp-eyed masthead lookout searching for the topsails of another frigate swept some 2,826 square miles of sea from his point at the center of a circle 60 miles in diameter. Fictional Ishmael and real-life navy lookout saw many nautical miles indeed.

Six thousand and eighty feet comprise the old Admiralty or nautical mile, the unit of marine distance abandoned in 1954 for a new, precise measurement—6,076 feet—that brings landsmen no nearer the horizon. Nautical measurement often infuriates landsmen wedded to the 5,280-foot mile, unwilling to hear definitions of *fathom, knot,* and other seagoing terms applicable to watery vastness, and already disconcerted by an amorphous horizon much like the horizon of one eighteenth-century philosopher, John Locke, "which sets the bounds between the enlightened and dark parts of things."[3] Of vaguest possible distance, shimmering on a curve, the horizon often seems a fiction, something that jars schoolroom memory.

Curvature of planet disconcerts many novice observers, too. Clearly, obviously, the tall sailing vessel "hull-down" against the sky is sailing far enough offshore to exhibit what every schoolchild knows Columbus trusted—visual evidence of a spherical planet. Summer visi-

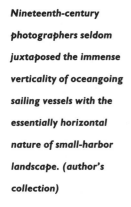

Nineteenth-century photographers seldom juxtaposed the immense verticality of oceangoing sailing vessels with the essentially horizontal nature of small-harbor landscape. (author's collection)

tors, even first-time beachgoers, recall their grade-school education with sudden interest, and study the sloops and other small craft sailing before them, hoping to see one hull-down, proving what Columbus hoped to extrapolate. Novice observers, especially those armed with binoculars, struggle to ascertain distances, unthinkingly believing that what they see is real.

But ocean vastness plays tricks, even on the educated eye, and loomings play the best tricks of all. Rare, remarked upon even by locals practiced in the ocean-sweeping scan that accompanies intimacy with beaches and ocean views, and long ago a part of every alongshore formal education, loomings seem even rarer now, perhaps because so few watch for what fascinated Jefferson, Melville, and other delighters in long-distance meteorology, savants who knew the meaning of the phrase landsmen use so carelessly, "looming on the horizon."

"The seamen call it *looming*. Philosophy is as yet in the rear of seamen, for so far from having accounted for it, she has not given it a name," mused Jefferson in 1785 in his *Notes on the State of Virginia* on a phenomenon whose "principal effect is to make distant objects appear larger, in opposition to the general law of vision, by which they are diminished." Jefferson admitted that he was "little acquainted with the phenomenon as it shows itself at sea," but he had seen it at Yorktown, "from which the water prospect eastwardly is without termination," where a canoe with three men aboard was "taken for a ship with its three masts." From Monticello, Jefferson watched the looming or shape-shifting of a far-off mountain, determining that "refraction will not account for this metamorphosis," since "by none of its laws, as yet developed, will it make a circle appear a square, or a cone a sphere." What puzzled him puzzled others, as he noted in his text, citing Thomas Shaw's *Travels or Observations Relating to Several Parts of Barbary and the Levant* (1738), a book that mattered much to the president of a new nation troubled by Barbary pirates. "It may be further observed," Shaw remarked about the visual effects of the Mediterranean storms called Levanters, "that vessels or other objects which are seen at a distance, appear to be vastly magnified, or to *loom*, in the mariner's expression."[4] How could a Virginia mountain some forty miles off change shape from dawn to dusk unless some meteorological phenomenon worked visual wonders, wonders perhaps offering Enlightenment insight into North African Old Testament miracles?

Melville knew the phenomenon, knew it in a deeper way than Jefferson, the Enlightenment landsman concerned with a mountain seen across land. Melville knew the principal effect, and others, too, across sea. In "Loomings," the first chapter of *Moby-Dick*, Melville sketches out the broad themes that form the focus of the novel. But first he touches on the near-mystical draw of the sea view. On a Sunday afternoon, he remarks, "stand thousands upon thousands

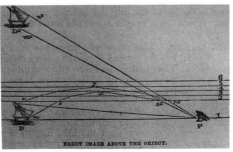

ERECT IMAGE ABOVE THE OBJECT.

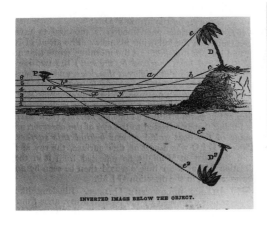

INVERTED IMAGE BELOW THE OBJECT.

of mortal men fixed in ocean reveries." These thousands of landsmen intrigue him. "Some leaning against the spiles; some seated upon the pier-heads; some looking over the bulwarks of ships from China; some high aloft in the rigging, as if striving to get a still better seaward peep."[5] The lookers have scant interest in the workings of the port, for the sabbath stills such activities. Nor are they attracted to seamen or even to the ships—after all, they look over the bulwarks of ships in from China, not at them. They seek only a view of the sea, what Jefferson called the water prospect without termination. Somehow that openness, at least to people usually "pent up in lath and plaster," opens something else, certainly reverie, possibly looming as seamen knew it, a glimpse *over* the horizon.

Two years after Melville coyly implied that seamen and seers can sometimes see over the horizon, beyond the curve of space if not of time, John Brocklesby explained how. His *Elements of Meteorology* aimed at advanced schoolchildren and focused on far more than prophesying weather changes. Brocklesby gloried in explaining "optical phenomena" and painstakingly distinguished simple looming from complex. By simple looming, he understood what Jefferson seems to have remarked, "displacement of terrestrial objects." Such displacement "is *ordinarily* seen in the slight elevation of coasts and ships, when viewed across the sea, and is then called *looming*." Local changes in surface temperature vary the density of the atmosphere, and consequently cause irregularities in refraction. But sometimes looming is anything but ordinary, and Brocklesby designates it by the landsman's term, *mirage*.

Using simple geometry to illustrate his historical examples, Brocklesby makes clear that extraordinary looming, while nothing more than an explainable atmospheric phenomenon, strikes mariners as visual magic. "Images of ships erect and inverted are seen in the air."

But the images are genuine representations of existing vessels, not delusions. In 1822, for example, one captain "recognized his father's ship, the *Fame*, by an inverted image of the vessel in the air, though it was subsequently found to have been at that time *thirty miles* distant, and *seventeen miles beyond the horizon.*" Such happenings prompted science-minded masters to record temperature and other meteorological information on the spot, and Brocklesby points out that such information, correctly interpreted, goes far toward explaining complex looming. On one occasion, "the temperature on deck was 54 degrees Fah., that at the mast-head 62 degrees Fah." Such variation in temperature might make loomings very common, but of course something had to be over the horizon to loom—otherwise mariners saw only a slightly different, indistinguishable horizon line quite close in the mist or fog.

On July 26, 1798, near Hastings on the Sussex coast, the cliffs of the French coast some fifty miles distant, and ordinarily below the horizon, suddenly became visible, and seemed to be only "*a few miles off.*" An alert landsman studied the phenomenon, and Brocklesby explains how "the sailors who accompanied him, pointed out and named the different places on the opposite coast which they were accustomed to visit. By the aid of a telescope, small vessels were plainly seen at anchor in the French harbors and the buildings on the heights beyond were distinctly visible." Witnessed by hundreds, recorded by a thoughtful observer canny enough to ask local mariners to describe the places suddenly in view, and widely reported, the Hastings looming provided Brocklesby with much material for explaining similar visual effects that badly disconcerted everyone not familiar with them.[6]

Explanation demands rigorous understanding of both Euclidian geometry and Newtonian optics, and Brocklesby provides both, ruthlessly demolishing all doubt that simple and complex looming occur, then revealing how slightly different atmospheric conditions cause refraction and reflection to join in definable ways, then demonstrating how to create miniature loomings in laboratories and drawing rooms. Brocklesby's lines and angles make clear that Columbus gambled indeed, for all that astronomers' mathematics had essentially proven a spherical earth. Not every ship sailing far out to sea vanishes in stages, appearing to beach-bound observers to sail over a curved sea, its hull disappearing before its lower yards, then its lower yards vanishing while its topsails remain visible. Sometimes looming makes a ship appear to topple, its suddenly inverted image confirming theories that the world ends at a definite edge. Then, too, looming sometimes lets beach-bound observers see beyond the horizon without knowing, see beyond-the-horizon ships distinctly, then see such ships vanish instantly and utterly.

So the five-foot, three-inch woman who shields her eyes against the glare and stares out-ward from the wavelets that slap her feet sees into a vastness for which few landsmen are prepared, either by experience or by education. Little in fact prepares landsmen for their first glimpse of the sea, let alone their first long stare. Sea, and sometimes shore, lacks scale, and without scale becomes not visually stimulating but almost deadening. "Men and boys would have appeared alike at a little distance, there being no object by which to measure them," mused Thoreau while trudging across the sands of Cape Cod. "Indeed, to an inlander, the Cape landscape is a constant mirage." But the mirage defies study, for unlike some desert oasis floating above some thirst-stricken wayfarer, the coastal mirage or looming—if identi-fied as such at all—shimmers above a moving sea. "On the beach there is a ceaseless activity, always something going on, in storm and in calm, winter and summer, night and day," he continued. "Even the sedentary man here enjoys a breadth of view which is almost equivalent to motion." No wonder so many beachgoers read or watch people or orient themselves to look along the beach, not out to sea. Or they close their eyes. To look outward for long, on all but the rarest of calm days, is to look at innumerable waves that mock any effort to distinguish looming—nowadays termed "abnormal refraction" in sailing manuals—from ordinary view.

And however much the rollers and waves distort the long-distance horizon-scraping view, the waves make continuous noise, the rote that distracts the acutest seeing. "If he is too lazy to look after all," Thoreau decided, "he can hardly help *hearing* the ceaseless dash and roar of the breakers."[7] Before enclosed automobiles and portable radios, especially before earphone-equipped miniature radios and stereo tape players, the rote transfixed, soothed, and irritated newcomers to the shore. "But it was the ever-present sound of the sea which made the great-est impression upon my bucolic mind," argues a character in George S. Wasson's *Home from Sea* (1908), one of the first American novels aimed directly at recording seashore experience. "Day and night, summer and winter, always the ceaseless rote of the sea, like the breath-ing of some great monster it seemed to me; sometimes very low and faint in the village, but still always noticeable in some degree, and at times jarring every window in the town with its thunderous rumble."[8] It is the rote that makes stupid any comparison of the sea or coastal realm with the great prairies, for however much the prairies appear limitless, however much the grassland wind sometimes sounds like the onshore wind, the prairies lack the rote, the ever-changing but ceaseless sound of sea clutching land.

Calenture—or a trace of it—touched William Ellery Channing as he trudged the Cape Cod sands beside his friend Thoreau, wondering at the young locals so accustomed to the roaring

rote that they hardly heard it, and struggling to fit ocean vastness into some other topographical terminology. "My companion compared the sea to the prairies," Thoreau remarked, of the comparison he made himself, noting in his journal that "the sea has that same streaked look that our meadows have in a gale."[9] Many landsmen make the comparison still, ignorant of long-lost symptomatic importances. Seeing the sea as prairie, or as farm fields spiked with trees, once indicated dangerous illness, as schooner-master-turned-poet Philip Freneau understood in 1786 in his gothic poem "The House of Night":

> He mention'd, too, the guileful *calenture*,
> Tempting the sailor on the deep sea main,
> That paints gay groves upon the ocean floor,
> Beckoning her victim to the faithless scene.

Like *looming, calenture* flourished as an oceangoing term alien to landsmen, and Freneau felt obligated to define it in a footnote: "an inflammatory fever, attended with a delirium, com-

As this postcard view suggests, turn-of-the-century amateur painters at the seashore often wound up learning techniques of portraiture rather than how to catch the glim on canvas. (author's collection)

mon in long voyages at sea, in which the diseased persons fancy the sea to be green fields and meadows, and, if they are not hindered, will leap overboard."[10] Calenture often struck mariners and passengers already ill—in his *Account of Two Voyages to New England* (1674), John Josselyn recounted the nastiness of sailing with passengers ill with both smallpox and calenture—and healthy onlookers sometimes remarked it as following sunstroke, or perhaps eyestrain.[11] Usually accompanied by high fever and the delirium-induced delusions seamen distinguished from proper looming-seeing, calenture disappeared in the oceanliner era, when passengers no longer knew the monotonous dreariness of months-long passages over ocean vastness, nor the ceaseless slapping of waves against wooden hulls. But calenture lingers as a flitting disease of first-time seashore visitors desperate to make topographical sense of the sea and willing to forget that no rote smites the ears of prairie travelers.

All alongshore everyone knows the rote, knows what knowing by rote means, getting something by heart, by heartbeat. For some newcomers, for the summer visitor walking seaward through the dunes for the first time, the rote of heavy surf is oddly bewildering, a sort of noise unknown, perhaps vaguely like that of Niagara, but rhythmic. Mysterious, throbbing, sometimes shaking not only the sand under foot but the asphalt-topped parking lot behind the dunes, the rote makes sense only when the visitor sees the surf, and even then the visitor may wonder how such massive waves enliven a windless day.

Along this stretch of coast the rote sounds always from the East, usually swept stronger and further inland with the East wind, exactly as John Greenleaf Whittier describes in "Snow-Bound":

> The wind blew east; we heard the roar
> Of Ocean on his wintry shore,
> And felt the strong pulse throbbing there
> Beat with low rhythm our inland air.

Massive winter gales, the "three-day no'theasters," throw up gigantic surf that pounds and pounds, not only damaging coastal structures and any vessel luckless enough to sail into it but slamming hearing into dullness, then temporary deafness, making any place indoors a refuge not just from cold, wind-driven rain and snow but from rhythmic, battering noise.

And the summer-day-at-the-beach rote, often together with the breeze, sometimes drowns out all landward sounds. Whittier catches that, too, in "The Tent on the Beach," a poem about vacationers escaping a heat wave. "Sometimes, when the wind was light / And dull the thunder of the beach," the campers hear church bells ring far off.[12] But most of the time the campers live in auditory isolation, as cut off as the inhabitants of the snowbound house. In the middle of the nineteenth century, no human sound, not even the whistle of the railroad locomotive, not even the shriek of the coastal steamer, carried far into the roar of the rote. However much the locomotive whistle disturbed the quiet of Thoreau's Concord woods and Walden Pond, alongshore the rote often reduced the blast to a feeble chirp that scarcely penetrated the nerve-numbing cacophony of ceaseless crashes. Even today no sound, not police-car siren or fire-truck woofer, reaches far into the sound of surf.

So the woman shielding her eyes, looking for looming, watching for some indication that vastness might be measured, not only hears the rote but, when the surf runs high, feels it, too. The beach throbs almost as though the sand has indeed become quick, and the sand and small cobbles hit by the waves move seaward, often with a gurgling, sucking noise wholly animate. "Though for some time I have not spoken of the roaring of the breakers, and the ceaseless flux and reflux of the waves, yet they did not for a moment cease to dash and roar, with such a tumult that if you had been there, you could scarcely have heard my voice the while," Thoreau wonders in his *Cape Cod*, "and they are dashing and roaring this very moment, though it may be with less din and violence, for there the sea never rests." To scrutinize anything under such conditions, even on an otherwise clear day, is to test and test again the resources of human vision, as Thoreau learned, as Anne Morrow Lindbergh learned, as anyone learns who tries as the local woman tries.

For that woman looks out upon waves. Few beachgoers stare long at waves, and even fewer note that waves come in groups, often five of the same size, then one slightly smaller or perhaps angling in from a different direction, then five again. Thoreau understood that American steamships could be tracked from seventy miles off simply by the massive plumes of black smoke produced by their bituminous-coal fires, but such long-distance envisioning, let alone the complications of looming, scarcely assist the woman watching a particular wave make the shore.[13] Following a wave to shore proves confoundedly difficult, almost as difficult as watching patterns of them, or learning how the waves make the rote. Is it the actual crash on shore that makes the sound, or the earlier crash on the outer sandbar?

Wave-watching hypnotizes, or nearly does. Kate Chopin addressed the effect obliquely in her novel *The Awakening* (1899), exploring the coastal realm as subtle stimulus to social experimentation, to freeing inhibitions. "The voice of the sea is seductive; never ceasing, whispering, clamoring, murmuring, inviting the soul to wander for a spell in abysses of solitude; to lose itself in mazes of inward contemplation. The voice of the sea speaks to the soul," she writes early in the novel, and she repeats these sentences word for word in its last paragraphs. Perhaps more than any other American author, certainly more than any other in her time, Chopin probed not so much at the soul swept by the sea but at the pysche released from its moorings by the all-encompassing seashore.

Her chief character, summertime beachgoer Edna Pontellier, drifts free long before she learns to swim, long before she realizes her affair with a young Creole. She drifts free when she loses herself in gazing "out as far as the blue sky went," and finds a sort of amoral listlessness in the gazing. " 'The sight of the water stretching so far away, those motionless sails against the blue sky, made a delicious picture that I just wanted to sit and look at,' " she says to a concerned friend. Of course, "delicious" gives away Chopin's intent. Something sensuous beyond speech suffuses Edna's being, and in time Chopin makes clear the sexual nature of the waves and vastness: "The sea was quiet now, and swelled lazily in broad billows that melted into one another and did not break except upon the beach in little foamy crests that coiled back like slow, white serpents."[14] For such prose *The Awakening* was banned from libraries and generally condemned, but Chopin's words endure as one entry into the mystery of beach-going, of wave-watching as narcotic, of the dangers of doing as Lindbergh recommends, of opening oneself to the sea.

Empty of history, utterly without past, looking as it might have looked nine thousand or nine million years ago, the sea flaunts its presentness to anyone determined to scrutinize its surface, to number its waves, to grasp the portent of Luke 21, to imagine "nations confused by the roaring of the sea and the waves." But more than wave rhythm governs the sea edge. Another measure of time stirs the sand and gropes among the rocks, a measure as utterly absent as the rote from vast prairies and deserts—and rural meadows. The five-foot, three-inch-tall woman knows sea level as dependent on the tides, and the tides have a rhythm of their own.

Generations enamored of Newtonian physics studied the tides as determinedly as they studied Newtonian optics, fascinated by a wholly predictable phenomenon of extraordinary importance to mariners, of course, but also to anyone bound for a walk down a low-lying beach or out to a nearby sandbar or toward some salt-marsh island. Let the tide come in, let it encircle the thoughtless walker, and only swimming brings the walker back to land free of the tide. But immediate effects aside, the tides once mattered for reasons beyond wet feet.

Tides beg the question of cause. Any acute observer, say the summer visitor spending two weeks in a seashore cottage, wonders after a day or two what causes the sequential rhythm of high tide and low. "The moon" suffices only for young children who ask about the tides, and "lunar cycle" satisfies only the most incurious adults. To anyone who looks over time, often over time longer than the typical summer vacation, some high tides are higher than others, often much higher. And even a careful one-day visitor may notice the wreckage high up on beaches or inland on salt marshes, and wonder if storm waves and winds alone accomplished the placement. Any decent explanation of the tides, the sort a lobsterman patiently makes to summer visitors buying dockside lobsters at retail prices, immediately encompasses not only the phases of the moon but the juxtaposition of moon and sun. Anything more complicated leads into the complexities of lunar cycles, of neap tides and onshore winds, of heavenly influence. And no one alongshore dismisses astrology out of hand, for heavenly bodies make the tides, and the lunar cycle, so frequently missed by inlanders not enamored of moon-watching, may have other effects, say on menstrual cycles, or on the willingness of fish to take bait, effects as unknown to contemporary science as complex looming was to Enlightenment philosophy.

But visually, tides change the appearance of the shore by altering its expanse. The novice visitor arrives some Saturday morning to find a gorgeous barrier beach reaching seaward half a mile from the dunes. A week later the visitor returns with friends, strides through the path between the dunes, and finds only a narrow strip of sand washed gently by tiny waves. When George Howard Darwin published his *Tides and Kindred Phenomena in the Solar System* in 1898, having already published parts of the book as articles in such mass-circulation magazines as *Harper's*, *Atlantic Monthly*, and *Century*, he assumed a readership familiar not only with the vastness of the sea but with its immediacies. "At the seashore it would be impossible to avoid noticing that some rocks or shoals which are continuously covered by the sea at one part of a fortnight are laid bare at others," he remarked in his introduction. "It is, in fact, about full and new moon that the range from low to high water is greatest, and at the moon's first or third quarter that the range is least."[15] In other words, Darwin assumed beachgoers who knew not only the cycle of tides but variations in the cycles, the last revealed by visible changes in the seashore environment. He assumed a readership of lookers.

Today, of course, many people use beaches far more actively than they did at the turn of the century, and people actively using the shore may observe less closely than their parasoled and straw-hatted predecessors, the beachgoers Winslow Homer painted, the beachgoers Darwin presumed to see accurately. But even a century or more ago, strangers to the beach did not scrutinize as closely as they might. "When we have returned from the seaside, we sometimes ask ourselves why we did not spend more time in gazing at the sea," Thoreau remarks near the middle of *Cape Cod*, "but very soon the traveller does not look at the sea more than at the heavens." Implicit in this remark lies a key into the character of Thoreau's book, and of other books by other beach visitors, and perhaps into contemporary beach-going-as-vacation, too. Thoreau wrote *Cape Cod* out of his travel experiences, not out of his experiences living on the beach, in the coastal realm. Everywhere in his book lurks the viewpoint of a man walking the beach toward Provincetown, a man remarking on the feel of sand under his feet—and in his shoes.[16] *Cape Cod* is fundamentally different from Chopin's *Awakening*, for example, because Chopin studies a woman spending a summer in the coastal realm, a woman learning to swim to be sure, but a woman nonetheless in one place, watching, and perhaps succumbing to rule. So many books about beaches follow Thoreau's model, not Chopin's, that contemporary readers may miss another, perhaps chief significance of the five-foot, three-inch woman who stares seaward.

That woman has achieved—almost—the sensual freedom of Chopin's thwarted heroine, who swims naked only once, when she drowns herself at the close of the novel. Nowadays Americans can wear almost nothing at the beach, and so they experience the environment directly. The five-foot, three-inch woman stands exposed to the elements, shielding her eyes behind sunglass lenses scarcely smaller than the wisps of nylon comprising her bikini, protecting her skin with sunscreen but otherwise confronting the seashore environment unencumbered with such stupid accessories as shoes. She experiences the shore as something beyond the visual. It becomes tactile.

Nowadays the ocean beach exists as the last place where adult Americans habitually walk barefoot.

Barefoot walking scarcely resembles the shod trudging of Thoreau the landsman, the landsman who insisted on shoes even in October. To walk the beach barefoot—and here on sunny January days regular beachgoers often walk barefoot, knowing that the dry sand is comfortably warm, and almost always warm enough to make barefoot walking less tiring than walking in shoes—is to walk in direct contact with an essentially natural environment, the one least touched by the human hand. Barefoot walking informs directly, immediately, of the varying sorts of sand or cobbles beneath one's feet, of the varying coolness and wetness of the sand, both often invisible even to the experienced eye, of the temperature of the water. Barefoot walking lets the walker feel the slightest tremors of the rote, tremors that hiking boots mask.

Summertime near-nakedness provides opportunities even greater than barefoot strolling, however, opportunities so blatantly obvious as to pass unremarked now but missed by past generations. The bikinied woman in her waterproof sunscreen can dash at will into the surf and retrieve the seaweed or half-submerged wreckage that Thoreau longed to recover. She can move from shore into waist-deep water and on to an exposed sandbar, and she can stand in the spray of massive surf and test her powers of perception by ascertaining the distance— in nautical or land miles—of the steamship this side of the horizon. Finally, of course, she can swim out at will, turn, and look at the beach from the sea, easily accomplishing something unknown to most Americans until the first third of the twentieth century.

And caught in tide, in wind, in the glare of reflected and refracted light, standing against wavy vastness, the woman stands as more than experienced observer. She becomes something else, a vertical element on the beach, an object of view, a seamark, a landmark.

On a near-limitless barrier beach reaching in both directions into the haze, that woman-as-figure becomes scaleless. "Men and boys would have appeared alike at a little distance, there being no object by which to measure them," Thoreau mused in a passage worth repeating.[17] Thoreau worried only about distinguishing height, for in his time men and boys dressed differently from women. But nowadays no voluminous skirts and other drapery define female silhouette along the beaches, and even the most sharp-eyed, practiced observer learns how difficult it is to discern anything precise about the human form in the distant haze. How far away is the person? Approaching or retreating? Is the person adult or child? Male or female? Naked, nearly naked, or fully clothed? Running or walking?

Given the circumstance, given someone moving in the sparkling, moisture-laden, far-off haze, such questions vex accomplished observers of inland landscape. Dunes may be high or low, and lacking anything remotely like built form—say utility poles stretching into the haze—the dunes become as scaleless as the person. Does anything suggest movement toward or away from the expert observer? How far away can gender be guessed at, let alone known with certainty? After all, people walk differently when they walk barefoot in the soft sand, and the "characteristic walk" so carelessly displayed on urban sidewalks disappears along the beach. Naked, or almost? Perhaps the metallic bikini provides a flash or two of light, but then again, perhaps the swimsuit is skin-toned, and so almost indistinguishable at a thousand yards. Is the figure a man, clad in almost nothing and striding almost effortlessly over damp sand, or is the figure a woman, loping carefully along the dry sand in an effort to build calf and thigh muscles?

On the beach, at right angles to the ocean vastness, people—and occasional dogs—are

often the only intruders in an otherwise natural place. Given the waves hitting the sand, usually at an oblique angle, to seaward, and a line of dunes to landward, the observer tired of staring out at ocean vastness suddenly discovers the equally disconcerting problem of making sense of near-limitless alongshore views that end in haze.

Never mind distinguishing man from woman. For many beachgoers, particularly those walking the beach early in the morning, the effort becomes one of determining if the human form half-obscured in haze walks on the sand or several feet above it.

Now and then the scalelessness of the barrier beach puzzled even accomplished painters. When Elihu Vedder painted his *Lair of the Sea Serpent* in 1864, for example, he apparently intended to convey the awesomeness of a massive sea serpent warming itself among great dunes of sand. But the painting is oddly disappointing to anyone familiar with beaches, and perhaps to anyone who studies it for long. The sea serpent suddenly shrinks, becoming nothing more than an eel, and the dunes shrivel, too, even as the grass becomes too tall. Something is wrong indeed, and perhaps the error lies in Vedder's effort to depict wind-stunted trees that under scrutiny seem like bits of rockweed holdfast cast ashore and half-buried in sand. The painting, however competently executed, is almost filmic to today's observers, who find their eyes shifting constantly from one out-of-proportion element to another and then back again. To look long at Vedder's painting is to experience something of the visual discomfort that accompanies casual beach-watching.

So the beach disconcerts, then disappoints many visitors anxious to photograph it, or paint it. Amateur photographers and painters quickly dismiss it, settling instead on harbors, stranded sailboats, or even a washed-up lobster pot or two. For many artists, the long view over sea or along beach is simply not rewarding, no more valuable for them than it is for less visually inclined summer visitors. The artists turn toward familiar, usually built things, especially quaint things, abandoning the long beach views to the locals who somehow enjoy their empty trickiness.

The local woman, whose eyes almost exactly five feet above the sand make her a perfect instrument for testing R. C. Carrington's findings concerning over-sea perception, sees further into empty trickiness, into what locals still call the "glim" or the "sea-glin," the staggering openness in which the eye reels while the rote booms. In the words of Wallace Stevens, she "beholds / Nothing that is not there and the nothing that is," unless, of course, she glimpses a genuine mirage, not a rare looming.[18]

She glimpses differently, her local gaze drifting across visual vastness, her practiced eye shielded by sunglasses or slight squint. Just as local lobstermen and fishermen find the fog-bound harbor channel by unconsciously distinguishing the sounds of the rote to its north and south, the local who looks seaward or "up and down beach" looks in a quasiconscious way.[19] " 'There's a knack to how you look. You got to . . . unfocus your eyes. What you want to do is let as much come in as you can,' " says a fisherman in John Casey's *Spartina*.[20] At the beach, where visitors tend to worry about how their bodies look to others, about their posture while walking through sand, "how you look" rarely means how one sees. Much passes unseen, unlooked-for, even ignored. Mirage is as likely as looming, for mirage and looming are the creatures of wilderness, and the sea, untamable and unbuildable, remains the wildest of wildernesses. The coast between Gurnet Light and Minot's Ledge endures as frontier, as marginal zone between the civilized landscape and rawest wildness, between simple seeing and complex gazing.

Painters and photographers shun much of it. Even the most experienced of visual artists learn the perils of seascape-making, and others either devote themselves to recording fragments of the larger alongshore realm or flee toward manageable inland scenery.

All alongshore, a moisture-laden atmosphere distorts color as easily as it distorts light, and often the distorted, distended colors preoccupy painters determined to record their transitory appearance. The rendition of alongshore color frequently forms the chief subject of many seascapes, and black-and-white reproduction does scant justice to artists who succeed in a most difficult endeavor. This book in fact includes few reproduced oils and watercolors, not only because paintings often reproduce badly but because black-and-white reproduction subtly implies that color serves purposes less important than those served by line and shadow. Any sustained observation of seascape painting indicates that many painters not only depict particular locales, types of vessels, and alongshore enterprises with painstaking accuracy but devote equal or more attention to rendering the colors that suffuse—indeed make—the scenes. Consider, for example, *Boston Harbor at Low Tide*, by Hendricks A. Hallett, painted about 1890 and a near-perfect illustration of the topic of this book. In the middle distance, a schooner passes in light airs and men repair the hull of a grounded brigantine. Ebb tide, "probably dead low tide," given the proximity of the channel marker to the exposed sand, reveals the shadows of the wharf pilings and strands a pulling boat. The schooner and the brig direct attention toward their bows, at the city of covered wharves, docked square-rigged ships, and smoking stacks, at activity more hurried than Hallett places in the middle distance. Are such the subject of *Boston Harbor at Low Tide?* Does it matter that Hallett infuses his canvas with rose-toned colors, that the buildings are essentially pink?[21] Nautical experts might quibble about his details of rigging, and urban historians might wonder about fidelity to skyline, but any observer must be stunned by Hallett's masterful use of rose to render his scene almost perfectly still, almost magical.

Paintings like *Boston Harbor at Low Tide* receive more and more attention nowadays, perhaps because historians of painting have discovered the extraordinary energy with which the Luminists infused American painting.[22] But any history of seascape painting, marine painting, alongshore painting misses in a odd way the once-powerful role of color in alongshore looking.

Until quite recently, perhaps until the era following the popularization of Kodak Kodachrome color-transparency film in the 1930s, astute observers of the coastal zone routinely recorded the colors about them, including the colors above the sea and the land. But most recorded the colors in words alone.

In retrospect, Thoreau's writing epitomizes a centuries-old verbal tradition. "Today it was the Purple Sea, an epithet which I should not before have accepted," he notes in *Cape Cod*. "There were distinct patches of the color of a purple grape with the bloom rubbed off." No painterly vocabulary informs this passage, but the landsman's effort to match a fleeting ocean color to that of a concord grape works perfectly. Of course, Thoreau admits that "first and

last the sea is of all colors," and he knows that in calm weather, "where the bottom tinges it, the sea is green, or greenish, as are some ponds." And he follows instruction, having read William Gilpin on scenery closely enough to quote the British coast-walker disparaging mountaintop colors as "mere coruscations compared with these marine colors, which are continually varying and shifting into each other in all the vivid splendor of the rainbow, through the space often of several leagues."[23] And since color change results directly from the most fleeting and subtle changes in weather, Thoreau records the atmospheric conditions that "make" the day. "In a journal it is important in a few words to describe the weather, or character of the day, as it affects our feelings," his own journal records in an entry of February 4, 1855. "That which was so important at the time cannot be unimportant to remember."[24] Nothing original informs either his *Cape Cod* remarks on light and color or his journal entry, however, but rather something especially well done.

In the middle of the nineteenth century, educated, thoughtful people struggled to see color well, to understand as accurately as possible Gilpin's association of changing sea colors with the changing colors of clouds and rainbows. "When the rays of the sun fell upon the fragments of vapor floating in the eastern quarter of the heavens, their jutting heads and broken edges gleamed with a flame-like hue; while, between the masses, the sky appeared of the deepest indigo," Brocklesby noted in *Elements of Meteorology*, explaining the union of trained scrutiny and aesthetic appreciation. "As the evening advanced, portions of the western stratum assumed the tints of *lead, lake pink, green, purple, violet, orange* and *crimson.*" As time passed, "the heavens seemed as if covered with a delicate *lace-work woven of prismatic rays,* and this phenomenon was succeeded by *green, purple,* and *violet* clouds in the west." In guiding his readers as surely among atmospheric colors as among the geometric intricacies of complex looming, Brocklesby insists on an understanding of phrases like "the intermingling colors consisting of *bronze, orange,* a *vivid grass green,* and a *golden yellow*" as an accomplishment of self-disciplined looking.[25] Weather, light, and color combine to interest the educated looker and to test the ability to put words to visual effects.[26] And nowhere else than at the edge of the land did the play of weather, light, and color test more rigorously. On sandy beaches, on cliffs overlooking bays, in the middle of salt marshes, on fogbound wharves suddenly illuminated by shafts of yellow light, nineteenth-century writers struggled in the eighteenth-century tradition. They struggled differently than painters struggled perhaps, but they struggled equally hard to record fleeting, ever-changing light and color.

What weather suffuses Hallett's rose-suffused painting? How is it that the stranded brig dries its sails in the sun while the schooner finds enough of a breeze to move? And how is it that the schooner sails against a breeze that blows the factory smoke but leaves untouched the drying sails? What weather makes the light, the color? What makes the character of the evening, the essence of the painting?

Questions like these still come to the minds of alongshore people who scrutinize paintings from another era. Now and then a well-trained aesthete smiles or smirks, overhearing another looker remarking on the rigging of some sailing vessel or the state of the tide or the placement of a channel buoy. Such talk supposedly smacks of simplicity, of small-mindedness, of angles of vision too narrow to encompass Cubism or postmodern abstraction. But it also smacks of the sea, and although the talk sometimes raises details that many aesthetes find annoyingly out of place—say the *red* lantern on Hallett's channel marker—often it focuses on color as something crucial to the painters and their long-dead original viewers.

But since the 1930s such questions have faded. The Depression-era popularization of Kodachrome film seduced many coast visitors away from disciplined scrutiny and disciplined recording in paint or in words. Color photography quickly became the color snapshot, a quick and easy replacement for looking and for time-consuming, suddenly old-fashioned recording. Armed with newly developed 35-millimeter cameras boasting high-speed shutters and easily adjustable f-stop rings, beachgoers learned to make color photographs far more quickly than the view-camera-toting, black-and-white photographers of the nineteenth century. And by the 1960s the old insistence on remarking and recording coastal zone color had almost disappeared from alongshore prose.

Only a handful of photographers know the warning issued by Eastman Kodak in 1985, in Reference Sheet E-73, a bulletin that reads remarkably like the Coast Guard *Notice to Mariners*. Sheet E-73 warns of the limitations of color-film chemistry. "Other colors—such as shades of chartreuse, lime, pink, and orange—may not reproduce so well," the company cautions. "It would be possible to design a film that would reproduce these colors better, but then some of the more important colors would suffer." Eastman Kodak designs color film first for its chief mission, the accurate recording of skin colors, then for its secondary mission, the accurate recording of "common 'memory' colors such as those of the sky, grass, sand, etc."[27] After that, color film may or may not do very well in recording, say, the chartreuse of so many salt-marsh grasses or the chartreuse bikini of the woman watching the horizon.

How does anyone anywhere alongshore precisely note, let alone precisely record, the colors that shift with season, weather, time of day? The barefoot historian, not wishing to burden himself with color wheels and heavy, perfect-bound books identifying color chips by number and by names in five languages, experiments with fold-out charts. It is useful to know that some marsh grasses sometimes appear a perfect match for "#215 Cadmium Green" and that sometimes the inshore waves appear an almost perfect "#149 French Ultramarine," but the big British fold-out chart fits awkwardly in his pocket, and nowhere in his swimsuit.[28] His local paint manufacturer, long renowned along the coast for house paints "adapted to the New England climate," as well as marine paints and varnishes adapted to New England coastal waters, offers chip-cards in a more pocketable size. But although the cards discriminate finely among gradations and hues, and are replaceable at no charge—a useful attribute considering the frequent soakings that accompany coastal-realm research—they name colors in very idiosyncratic ways. Color cards immediately useful in the springtime salt marshes carry chips labeled "green ribbon," "green berry," "monsoon," "periwig," and "bistro green," all useful remembrancers to the historian leaning on his oars and scribbling in his notebook, but of no value to anyone lacking the cards. What good does it do anyone to know that a vast mat of submerged seaweed had turned the ocean "truck garden" or that a gathering gale makes the inshore water "Baltic blue"?[29]

And photography proves of little use, too, especially along an ordinary seacoast. In Cape Cod, Thoreau remarks that he and Ellery Channing were mistaken for bank robbers merely for their hiking along the beaches rather than along the main roads.[30] Any photographer along this stretch of coast is mistaken for a voyeur, if not worse.

Until well into the 1920s, photographers moved alongshore with increasing freedom. After about 1900, they savored the joys of compact cameras, especially the "post-card" format cameras B. F. Langland recommended in his article of 1918 in *Photo-Era*, "Alongshore with a Camera." Since the "photographer of alongshore scenes frequently has to ferry everything," the hand-held cameras liberate him from the drudgery of carrying large-format, tripod-mounted view cameras over difficult terrain.[31] But as late as 1920, when Raymond E. Hanson and Herbert B. Turner published "A Provincetown Pilgrimage" in *Photo-Era*, many serious amateur photographers still worked with both hand-held and tripod-mounted cameras. The two photographers found much of the sandy shore "windswept, cold and desolate," and while Provincetown itself struck them as "quaint and curious," they needed all their equipment to work in such difficult lighting conditions. Having had the foresight to bring several types of film, the two men struggled to find appropriate combinations of camera and film, with haphazard success. "The panchromatic plate gave a more colorful rendering, but cut out more haze than was perhaps desirable," they concluded.[32] Lighting continuously challenged alongshore photographers, many of whom discovered that they had brought the wrong lenses and film to places difficult and time-consuming to reach. Another photographer, John S. Neary, warned readers of his "Sand-Dunes and the Camera" in 1920 to eschew all inland photographic wisdom concerning camera positioning, work only before ten in the morning and after three in the afternoon, and strive to find "light to transparent shadows." Otherwise, no matter how careful and experienced, the photographer will get only the most contrasty of negatives that make "hard" and useless prints no matter how much attention they receive in the darkroom.[33] Moving the camera into a rowboat or other small craft only worsened problems of lighting and glare, of course, and taxed—and overtaxed—the patience and skill of many photographers, amateur and professional, between 1880 and 1930. In "Vessels as Pictorial Subjects," William S. Davis warned that "much of the work will be done necessarily afloat" and only small cameras will do, although some diehards devoted to tripod cameras will find that "the end of a pier at a seaport or the bank of a commercially important river usually offers the best stationary standpoint from which to photograph passing shipping and anchored craft." But the best shots require a hand-held camera with a shutter set at one-hundredth of a second and a lens stopped down to at least f.11, especially on a sunny day, when the photographer is "not fortunate enough to be able to direct the movements of a private yacht or launch." In such cases, the small "speed" camera compensates for the inability of the photographer to row while composing an image.[34] The outpouring of alongshore photography articles in both photography journals and in such general interest magazines as *Country Life* reflects rapid advances in photographic technology but demonstrates also the rapid shift away from late-nineteenth-century large-format, tripod-camera work to late 1930s snapshot image-making.[35] And it seems that in the late 1920s, holders of the speed camera, or snapshot camera, as everyone called it, became suspect as they aimed their cameras less and less at sand dunes, quaint wharves, sunsets, and schooners and more and more at scantily clad women.

While great ports, resorts, beach clubs, and wilderness shores attract photographers interested in oil tankers, Victorian hotels, film stars, and mating walruses, stretches of ordinary seacoast fascinate few professional photographers and bedevil amateur ones. Most owners of 35-millimeter cameras learn soon that salt air and blowing sand wreak havoc with delicate mechanisms, no matter how carefully they enclose them in cases, plastic bags, and airtight boxes. The salt-laden moisture that makes spectacularly colorful cloud formations sparks corrosion first in electronic systems, then in mechanical parts; and sand, even a single grain of it, scratches film and tears shutters. After the late 1970s 35-millimeter-photography craze ended, almost no one brought a fine camera to the stretch of coast that figures in this book, and the barefoot historian is always conspicuous with his. And often people, even summer people leery of speaking with locals, warn him that he risks damaging his camera beyond repair.

He knows. Three fine cameras met their end in the service of this book. Two 35-millimeter cameras and a splendid medium-format camera, all veterans of nationwide landscape rambling, clicked for the last time at the uttermost edge of landscape. One wrecked in the middle of a winter gale, and the others silently failed on summer days. No wonder the few amateur beach photographers now favor the disposable cameras manufactured by Eastman Kodak and Fuji, especially the ones advertised as watertight down to ten feet below.

Black-and-white or color makes little difference in the end, however, for anyone, especially a man, carrying a camera badly upsets nearly everyone all alongshore. No matter what he photographs, no matter how harmless he looks, no matter that his wife and sons surround him and mark him definitely as a respectable family man, no matter that he is a known local—the historian feels the sting of angry looks. Women, even local women wholly comfortable in their

This **Harper's Weekly** *image of July 17, 1915, hints at how early twentieth-century women learned to fear— and exploit—the male photographer-as-rapist. (Harvard College Library)*

Although How to Stuff a Wild Bikini, *like other "bikini beach" movies of the 1960s, never approached fine art, it condemned the rudeness of the camera-equipped voyeur. (Museum of Modern Art, Film Stills Archive)*

exiguous bikinis, avoid the camera-carrying man, and parents grow protective when he walks by the toddlers who paddle naked in a tidal pool. He might as well be a pirate come to pillage or a shark come to snack. Fishermen and lobstermen scrutinize him, perhaps suspecting he is from some government agency, and harmless, law-abiding clam diggers wonder if they have done something wrong and pat their shorts or swimsuits for their clamming licenses. Boaters wonder if he records some failure to obey rules of navigation, or some illegal equipment or nasty behavior. The camera is impossible to hide, much less use, in a place where almost everyone is nearly naked, and once spotted, even in its case, even in its case wrapped in a plastic bag, the camera alters behavior. The surf fisher suddenly hides a can of beer under his iced bait, the teenage boys charging up the channel throttle back their engine and look innocent, the scampering Frisbee-catcher glances down at her breasts and tugs up her bandeau, the assistant harbormaster hurries a fishing pole out of sight. Everything is changed, especially when regular beachgoers notice the same photographer day after day, and notice that he never photographs the piping plover or least sandpiper, that he never aims his lens at flashy, orange-sailed fiberglass sailboats, and that most of the time he appears to photograph nothing, absolutely nothing, of any customarily accepted value.

So the historian, who knows that he can publish no photograph of anyone without obtaining a written release, stops carrying a camera and tries to notice, to realize, to scrutinize marginality, to see into the glim.

Nearly everything in the glim is faded, dull, soft in color and in tone. At exotic coastal places perhaps everything glistens in vibrant colors, in neon colors, in splendid pyramids of reflective glass and plastic. But here the sun and the wind and the surf conspire to corrode and pit more than cameras. They fade almost everything, often rapidly, and their force is so great that they actually destroy some things, like plastic decks and synthetic-fabric sails. In late May, the first of the locals show up in the glim as bright spots of neon, their new swimsuits glistening wet or dry, and boats slide into the water fresh-painted. But within two months, swimsuits and boats are softer, duller, far diminished in saturation and hue, and in the mist down the beach or just offshore, suddenly as scarcely visible as a long-faded orange life jacket. Washed out, the two-week tourist calls it, speaking of a scene that lacks bright colors and is suffused in too-bright sunlight. But the fadedness, the softness of color glowing but rarely glistening in the dazzle, in the sand-reflected, water-reflected light, deserves sustained scrutiny. What belongs alongshore, at least along the ordinary shore, is always faded, for the coastal realm fades everything, shades everything into the glim. "They were warmer colors than I had associated with the New England coast," Thoreau concludes of the September colors alongshore.[36]

And at first glance in late summer everything seems soft, gentled in color. By September the swimsuits have faded into a softness unimaginable in stores, and the boats have chalked into pastels. The whole marginal zone seems soft, gentle enough for an infant, soothing to any eye. But harshness cruises in the glim, a harshness never, ever soft, a harshness always waiting for any mariner closing the land.

To look from the shore far into the glim is to glimpse it, to recognize the harshness always inshore.

*In a depiction of the
intrepid yachtsman
running far offshore,
the October 1902 issue
of the Rudder illustrated
an experience many
amateur mariners hoped
never to encounter.
(Harvard College
Library)*

C ast and retrieved, then cast and retrieved again, the tallow-bottomed dipsey lead guided mariners into soundings, the shallow water abutting America. Like old Pew, the blind pirate tap-tapping his way along the coast road in Robert Louis Stevenson's *Treasure Island*, mariners felt their way inshore, exploring the edge of the New World, then fishing and bringing colonists, then arriving home from anywhere deeply laden with cargo and passengers.[1] The dipsey lead, a sixty- to seventy-pound weight attached to some one hundred fathom —six hundred feet—of line, tested more than depth. Packed into its concave end, tallow or soap gathered up traces of bottom—sand or pebbles or blue shells—the raw material of interpretation. Before shallops grounded on sand, before boots splashed through wavelets, before Europeans planted flags and claimed dominion, the dipsey lead touched American ground.

Making land began with sniffing the land breeze, the air full of earth smells and vegetable odors that capitivated sixteenth- and seventeenth-century adventurers as it warned them to try the dipsey lead. The massive lead—its name an ancient corruption of "deep sea"—dropped straight down through coastal currents and seaweed to provide definite information about shelving bottom.[2] In daylight and good weather the dipsey lead proved a comfort to masters conning clumsy vessels toward an unknown coast. At night or in fog it became a necessity. "This day we sounded divers times and found ourselves on another bank, at first forty fathom, after thirty-six, after thirty-three, after twenty-four. We thought it to have been the bank over against Cape Sable, but we were deceived, for we know not certainly where we were because of the fog," wrote clergyman-passenger Francis Higginson in a letter dated July 24, 1629. "But perceiving the bank to grow still the shallower we found it twenty-seven and twenty-four fathoms. Therefore being a fog and fearing we were too near land, we tacked about for sea room for two or three watches and steered south east." Volumes of exploration adventure and colonization terror lurk in Higginson's brief sentences. Landsmen learned what mariners knew—safety lay in the mid-Atlantic, however small the vessel. Once in soundings, once inshore, coastal currents and weather might combine to cast the ship onto ledges or rocks. By 1629, the master of the *Talbot* had somewhat useful charts and sailing directions for New England, but the actual landfall worried him as much as any adventurer a century before, and it worried Higginson more. Just as a crewman aloft at the masthead sighted land, Master Beecher sighted surf breaking on what he decided might be uncharted ledges. He ordered out the ship's boat, investigated, and found only a mass of floating seaweed.[3] In soundings every master had not only his usual duty to ship, crew, and passengers but an additional responsibility to study and chart every possible hazard for the safety of those who sailed after. Master Beecher did his duty, to the discomfort of Higginson and other passengers anxious to reach land and frightened of the shallow water in which the *Talbot* moved so hesitantly.

Its long line divided into units marked by strips of different-colored cloth and leather, the dipsey lead helped mariners *feel* their way inshore. As decades passed, word spread about the meaning of gravel or clay stuck to the bottom of leads retrieved from thirty fathoms, of sand and broken shells stuck to leads brought up from twenty-two fathoms. Slowly, fitfully emerged an image of the banks inland from the hundred-fathom line. Mariners groping landward trusted the dipsey lead for its very precision. A full-grown man hauling in its line pulled in a fathom each time he spread wide his arms, grasped the cord, and heaved; a fathom—six feet at a reach—helped him feel by day and by night what the line markings counted—exact depth. And more than one mariner tasted the bottom stuff clinging to the tallow, an ancient custom that endured into the twentieth century among New England fishermen retrieving leads by night.[4] Knowing the bottom helped sixteenth- and seventeenth-century fishermen of Bristol and other European cities fishing for cod on "the banks" of the New World, and it helped the explorers and mariners gunkholing their way to America.

Gunkholing is shoal-water seamanship, not blue-water navigation. The word itself derives from *gunk*, a slang word of Scottish and Irish origin meaning a hoax or a jilting.[5] Nowadays deep-sea seafarers—and yachtsmen—often know nothing of the term, and lexicographers intrigued with nautical usage miss it because they associate nautical language with deep-sea seafaring.[6] But fishers, the Coast Guard, and adventurous "small boaters" conning shoal-draft craft use it routinely to designate the mystery of entering gunkholes.[7] According to Gershom Bradford's *Mariner's Dictionary* of 1952, a gunkhole is "a small narrow channel dangerous to

When James Wilson Carmichael painted *Discovery of America: Columbus Landing on Watling Island in the Bahamas 12th October, 1492, early in the nineteenth century, gunkholing remained a routine activity of exploration. (Mary Anne Stets Photo, Mystic Seaport, Mystic, Conn.)*

navigate owing to current and to numerous rocks and ledges." Moreover, it is "a small anchorage, usually shallow. Gunkholing adds much interest and pleasure for shoal-draft boatmen."[8] This definition perhaps inspired the editors of *Webster's Third New International Dictionary* (1961) to include the word, defined as "a shallow cover or channel nearly unnavigable because of mud, rocks, or vegetation," but to associate it with the term *gunk*, "filthy, sticky, or greasy matter usually objectionably messy or smelly," not with the Scottish and Irish derivation given in the *Second New International Dictionary* of 1934.[9] Modern lexicographers dismiss *gunk* and *gunkhole* as marginal words indeed, and no longer see *gunkholing* as important.

But gunkholing lies at the heart of shallow-water seamanship, and it was the basis for all exploration inshore of dipsey-lead soundings. Once mariners found water too shallow for the dipsey they used the hand lead, a much lighter device—weighing from seven to fourteen pounds—for quickly sounding shallow depths. Sunshine helped lookouts distinguish shallow from shoal water, but on overcast days the hand lead helped explorers gunkhole along—or escape into deeper water. Of course, explorers ventured along the unknown American coast in very small vessels, precisely to lessen their danger from accidental grounding. In 1606 Samuel de Champlain went gunkholing along Cape Cod in a bark of only eighteen tons, and time and again he blundered into shoal water. Three miles from shore, the explorers found only three or four fathoms of water and then, in quickening terror, watched the water shoal into sandbanks. "On going a little further, the depth suddenly diminished to a fathom and a half and two fathoms, which alarmed us, since we saw the sea breaking all around, but no passage by which we could retrace our course, for the wind was directly contrary." The adventurers pushed on, feeling their way. "Accordingly, being shut in among the breakers and sand banks, we had to go at haphazard where there seemed to be the most water for our bark, which was at most only four feet." In hideous surf, with a contrary wind, going "at haphazard" proved almost fatal. In smashing over a sandbar they broke their rudder, and once deprived of steering ability had to find refuge immediately.

With the aid of Narragansett pilot they made an anchorage abreast a small inlet, in six fathoms of water and light surf, and over a bottom solid enough to hold anchor. Just inland, separated from their roadstead by a sandbar, lay a shallow harbor, and at low tide the next day, Champlain writes, "men were sent to set stakes at the end of a sand bank at the mouth of the harbor, where, the tide rising, we entered in two fathoms of water." Staking the sandbars at low tide with long sticks or saplings is boundary-marking not for ownership but for way-finding, for although the snug harbor made a fine place to secure the bark while men went ashore to build a forge and repair the rudder, its one fathom of water at low tide recommended it for little else. "It would be an excellent place to erect buildings and lay the foundations of a State, if the harbor were somewhat deeper and the entrance safer," Champlain decided. But the whole coast "though low is fine and good, yet difficult of access, there being no harbors, many reefs and shallow waters for the distance of nearly two leagues from land." [10] Champlain named the headlands and the worst of the sandbars, but he determined to build no colony around the harbor. Any place with sandbars and shoals from four to six miles from shore could amount to little, for no deeply laden cargo vessel could live long among such awesome hazards, and masters of cargo vessels had little expertise in gunkholing. Once in soundings they dreaded the shallows, and turned for sea room at the onset of dark, foul weather—or the jitters. Champlain knew the immense sandbars were no place for merchant vessels.

Fourteen years later the *Mayflower*, storm-damaged, overcrowded, and heavily laden, made soundings off the eastern tip of Cape Cod and, bound for "Hudson's River," swung south with a fair breeze, directly into the sandbanks that had terrified Champlain. Joyfulness vanished. "They fell amongst dangerous shoals and roaring breakers, and they were so far intangled therewith as they conceived themselves in great danger; and the wind shrinking upon them withall, they resolved to bear up again for the Cape, and thought themselves happy to get out of those dangers before night overtook them, as by God's providence they did," William Brad-

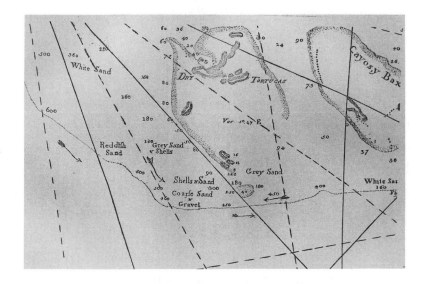

ford recorded. Only the "shrinking wind," God's providence in Bradford's thinking, saved the *Mayflower* from almost certain shipwreck, for sudden veering of the wind enabled its master to put about and sail away from the hazards into which the same northwest wind had driven the French explorers in their shoal-draft bark.

Bradford never forgot the wild moments that followed the first hours of joy, and with good reason. Six weeks after the *Mayflower* found safety in Provincetown harbor, his wife, Dorothy, apparently unwilling to confront further exploration let alone life in a hostile place, drowned herself. Bradford's history of the Plymouth Colony devotes a paragraph to the named shoals and headlands of Cape Cod, among them Point Care, Tucker's Terrour, and Malabar, the last named by Champlain for the "bad bar" in which his little bark damaged its rudder. Most of Bradford's chronicle emphasizes adventure in forest wilderness, but the inshore chaos of hazard-strewn ocean is its permanent backdrop, and a hinge in the narrative, for in that shallow-water wilderness Bradford's married life ended and his life as a colonist began. "And for the season it was winter, and they that know the winters of that country know them to be sharp and violent, and subject to cruel and fierce storms, dangerous to travel to known places, much more to search an unknown coast." That winter, while the *Mayflower* rode at anchor in Provincetown Harbor, the colonists went gunkholing in their shallop, a large rowing and sailing vessel more fitted to searching an unknown coast than the oceangoing merchant ship. Not again did the master of the *Mayflower*—and his crew—go gunkholing on behalf of the settlers. "What heard they daily from the master and company, but that with speed they

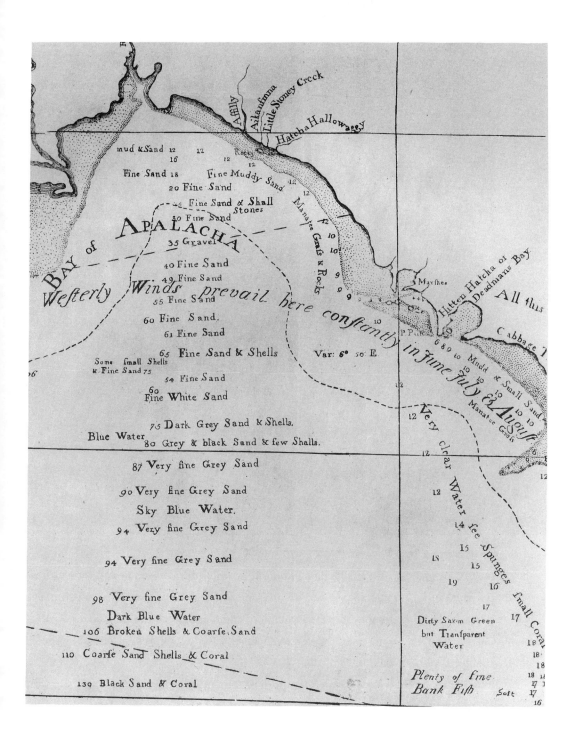

should look out a place with their shallop, where they would be at some near distance," remembered Bradford, "for the season was such as he would not stir from thence till a safe harbor was discovered by them where they would be, and he might go without danger."[11] Plymouth Colony, New England, indeed much of America is only land west of a big gunkhole, sandy, shallow Plymouth Bay.

Accurate charts might have mitigated the peril of the *Mayflower* by instructing its master in safe channels and anchorages, but Christopher Jones had no such charts. John Smith cautioned in 1616 that "the coast is yet still but even as a coast unknown and undiscovered," largely due to charts "so unlike each to other, and most so differing from any true propor-

tion, or resemblance of the country, as they did me no more good, than so much waste paper, though they cost me more." [12] Much of the chart-making problem originated in the simple fact that square-rigged vessels big enough to make the Atlantic passage were too unwieldly to explore shallow channels. Relief and joy at landfall routinely turned to horror. "Thus we parted from the land, which we had not so much before desired, and at the first sight rejoiced," wrote James Rosier in 1605 of a harrowing experience among shoals unwittingly entered, "as now we all joyfully praised God, that it had pleased Him to deliver us from so imminent danger." Rosier blamed his near-calamity on "sea charts very false, putting land where none is," but he might have phrased it differently—"putting deep water where none is." [13] But even the best gunkholers, even Champlain, who carried a shallop for probing the shallowest inlets, confronted something no one took seriously for decades—the moving of the coast.

South of present-day Scituate on Massachusetts Bay, the coast is predominantly sandy and liable to continuous modification by land-devouring storms. Although major headlands and other topographical features retained their shapes throughout the sixteenth and seventeenth centuries, lesser landforms like sandbars and barrier beaches—and now and then entire islands—changed location or disappeared entirely. Not even the "wandering wood" Edmund Spenser created in his *Faerie Queen* in 1596 caused such confusion as the southern New England coast, which wandered even as navigators and colonists wandered along it, charting places that moved from one season to the next.[14]

Most confusing and deadly were the shoals between Cape Cod and Martha's Vineyard and those east of Nantucket, the maze of ledges, sandbars, and occasional great rocks that un-nerved Gosnold, Waymouth, Champlain, and other gunkholing explorers, and among which the *Mayflower*, but for a fluke of wind, would have foundered. Each master sounded and sketched his course among the hazards, and returned to Europe with a crude chart ready for engraving. Later, at end of the eighteenth century, "pilots" brought their sketches to Boston for reproduction, hoping like their forebears to delineate the complexity of shoal-water danger and perhaps profit from their experience. But the ever-wandering sands always defeated the pilots. No matter how accurate their work with hand lead and compass, within a winter or two some sandbar had drifted into the channels delineated as safe, or at least navigable.

On November 16, 1701, for example, the sloop *Mary* anchored "against Webb's Island," just south of Cape Cod, about three leagues off present-day Chatham. For at least a century, mariners had worried about entering the general vicinity, and explorers had tried to chart and name its features. Champlain's expedition, having escaped the "bad bar" it christened "Malle Barre," steered into the shoals, and called the headland "Cape Batturier," meaning a bank battered by waves.[15] As colonists gunkholed about the immensity of shoals, Cape Batturier became known simply as Sandy Point—and known, too, for its slow movement as currents and storms washed sand to and from it. Between 1606 and 1701, an island near Sandy Point acquired the name Webb's Island. The island appears named—and distinctly located—on a chart in Cyprian Southack's *New England Coasting Pilot*, which appeared in the early 1730s, but is conspicuously absent from such detailed late seventeenth-century charts as Gulielmus Hack's depiction of New England waters.[16] Hack's omission of Webb's Island is curious. His chart locates and names such tiny islands as the Elizabeth Islands and Sturgion Isle, and

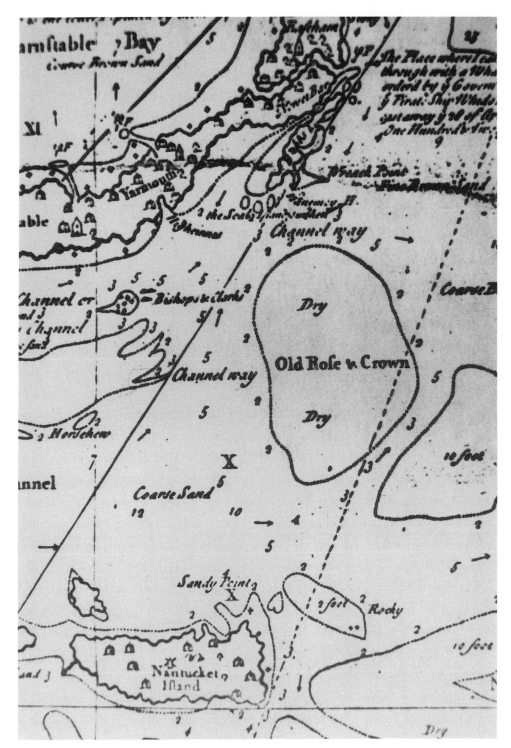

In the early 1730s, Cyprian Southack's **New England Coasting Pilot** located Webb's Island just north of Old Rose & Crown shoal. (Harvard College Library)

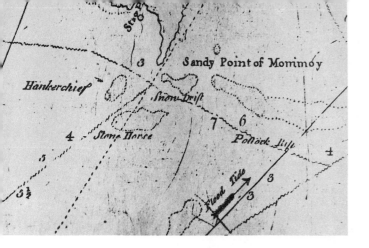

By 1791, Webb's Island
had become the Sandy
Point of Mommoy Paul
Pinkham detailed in
"A Chart of Nantucket
Shoals." (Harvard
College Library)

shoals like Middle Ground and Horseshoe. Along with so many other cartographers of his day, say, like the maker of the map affixed to William Wood's *New England's Prospect* (1634), Hack may simply have focused on the mainland and emphasized prosperous settlements to the omission of small coastal features.[17] But given Hack's attention to shoal-water detail, his omission of Webb's Island perhaps derives from another cause. If its subsequent shape-shifting indicates anything, it is that in Hack's time Webb's Island was simply a peninsula.

Continuous erosion and location-shifting landforms complicated chart-making and turned every coasting passage into potential gunkholing. By the late eighteenth century Webb's Island had utterly vanished. "When the English first settled upon the Cape, there was an island off Chatham, at three leagues distance, called Webb's Island, containing twenty acres, covered with red cedar or savin," remarked a writer in 1791 in an issue of *Massachusetts Magazine*. "This island has been wholly worn away for almost a century. A large rock that was upon the island, and which settled as the earth washed away, now marks the place."[18] Had the island vanished a century before this account, or only sixty years earlier, as Southack's *New England Coasting Pilot* suggests? Did it vanish by creeping toward shore and joining the mainland, or did it simply erode away, over the years, or overnight?

Wherever the *Mary* anchored that night no subsequent mariner could be sure, no matter what chart he carried, no matter what his local knowledge of changing hazards. Only north of Scituate did hazards tend to keep their place. And in many ways the fixed hazards did more damage than the shifting bars of sand.

Rocks stay put, and ledges, often merely exposed bedrock, stay put over centuries. But rocks and ledges forgive nothing. Unlike the sandbar that damaged Champlain's rudder, rocks and ledges destroy what brushes against them, and even the gentlest collision means scraping along barnacle-encrusted surfaces. Colonists quickly learned what mariners knew. Rocky hazards punctured ships instantly, and barnacles flayed skin from hapless swimmers who thrashed toward safety. Better, many decided, a sandbar in all its implacable but smooth solidity than sharp, barnacle-smothered ledges slicing first the timbers of the ship, then the flesh of the shipwrecked.

Rocky hazards combined with uncertain weather to make gunkholing incredibly dangerous, and later to make perilous even ordinary coastal trips. To be sure, explorers and colonists quickly confronted harsher winters than that of Gulf Stream–warmed England, as Bradford reminded his readers in his description of gunkholing toward Plymouth Bay. "The weather was very cold, and it froze so hard as the spray of the sea lighting on their coats, they were as if they had been glazed."[19] But New World weather long surprised mariners and colonists, who struggled for decades to learn the signs, especially the signs far out in the glim, that might lead to reliable forecasting for seafaring and farming. Only as generations passed did mariners and landsmen begin to understand that in New England, and especially along the coast, weather is impossible to predict accurately. Now and then it astonished even the Indians.

Consider the August storm of 1635. Spanish settlers in the Caribbean had learned firsthand of storms unlike any in Spain, but into the seventeenth century many Britons did not even know the Spanish term for wind beyond imagination. And those few who did know it, as Shakespeare did, knew it wrongly, assuming it meant a sort of waterspout.[20] Not for generations after colonization did the English learn something of the diabolism implicit in the Spanish.

Hurricane derives not from the Old Swedish *hurra*, whirlwind, the root word of English terms like *hurry* and *hurly-burly*, but from *huracan*, a Spanish word English mariners now and then corrupted into *hurricano*, about as often as they pronounced it correctly. Spanish conquistadores yanked the word whole from the Quiche, one of the Mayan tribes they overran in 1524, in the territory the newcomers called Guatemala. The Quiche worshiped and feared Hurakan, god of thunder and lightning, of pure storm. Forever after Pedro de Alvarado conquered the Quiche, Hurakan intermittently wrought vengeance on the newcomers.[21] Slowly the English learned that a hurricane was not the puny waterspout of King Lear's raging soliloquy but a storm beyond European knowing, a storm born in tropical force and reaching to northern latitudes, a storm in which no Spanish galleon, no English vessel could live inshore, a storm that brought ocean inland.

John Winthrop, governor of the Puritan colony, had no warning of the hurricane of 1635, for the weather signs indicated nothing to the strangers only five years resident on the coast. Moreover, the vengeance of Hurakan came by night. "The wind having blown hard at S. and S.W. a week before, about midnight it came up at N.E. and blew with such violence, with abundance of rain, that it blew down many hundreds of trees, near the towns, overthrew some houses, drove the ships from their anchors," he recorded in his journal. Suddenly Boston Harbor became not a refuge but chaos itself. "The *Great Hope*, of Ipswich, being about four hundred tons, was driven on ground" and then flung from its momentary resting place across the harbor to Charlestown, where it wrecked again. Winthrop knew nothing of the "eye" of the storm, the momentary calm that Spaniards knew presaged a wind veering to the opposite direction, but by eight in the morning the blast had shifted nearly 180 degrees. So madly did the wind blow that the high tide suddenly dropped three feet, then began to flow again. "The sea was grown so high with the N.E. wind, that, meeting with the ebb, it forced it back again." [22] The tide forced back stunned Winthrop and the other colonists. In English folklore, even King Canute could not command such a thing. No chart, no dipsey lead, nothing human could interpret such storm, but not for days did landsmen learn the immensity of the horror elsewhere—in the shoal water just a league or so off their tumbled landscape.

Three days before the storm two families took passage in a pinnace, a small coasting vessel, bound north along the coast from Boston for Marblehead. Avary Thacher, a clergyman asked to lead the church at Marblehead, brought his family and his cousin Anthony, a tailor, with his wife and children, too—eighteen Thachers in all, plus a servant girl, the master of the pinnace, and four seamen. A few weeks after the storm, Anthony Thacher set down his experiences in a letter to a brother in England. Anthony Thacher knew a different storm than John Winthrop. He knew the storm inshore.

Around ten o'clock at night on August 14, the old sails of the pinnace split in "a fine fresh gale of wind," and the master determined to ride at anchor that night and bend on new sails at daybreak. "But before daylight it pleased God to send so mighty a storm as the like was never felt in New England since the English came there nor in the memory of any of the Indians." So "furious" was the storm that the anchor "came home" out of the sea, and then the anchor cable broke free of the vessel. "Then our sailors knew not what to do but were driven as pleased the storm and waves." Thacher looked about the cabin at "my cousin, his wife and children and maid servant, my wife and my tender babes" trying to comfort each other, then felt the vessel "lifted up upon a rock between two high rocks yet all was but one rock but ragged, with the stroke whereof the water came into the pinnace." Wave after immense wave smashed the pinnace "to pieces." Thacher slipped into the present tense, his prose demonstrating beyond doubt his reliving the horror.

"Now look with me upon our distresses and consider of my misery," he wrote, cataloguing the water "violently overwhelming us," their provisions afloat, and his wife and children about to be "swallowed up and dashed to pieces against the rocks by the merciless waves." Immediately following the strike, both masts went overboard, the forward half of the pinnace sheered off, and the children fell silent in terror as the adults determined to die in love, not panic. But panic did follow, as the seamen, then the master lurched overboard, as Thacher put on his greatcoat, then peeled it off to prop under his baby. The vessel shattered into pieces, and a great wave flung Thacher onto the rocks. From a small hollow in the top of the rock, Thacher, one of his daughters, his cousin, Avary, and his eldest son gathered their wits and tried to encourage the others to join them. Then a single "cruel" wave scattered the remnants of the pinnace. Thacher saw his wife swept past "on the greater part of the half deck," then saw his children drown, then saw the wave that cleaned the rock of everyone but himself— all in an instant. Thacher grabbed a "board or plank of the pinnace" and thereafter was flung from ledge to ledge "a great while," struggling to grab a better piece of wreckage, feeling the rocks once with his right foot, perhaps trying to die. He came ashore nearly insensible, but almost within reach of his wife, whose legs were buried beneath wreckage. "When we were come each to other we went up into the land and sat us down under a cedar tree, which the winds had thrown down, where we sat about an hour, even dead with cold." Then Thacher staggered along the rocks, searching for the others. He found no one.

"Oh I yet see their cheeks, poor, silent lambs, pleading pity and help at my hands. My little babe (ah, poor Peter) sitting in his sister Edith's arms, who to the utmost of her power sheltered him out of the waters, my poor William standing close unto her, all three of them looking ruefully on me on the rock, their very countenance calling unto me to help them, whom I could not go unto, neither could they come unto me." Separated by "merciless waves," locked in each other's eyes, the Thachers died. Thacher's prose quivers with guilt when he writes of "how I had occasioned them out of their native land, who might have left them there," all thoughts that "do press down my heavy heart very much." His is no chronicler's prose, no document written for eyes other than his brother home in England, a brother to whom duty demanded explanation. His letter is as much an effort to ease the pressing on his heart as an effort to inform, however, and its final paragraph is timeless.

Thacher and his wife subsisted on a drowned goat, warmed themselves by a fire, and were

Flood tides carried unwary mariners into the dreaded Horse Shoe shoal south of Cape Cod. In 1786, Thomas Jones warned readers of his "Pilot's Chart of Nantucket Shoals" that tidal currents might lead to disaster. (Harvard College Library)

taken off in a small boat on Monday afternoon. "We went off that desolate island, which I named after my own name, 'Thacher's Woe,' and the rock I named "Avary his Fall,' to the end their fall and loss and mine own might be had in perpetual remembrance," he concluded. "In the island lies buried the body of my cousin's eldest daughter, whom I found dead on the shore."[23] Hurakan had triumphed mightily, close inshore.

Hazards are grave markers, gigantic headstones remembering awesome horror, sudden or drawn-out death. Many tourists lathering on sunscreen while staring into the glim scarcely discern the brownish-green ridges that jut from the blue swells, marked on windy days by far-off spray. But to lobstermen, to fishermen, to locals accustomed to looking into the glim, each hazard has a name peculiar to it, and any one name has a history, always of death. Collamore's Ledge, for example, acquired its name in 1693, when a Scituate coaster commanded by Anthony Collamore struck the ledge and sank off its home-port shore during a brief but sudden December snowsquall. Collamore's clergyman, Deodat Lawson, composed a long, lugubrious poem commemorating the calamity and investing it with providential significance. According to "Threnodia," the storm came up out of a clear sky:

> But dangers great did quickly him surprise,
> The clouds did gather and obscure the sun;
> Winds whistled, snow came thick and sea did rise;
> And he was at a loss which way to run.

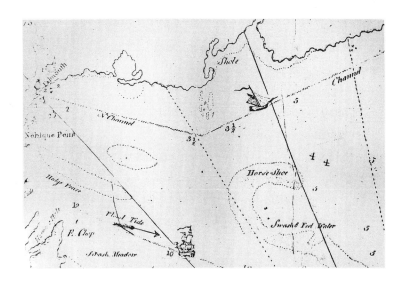

Lawson imagined little. The shipwreck occurred just beyond farms, in full view, apparently, of Scituate townspeople—and Collamore's wife.[24] Inshore, shipwreck happened when mariners could find no sea room, no way to navigate to deep, safe water no dipsey lead could sound.

Thacher's children and Collamore died within reach of land, literally on land, for the great seas smashed their vessels again and again on exposed rocks. Thacher stared at his children as they drowned wide-eyed a few feet away; Collamore must have stared at the dry land so close by. What characterizes inshore shipwreck, of course, is not only the wilderness of storm and sea but wilderness horror in full, close view of onshore watchers perfectly safe from harm. Perhaps it is that heartrending safety that keeps *seascape* a suspect word. Surely it is the summertime sunny-day view that makes so many millions immune to the hazards just offshore, half-hidden in the glim. And the names grow ever more meaningless now, as they started to in steamship days.

Great ships stay offshore, and move toward deep harbors along specific channels. Inshore meant only death to a late nineteenth-century Atlantic steamer, means only catastrophe to the contemporary freighter, tanker, or aircraft carrier. Inshore is a zone abandoned to lobstermen and the Coast Guard, the last dismissed as "shallow-water service" by the blue-water Navy, which gratefully abandoned any responsibility for probing, let alone patrolling it. But among the lobstermen, among the small boaters happily or fearfully gunkholing along with and without charts, the names rise up from charts. The Graves, Roaring Bulls, Hypocrite Channel, and Northeast Grave speak of destruction and death. Old Rose and Crown, named for the beer-mug froth of a seventeenth-century London pub, still grabs mariners, as does Norman's Woe.[25] Lobstermen bark into radio microphones, shouting for Coast Guard rescue of uninformed, too-bold summer sailors bound for drowning in fiberglass sailboats. The word echoes in the farmhouse kitchen. "Off the Salvages, I tell you. . . ."

Salvages, what the firstcomers called the natives, the older form of the word *savage*. Now archaic, now forgotten, the relation between *salvage* and *savage* lingers in the poetry of T. S. Eliot's *Four Quartets* and in the everyday speech of fishermen, lobstermen, and summertime gunkholers. Eliot prefaces "The Dry Salvages" with a hint of explanation—"presumably *les trois sauvages*—is a small group of rocks, with a beacon, off the N.E. coast of Cape Ann, Massachusetts. *Salvages* is pronounced to rhyme with *assuages*. *Groaner:* a whistling buoy." The lines that follow his preface defy explanation, but the relation of *salvage*, something saved from the sea, from shipwreck, and *savage*, something wild, echoes clearly enough: "the whine in the rigging,/The menace and caress of wave that breaks on water,/The distant rote in the

granite teeth." Surely the inshore region is dangerous, so dangerous that an anchored buoy, an aid to navigation, warns mariners.

The groaner heaves continuously, every swell forcing air upward through its whistle, announcing its presence "under the oppression of the silent fog," together with "the tolling bell" of another buoy warning off seafarers while giving some indication of position. At the Salvages, a mile or so from Thacher's Island,

> The tolling bell
> Measures time not our time, rung by the unhurried
> Ground swell, a time
> Older than the time of chronometers.[26]

The rote of the sea, the ceaseless repetition, changes as it strikes the rocks, the granite teeth that snare and eat ships and humans. In 1943, with lighthouses blacked out and enemy submarines patrolling close inshore, the Salvages perhaps reminded Eliot of a world on the edge of destruction. But at any time mariners hear a groaner or bell buoy in the fog, hear the

Given the right conditions of cloud and background, a framework buoy can blend into its surroundings and, when fog rolls in, vanish almost instantly. (JRS)

On an overcast day, even a landlubber can find a can buoy marked "3," and even a novice small boater knows it as a warning not to be slighted. (JRS)

lighthouse foghorn boom out, hear the sound of surf in the rain-swept dark, they know immediately of the savaging of ships, savaging ending in bits of wreckage beachcombers salvage from the waves.

Some yachting and all small boating involve intermittent gunkholing and continuous brushes with inshore hazards. Avery's Ledge, Collamore's Ledge, the Salvages mean much to amateur mariners simply because little vessels stick close inshore, very near the rote of waves on sand or granite teeth. On a summer Saturday, fleets of little boats venture beyond harbors, fanning out to polka-dot the ocean with flashes of red fiberglass hulls and white and yellow sails. But prolonged observation from the edge of cliffs or even atop dunes confirms what even a casual cruise among the fair-weather seagoers makes comically obvious. The pleasure boaters rarely set course for other harbors north or south, but instead cruise about, the sailboats tacking and falling off with the wind, the engine boats now and then speeding up to slam over the rollers, then slowing again to turn back, parallel to the coast. Almost every boat stays beyond the outermost hazards, away from any sandbar or rocky obstruction that might sheer off outboard engine or puncture precious hull. But the boats steer clear of the horizon, too, and rarely move beyond sight of land. When gas runs low, the engine boats turn shoreward, and by four or five in the evening the fleet is heading home, converging from a wide arc of sea into the narrow, familiar channel that leads into the harbor, toward moorings, gasoline docks, boat trailers, chowder. Rising wind or thickening fog hastens the exit from the sea, of course, and calm or evening sunshine slows it, but as darkness falls or the wind rises, the barefoot historian clutching his old telescope always descries one or two latecomers making haste, scurrying across the emptying sea, skirting ledges or sandbanks a little closer than prudence dictates.

Cruising yachtsmen, those bound from Key Biscayne to Bar Harbor perhaps, bring their massive ketches and other sailing vessels into harbor early, almost always under power, not sail. For them the harbor is motel only, a stop along a longer passage, a stop promising a yacht-club mooring, a yacht-club hot shower, a yacht-club dinner, a stop suggesting always the possibility of gunkholing for a few minutes if the government chart is slightly out of date, if a sandbar has crept a few yards further out, if near collision mandates steering out of the channel. But the experienced yachtsman versed in alongshore navigation and the ways of careless motorboaters worries too about the crush of traffic at end-of-day, about the dozens of little motorboats darting landward. So sometimes he or she foregoes the easy if congested safety of the late-afternoon harbor and steers onward in the gathering dark, aiming for landfall after the evening traffic jam. And sometimes, in reasonably gentle weather, the most experienced of yachtsmen fall into the chaos of the hazardous zone.

Contemporary written accounts of gunkholing into terror are few, although tales are common all along the coast. Only the rare mariner writes a permanent memorial of personal stupidity or carelessness, but now and then one writes of bad luck. Such is the account of the last passage of the schooner *Tyehee*, a thirty-six-foot vessel crewed by three men who encountered uncanny confusion close inshore.

John H. Gilchrist's "Wreck at Cape Ann" is as graceful and lively as *Tyehee* on its last passage, and its appearance in *WoodenBoat Magazine* in 1988 derives from its author's endeavor to give meaning to the loss of the schooner and from the magazine editor's fervent desire to instruct. "What is most striking about this whole tale is how easily any of us might find ourselves in similar circumstances," notes the editor, Jon Wilson, "and how little we know about what we might do if faced with some of the same decisions." Gilchrist's narrative is as honest as Thacher's, and perhaps more worrisome, for Gilchrist wrecked on a charted coast, one punctuated with buoys and other contemporary aids to navigation beyond the imagination of the Marblehead-bound tailor. And then again, *Tyehee* wrecked almost exactly where Thacher's little pinnace slammed ashore centuries earlier.

Tyehee foundered on a routine passage between New London, Connecticut, and Gloucester, Massachusetts, a passage wholly uneventful for almost three days, a passage that veered into uncertainty at 1:20 P.M. on the third day, as *Tyehee* skirted Farnham Rock off Plymouth, the dividing point between Cape Cod Bay and Massachusetts Bay. "Our choice was clear: eschew all derring-do by hugging the shore and making for Scituate, barely 10 miles away, or take advantage of a durable prevailing wind and stand for Gloucester 30 nautical miles practically due north." Deciding for "the exhilaration of an extended offshore sail" and planning on arriving an hour before dark, the men disregarded the weather forecast of possible evening thunderstorms, bent on more sail, and set off, steering a compass course beyond sight of land. Things went well, and around 4:30 they "picked up" a buoy and tried to match its identification numbers with those printed on their chart. The buoy, a massive, open-frame "hotel" buoy meant mostly for seagoing ships, proved impossible to identify—rust obscured one of its painted letters, and Gilchrist could make out only an "H," a letter absent from any buoy listed on his charts. The men knew that such aids are frequently replaced and relettered, and that their five-year-old charts meant not much could be done in the way of reliable identification. "Our failure to clearly identify the 'H' buoy unsettled me; running across an uncharted buoy, especially one so imposingly official, is like finding a road sign in the wilderness proclaiming a village that is nowhere to be seen," Gilchrist notes, sliding unconsciously into seventeenth-century idiom. Suddenly, wilderness intrudes, and with it incipient bewilderment. "Of course, without a clear identification, the marker told us nothing," and since the men had not calibrated their electronic fathometer, today's substitute for the dipsey lead, they could not locate their position by ascertaining the depth of water beneath *Tyehee*. Uncertainty had become unease.

And windy weather had turned rainy, and stormier by the hour. With one man hurt by a jibing spar, no foul-weather clothing, and growing doubts over the accuracy of their course, the men discovered that the electric system had failed. *Tyehee* carried emergency oil-burning lamps, great red and green navigation lanterns straight out of the nineteenth century, and an emergency binnacle lantern, too, but the men had not unpacked them, let alone prepared them for use. So, holding a flashlight over their compass, and running totally dark and thus liable to collision, *Tyehee* forged ahead, quite possibly steering straight out to open ocean, bound for Iceland. Then, "heaven be praised"—by this time Gilchrist had begun to think of divinity—"a red flash pierced the blackness like a stage light dimly pulsing behind the scrim. Other lights, probably street or house lights, not as high and not as bright, materialized," and overjoyed, the men took in sail, "fired up the engine," and made for Gloucester Harbor.

Closer and closer *Tyehee* motored, caught in darkness and rain, its crew indulging in "a spell of self-congratulation." Thacher's Island lighthouse appeared on their chart, and in their mind's eye, one of its twin 160-foot-high lighthouses flashed regularly. Everything seemed correct. As Gilchrist makes clear sometimes, and as pulses subtly through his narrative elsewhere, the men wanted everything to be correct, wanted correctness so badly in their fatigue that they disregarded obvious evidence of compounding error. "That the breakwater tower and smaller buoys were nowhere in sight seemed more puzzling than alarming, when in truth it ought to have been more alarming than puzzling." When they could see none of the many lights that clearly illuminate one of the busiest fishing-boat harbors in the nation, they determined that some "evening mist" had blanketed the inner harbor. They had sailed from Farnham Rock for Gloucester, and Gloucester they expected, desired, and had found, but only in their imagination.

"With agonizing suddenness, a fearful crack seized *Tyehee* from below—she shuddered, balancing for an instant upon some unseen object, then nosed down, stern lifted high." Disbelief became horror. "In a flash, *Tyehee's* stately progress in the water degenerated into a senseless pirouette, like a spinning top that has lost its balance." But balance had been lost long before, the whole balance essential to alongshore navigation, to approaching a treacherous coast, to gunkholing at night—which is what navigation with outdated charts, flashlights, and lack of local knowledge becomes. The schooner broke free of the ledge, blundered onward, and struck again, "there to thrash about as if to escape the clutches of a giant hand reaching out of the depths." Within seconds, Gilchrist reports in brutal honesty, "her crew was a study in distraction, clinging to its wits, fending off waves of shock and horror and disbelief, blindly groping for a device, an idea, a plan, anything that had the slightest chance of averting almost certain calamity." He admits he "was transported with fright." The "low granite ledge"—the teeth of Eliot's poem focused on a group of rocks adjacent to the ledges that eviscerated *Tyehee*—destroyed the schooner along with the pride of the men hired to sail it to its owners. Shipwreck here is more than physical. It is the wreck of reason in the clutches of primal force. The men cling to their wits, fight off waves of shock, grope for a plan, a device. Everything constructed, schooner and plan, comes apart. Even the emergency life raft inflates instantly with a tug of its ripcord, then slowly deflates, perhaps punctured by a sharp object on deck, perhaps defective from the beginning.

Tyehee breaks into kindling, vomiting its crew into the barnacle-encrusted ledges. Gilchrist had turned down a life jacket, thinking he could swim better without it. "Such irrationality, as it plainly seems on tranquil reflection, denies the transparent advantages of a life jacket—extra flotation and, in this situation, protection for a torso caught between wave and barnacled ledge." The tale devolves into a nightmare of getting ashore, of finding the tiny island occupied only by an abandoned lighthouse somehow the property of the Audubon Society, and of cold, despairing thoughts of being marooned without food, water, and dry, warm clothing in a windswept bird sanctuary, a deliberate, planned wilderness environment. Later, having finally attracted the attention of a lobsterman pulling pots close among the rocks, the men are rescued from the now fogbound island. "It all seemed beyond imagining, like wakening from a dream so starkly hideous that it displaces all capacity for normal thought." The inshore zone had grabbed another pleasure vessel, wrenching it from the world of gentle cruising into the realm of chaos.

What error led to destruction, to near death? Some hubris in sailing with outdated charts, some twilight-zone carelessness in deserting the rusted "H" buoy? Did poorly marked charts lead the men astray, as they later argued? Or was it desire only? Was it the warping power of the will that made the *Tyehee* crew see the flashing red beacon as the Gloucester Harbor lighthouse? Did they want their destination, their safe harbor, so badly that they saw the flashing red lighthouse as the Gloucester Harbor lighthouse marked on the flashlight-illuminated chart under their fingers? Perhaps so. Perhaps wanting made them see in the red flashes what they wanted to see, though it was not there.

Gloucester Light flashes white. Not red.

Pilotage is a most ancient art, one conducted only by locals, say by the Narrangansett who guided Champlain's damaged bark into the shallow-water harbor. Without a pilot, the mariner bound close inshore, the amateur crew of a sixty-year-old schooner, the fogbound teenagers probing carefully in their parents' fiberglass speedboat—all become gunkholers, for better or worse. Alongshore or coastwise navigation is a gigantic connect-the-dots puzzle, a cruising from one seamark to another, from one known headland to Farnham Rock, from Farnham to Gurnet Light, from Gurnet Light to what? To Scituate Light, or to some mysterious "H" buoy? As the mariner or the summer sailor edges closer toward the rocks and ledges and safe harbors, coastwise navigation becomes inshore navigation on sunny days and clear nights. And in fog or drizzle or driving rain it becomes gunkholing.

Coastal wilderness moves around. Unlike the forest wilderness in which even the tenderest

of lost tenderfoot hikers can stop, collect their wits, blaze a trail—and expect to find the trees stationary—the coastal wilderness incarnates motion. The seafarer moves constantly, sometimes purposefully, deliberately, sometimes in Champlain's haphazard, willy-nilly way. Now and then seamarks shift, go adrift with the wind and tide, and make mysteries of sandbars appearing from blue water. How to connect the dots, when one or more dots move? What of the buoy out of place, the hotel marker free of its anchor and drifting with current and wind, a perfect Judas goat? What of Thacher's unnamed pinnace, what of *Tyehee*, what of any little boat slamming up and down in the rollers, plunging into troughs too deep to give its crew any view of anything but water, charging upward to the crests, there to linger for an instant while its occupants stare madly about?

Madness, unreason, bewilderment in the old sense of *be-wildered*, of lost in wildness, such are the results of navigational failure. "It is of great use to the sailor to know the length of his line, though he cannot with it fathom all the depths of the ocean," mused John Locke in *An Essay Concerning Human Understanding*, his monumental effort to decipher what the mind can know of things that concern everyday conduct. "It is well he knows that it is long enough to reach the bottom, at such places as are necessary to direct his voyage, and caution him against running upon shoals that may ruin him."[28] No wonder *Mayflower* swung back, toward deep water, toward a wilderness of waves perhaps, but of waves only, not other hazards closer than the dipsey lead need tell. There, at least, its master had time to think, to worry about the seamanship required to meet each roller, not the navigation required to close the coast again.

Away from the hazards, in deeper water, mariners know and relish the safety of the offing, the refuge of the *Mayflower* master and all other deep-water seafarers. *Offing*, explained William Falconer in his *Universal Dictionary of the Marine* in 1769, "implies out at sea; or at a competent distance from the shore, and generally out of anchor ground." A prudent mariner keeps not only some distance from inshore hazards but a competent distance, the distance

Hurricanes and northeast gales play hob even in harbors, slamming small boats into unlikely places and reminding locals of the force of millennium storms, those that come once in a thousand years. (JRS)

that permits a wide range of maneuvers under any conditions of current, darkness, snow-squall, fog, and tempest, the range of response that insures safety. While some ships of Falconer's era, perhaps warships well provided with cables and courage, might anchor in ninety or a hundred fathoms of calm sea, Falconer knew that such anchoring posed technical difficulties beyond the capacities of most merchantmen. Few eighteenth-century vessels voyaged with six hundred feet of prime anchor cable, and fewer still carried crew enough to splice and set the hundred-foot lengths of cable anyway.[29] Ships did not anchor out in the offing in 1769 or 1869, and almost never do they anchor in the offing today. The offing exists as sea room, maneuvering space, as refuge from the inshore hazards and from the steep waves passing over any ground reached by the dipsey lead. In the offing mariners feel safe, even in great storms.

Inshore they feel otherwise. Inshore lies shipwreck, the end of the ship, the beginning of the small-boat work long-distance mariners traditionally loathe. "Judge, then, what promise of salvation for us, had we shipwrecked; yet in this state, one merchant ship out of three, keeps its boats," Melville wrote in *Redburn* (1849). "Like most merchant ships, we had but two boats: the long-boat and the jolly-boat. The long-boat, by far the largest and stoutest of the two, was permanently bolted down to the deck, by iron bars attached to its sides. It was almost as much of a fixture as the vessel's keel. It was filled with pigs, fowls, firewood, and coals. Over this the jolly-boat was capsized without a *thole-pin* in the gunwales; its bottom bleaching and cracking in the sun."[30] In Melville's novel, a callous captain determines to do nothing to rescue a seaman fallen overboard, since the jolly-boat would have taken some fifteen minutes to launch. In outraged prose, Melville moves from fiction to diatribe, emphasizing the awesome peril of any merchant ship working inshore, its boats as unready for emergency as its officers and crew are unwilling to consider emergency, its passengers helpless from the moment the ship strikes a hazard, splinters, and begins to fill.

What Melville feared, Thoreau encountered and memorialized in the first pages of *Cape Cod* in a description of the wreck of the immigrant brig *St. John*.[31] The bodies tossed on the sand, the bits of wreckage floating among the rockweed, the hulk in the surf; such are the fruits of gunkholing gone wrong.

But landsmen still think safety lies inshore, or just beyond, in some snug harbor insulated from the gale, in some marina near a good restaurant and a warm inn with showers.

Gunkholing honors its ancient origins. To gunkhole even on a sunny, windless day means getting jilted or hoaxed by what seem fair channels and safe anchorages, and on rainy nights being hoaxed by Judas buoys or some fervent desire for a white-flashing lighthouse. It means venturing inshore, exploring in the seventeenth-century style, threading guzzles, closing the beach.

3 GUZZLE

S and moves. It blows in the wind, shifts with every wave and current, compacts or slides under foot. Sometimes it collapses, as when part of a sandy cliff gives way and cascades down toward the waves gnawing its base. Sand is unstable, even when wet and glistening and shaped into a castle, and its lack of permanence, of predictable behavior, makes novice beachgoers uncomfortable, uneasy.

Discomfort begins in walking. Sand quickly fills shoes, or abrades skin rubbing on sandal thongs. Walking well requires removing shoes, the antithesis of boulevard or backpacking behavior. But then the sand is so often hot, so searing to delicate soles, that midday visitors hop across the dry, shifting white sand toward the cooler sand near the waves, always unmindful that the black-sand beaches of the Pacific burn even hotter. And on days cloudy and sunny, the dry sand, white or black, moves under shod or bare foot, sliding sideways somehow, throwing the walker forward on toes, tiring those unlearned in beach-walking, making many feel ridiculous or fat or old.

Contact with sand is direct contact, often the only direct tactile contact with the earth's surface many contemporary Americans enjoy or endure. Bereft of shoes and almost nude, protected from sun by sunscreen and sunglasses perhaps, but otherwise almost wholly exposed to air and water, the novice summertime beachgoer confronts the sand as both expected and utterly novel. Walking is hard. Sitting is harder. The barefoot historian studying beach-walking does as all other locals do. He scoops a shallow hole in the sand and sits in it, avoiding both hot sand and thigh and back contortions, then watches newcomers spread towels or blankets on the sand and sit hunched against enjoyment. Newcomers, casual visitors, even determined summer-long residents rarely sit on sand—most bathing suits and nylon shorts

As early as the 1920s, farmers and other residents near sand dunes experimented with snow fences in a determined attempt to stabilize dunes bereft of sand-holding grass. (author's collection)

offer too little protection from the hot, gritty grains, for to sit without the so-comfortable shallow depression is to sit and squirm, and squirming works sand between skin and swimsuit. Sand becomes irritant, something clean but infuriating—something invasive.

Wind abets invasion. Walking and running kick up sand that floats along, lighting on recumbent sunscreen-slathered bodies nearby, blowing past sunglasses into eyes that search some fickle horizon, settling into sandwiches and salads. Intense heat conjures up miniature whirlwinds, the sanddevils that dart along the beach for a few moments, enthralling children and angering adults momentarily blinded by a vortex that fills eyes and hair with sand. If the welcome zephyr freshens into breeze, the sand moves faster, almost malevolently. Blanket-types scurry about for stones to make fast the four corners of their civilized patch, then screech as the blanket lifts, blasting sand into clothes bag and picnic basket. Parasols topple and folding canvas chairs tip, minor problems but for the sand riding the wind into the eyes, hair, and ears of would-be rescuers.

And sand settles, too. When the breeze drops suddenly, more than immense box kites fall. Suddenly sand plummets almost straight down, onto the top of the soft-drink can, even through its little hole. It falls into shirt pockets, into camera bags, into ice-cube boxes, into sunglass cases. Its trajectory sharpened by the falling wind, it falls invisibly into the tiniest of crevices.

Sunscreen makes sand stick in ways unknown a few years ago. No matter how blatant the advertisements for greaselessness, freshly applied sunscreen traps sand. Summer visitors not yet accustomed to the beach environment, perhaps accustomed to no natural environment at all, fail to apply sunscreen at home, the way locals do. Instead they peel off pants and shirts, then lather themselves against skin cancer. On windless days their immaculate blankets protect them from the sand, at least until a child runs past. But few days are windless. Grains of sand touch the sunscreen and stick, apparently unnoticed at first, but then felt as irritant— and brushed, but not off. The grains dig and become even stickier, distracting, then infuriating the mind within the lathered, crusty body.

Men and women leap to their feet, swiping at arms and legs as though beset by midges or even hornets, slapping around—but rarely off—the grains of sand. Sometimes the wind rises just as a freshly smeared adult sets off for a walk, driving the sand against skin as sticky as flypaper. Men find the sand moving between skin and the inch-wide waistbands of shorts, and pull aside waistbands momentarily, only to realize that some of the grains have dropped further, into more tender zones. One-piece-suited women brush sand from the tops of breasts,

find that it cascades down cleavage to lodge against stomachs forever rubbing against nylon, and discover what local women know—the briefest bikini offers the greatest freedom from trapped sand. When slapping and brushing fail, some men and women flee the sand, running into the surf and splashing about while rearranging swimsuits.

But surf roils the sand. Most "swimmers" merely wade and splash about, and rarely submerge their faces, let alone open their eyes in the gray murk of sand roiled by gentle waves. The barefoot historian watches many sunscreened summer visitors literally laving themselves in the sea, gently washing off the sand, and apparently eroding the sunscreen, too, for they stride back to dry land, slather on more sunscreen, and begin anew their battle with sand.

Such antics may momentarily amuse those experienced enough to apply sunscreen before arriving at the beach, to wear the briefest swimsuits an easygoing law allows, and to scoop a hole in the hot sand before sitting, but they open on the larger world of quicksand, the sand that is quick, living. In the old prayer-book sense of "the quick and the dead," seashore sand very definitely is quick—with some quicker than other—and inland soil and rock are dead, unmoving, lifeless. Of course, beach sand is almost sterile, unfertile for agricultural planting, and almost unable to sustain the poverty grass and other sand dune–stabilizing vegetation that sometimes hold living sand in place, at least briefly. Inland soil, say loam, sustains life, is fertile to farmer, gardener, and forester, but never does anyone designate it "quick." It stays put, except when extraordinary wind or rain erodes it.

Land that moves vexes more than sunscreened beach walkers, frightens more than mariners closing inshore. It challenges anyone who writes and talks of it, who tries to pin it down in some landsman's way. It makes anyone, especially the summer visitor who has walked far, far down the barrier beach in the hot sun and knows how far he or she must walk back, think of deserts.

Only *desert* properly designates the sand, the dunes, the lack of drinkable water. Drifting, lifeless sand that makes walking and sometimes even seeing difficult, that cannot be shaped into a deep hole or anything else useful as shelter, that drifts inland in great dunes, burying fields and marsh and channels almost as quickly as it buries momentarily discarded toys, fishing gear, and swimsuits, that supports nothing that affords shade to the parched and sunburnt trudger must make anyone—especially the trudger—think momentarily of camels, oases, caravans, the French Foreign Legion. All the sea, all the undrinkable water making the rote, only intensifies the image.

But the image is not particularly old. In the United States, it dates only to the late eighteenth century, when many Americans had forgotten the colonial-era, coast-as-wilderness image unconsciously constructed by would-be settlers desperately weary of the Atlantic passage. Two postrevolutionary events combined to reshape American imagery. On the one hand, the Louisiana Purchase put many educated Americans in mind of the Great American Desert rumored to lay athwart any effort toward Manifest Destiny—or at least across what might be the path to Santa Fe and other settlements in northern New Spain. On the other, growing difficulties with the Barbary corsairs directed attention at the coast of North Africa, and especially at sand as unruly—and unruled—as the pirates who preyed on American merchant vessels. Jefferson's interest in foreign affairs perhaps led him to read Shaw's description of Barbary Coast looming, but many other Americans poised at the margin of the nineteenth century harbored deep interests in North Africa, too, and saw American beaches as the nearest examples of the great deserts across the sea, especially when they walked or rode horseback along them.

"The impression made by this landscape cannot be realized without experience. It was a compound of wildness, gloom, and solitude," grumbled Timothy Dwight in 1802 as he plodded through the sandy wastes of Cape Cod beaches. "I felt myself transported to the borders of Nubia," he continued before cataloguing desert explorers whose books he knew. "A troop of Bedouins would have finished the picture, banished every thought of our own country, and

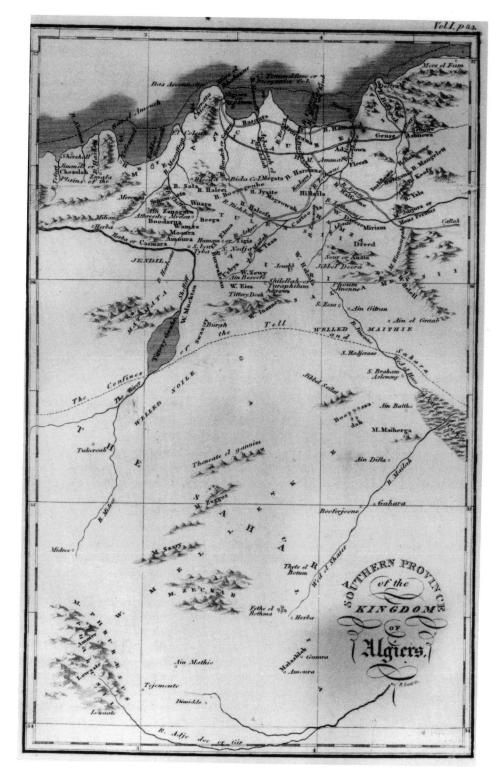

In *Travels or Observations Relating to Several Parts of Barbary and the Levant (1738)*, Thomas Shaw made clear the intimate relation of the African desert and the desert of the North African coastal realm. His point of view shaped early nineteenth-century American thinking. (Harvard College Library)

set us down in an African waste." As the day grew hotter Dwight lost all patience with the sand, condemning it as an "evil" that "spreads a perfect sterility in its progress and entirely desolates the ground on which it falls," and learning that the rising wind brings no relief. With horses "obliged to move slowly and with extreme difficulty," Dwight pushed on, his thoughts turning from African deserts to the desolate areas of Scripture. "Nothing could better elucidate the strength and beauty of that fine image of Isaiah, 'a weary land'; and to us 'the shadow of a great rock' would have been inexpressibly delightful." But on the beach and the parallel road fetlock-deep or deeper in drifted sand Dwight found no great rock affording shade. Indeed, he found no rocks at all, but only sand shimmering in the heat, sand endless in the glare, sand most un-American.

Dwight brought baggage to the barrier beaches, but not simply the baggage in his chaise. He brought all the baggage of an inland American, an American raised east of the high plains scarcely explored in his day, and still alien to most contemporary Americans. "In the mind of an American, frequent forests, and frequent as well as fine groves, are almost necessarily associated with all his ideas of fertility, warmth, agricultural prosperity, and beauty of landscape," he asserted elsewhere in his monumental four-volume *Travels in New England and New York*, an encyclopedic and usually open-minded record of his painstaking scrutiny of his part of the new nation. "Nor can he easily believe that a country destitute of trees is not destitute of fertility."[1] But no matter how he intellectualized the American equation of trees with beauty and agricultural fertility, the beach rammed home the utter practicality of trees. Trees afford not only shelter from wind—and so from blowing sand—but shade, too. Eliminate trees and almost all grass, make the sandy environment stretch on endlessly into the glimmering haze, and parch it with a searing sun, and Dwight shifted images immediately, thinking first—as befit anyone widely read—of African deserts, then, as the day grew hotter and the horses winded, of the desert of Scripture, the wilderness of testing.

Dwight's shift to desert metaphor originates in his reading of African explorers, especially the works of James Bruce, who published his *Travels to Discover the Source of the Nile* in 1790. The Scottish explorer had pushed further and further into the Egyptian desert, finally walking barefoot after his shoes had fallen into ruins, his feet "very much inflamed by the burning sand." North of Wadi el Halboub, he relates, he was "at once surprised and terrified by a sight surely one of the most magnificent in the world. In that vast expanse of desert, from W. and to N. W. of us, we saw a number of prodigious pillars of sand at different distances, at times moving with great celerity, at others stalking on with majestic slowness." Bruce tried to study the immense sanddevils, but emotion overcame him. The pillars moved off, "leaving an impression upon my mind to which I can give no name, though surely one ingredient in it was fear, with a considerable deal of wonder and astonishment." Several days later Bruce again encountered the phenomenon, at sunrise, when pillars appeared in the East, "rays shining through them for near an hour" giving them "an appearance of pillars of fire." As his Greek servants panicked in fear of "the day of judgment," his Arab guides worried that such a display at sunrise presaged a simoom, the violent dry wind of the desert. Bruce quieted his Europeans and trudged onward, searching for rocks to shelter his expedition from the imminent sandstorms.[2]

Shaw, Bruce, and other explorers of African coastal and inland sands discovered the necessity of the burnoose, the loose, cloaklike cloth wrapped about the face when confronting driving sand.[3] But few if any early nineteenth-century American beachgoers affected such sensible attire. Dwight, Thoreau, and other beachwalkers favored the hats and shirts of inlanders, as do most contemporary visitors. Perhaps sunglasses double as goggles nowadays, but when the wind rises on a hot summer day, only a few locals remove their shirts and fasten them about their faces. However much sand invades the eyes, it invades nose and mouth even more, and the local plodding homeward into the teeth of a rising wind values kerchief, skirt, or shirt as breathing filter. Blowing sand at first strikes bare skin as a thousand faint pinpricks, but as the skin numbs the sand becomes mere nuisance. No one, however, not even a local, can long endure it blasting into nose and mouth. The swimsuited walker-into-the-wind longs for a shirt or bandana, lowers his or her head toward the ground, and—if talking—starts to swallow sand—or else turns around and walks backward, staring at moving sand. On occasion the barefoot landscape historian, ensconced in a sheltering hollow at the base of a dune, watches a beach walker have enough of desertlike conditions and run dry-throated into the waves for momentary respite from blowing sand.

And to find quicksand in the shallows? Perhaps, if the tide is coming in and all other conditions are precisely right—or wrong.

Quicksand terrifies inlanders. More than sharks or kraken, jellyfish or undertow, quicksand strikes at the foundation of inland psyche. Land ought not move. Blowing sand, drifting sand, sand sliding about underfoot may irritate to distraction, but quicksand kills. Though it is very rare, it lurks—along with the Great American Desert—everywhere in the literature and oral tradition of westward exploration and is periodically pointed out by high plains and

Rocky Mountain writers intent on demonstrating the longevity of appalling wilderness. Bound up with moral and sexual overtones, quicksand lies just off the path to Manifest Destiny— and just off the interstate highway to Las Vegas.

Captain Randolph B. Marcy of the United States Army set out the reality of quicksand for westward-moving Americans in his *Prairie Traveller*, a detailed how-to-do-it guide of 1863 aimed at farmers intimately familiar with equipment and horses, but not with quicksand. "A man incurs no danger in walking over quicksand, provided he step rapidly, and he will soon detect the safest ground," Marcy asserted of the only sort of quicksand he knew, that caused by rising rivers. Although he only vaguely understood the physics of quicksand, Marcy knew enough to insist that before rivers rise perceptively, their gathering flow lifts sand from the bottom and pushes it up under otherwise dry sand. In such circumstances the dry sand cheats the viewer and can pull down horses and wagons, especially if the horses panic. But someone walking with springy step can probe the sand in safety, pound down stakes to guide the wagon train across, and can help keep the horses moving, especially if the horses drink their fill before they ford the river. Unlike many Americans of his era, especially Southerners, Marcy has little regard for mules, believing them more likely than horses to panic in quicksand: "Mules will often stop from fear, and when once embarrassed in the sand, they lie down, and will not use the slightest exertion to regain their footing."[4] At that point, he counsels, wise settlers will drag the mules out with ropes.

But lying down is exactly what an intelligent animals ought to do. Horses, cows, and other beasts less intelligent than the mule so often maligned by Northerners thrash in quicksand, working themselves deeper, exhausting themselves, and sometimes breaking legs—all the while panicking nearby livestock. In *Desert Solitaire: A Season in the Wilderness* (1968), Edward Abbey describes a modern cattle-driving experience that merely confirms most landsmen's worst fears about the sand that sucks, then kills. "Puddles of quicksand lay ahead of us," he writes of a day gone wrong. "The sand quivered like jelly beneath the cow's hooves, broke open, sucked at the plunged feet. Panicked, the cow struggled through, splashing mud and sand. Safe. As we went on I looked back and saw the holes the cow had made fill up and brim over with water, like suppurating sores." A short while later the same steer mires again, "bellydeep in the soup," no longer struggling but waiting to die. The cowboys rope it, and drag it to dry land, saving its life and providing Abbey the opportunity to describe quicksand.

His description resembles Marcy's, for it, too, derives from the oral tradition of the Great West. "Ordinarily it is possible for a man to walk across quicksand, if he keeps moving. But if he stops, funny things begin to happen," Abbey explains matter-of-factly. "The surface of the quicksand, which may look as firm as the wet sand on an ocean beach, begins to liquefy beneath his feet. He finds himself sinking slowly into a jelly-like substance, soft and quivering, which clasps itself around his ankles with the suction power of any viscous fluid." Abbey explains that it is almost impossible for people to work themselves more than waist-deep in the stuff, that four-wheel-drive vehicles bog down in it, and that the chief danger is dying of thirst and starvation, not drowning or suffocation, and he rounds out his analysis by describing how he rescued a friend caught in quicksand. "Although I hesitate to deprive quicksand of its sinister glamour I must confess that I have not yet heard of a case where a machine, an animal, or a man has actually sunk *completely* out of sight in the stuff."[5] On the surface, Abbey's description is as firm as his understanding of wet sand on an ocean beach.

Of course, an attentive reader might wonder why the steer moving across the quicksand mired not once, but twice. Abbey says nothing about the steer stopping for any reason, say to browse or to assess the situation underfoot. His description of rescuing his friend implies that the man sunk merely because he had a slight limp—after all, Abbey had walked across the same quicksand with no difficulty. Quicksand is a simple enough hazard, at least in Utah, very understandable and not especially dangerous. Only Hollywood makes it into a sort of demonic force, he implies, and although that force scares tourists, wise cowboys push on unafraid.

Coastal quicksand is not the puny hazard dismissed by Marcy, Abbey, and other Westerners but something utterly different for all that its physics remain the same. Unlike its inland relations, coastal quicksand comes and goes with the tides and with currents, and its fluidity rightly terrifies everyone, especially locals. Along this stretch of Atlantic shore so far east of Utah, quicksand lives in its own tradition, one rooted in the coast of the Old World, and steeped in old language indeed.

"There be also here uncertain sands not to be trusted, but ready to catch and swallow, they call them Quick-sands, so dangerous for travelers, while at a low water when tide is past they seek to go the nearest way, that they had need to take a very good heed lest in a going afoot they suffer not shipwreck and be cast away on the land," warned William Camden in his description of Britain in 1610. "But especially about the mouth of Cokar, where, as it were, in a field of syrts or Quick-sands," stood Cokar-sand Abbey, itself a victim of the syrts.[6] As did so many writers in that era of fluid, ever-changing English, Camden used the erudite name for the human-swallowing hazard, and the localism, too. Perhaps the tales about the Cokar sands made him especially anxious to get the designation right, but more likely he aimed his book at readers familiar only remotely, if at all, with the word *syrt*. Throughout the seventeenth century, English writers—and translators—routinely used *syrt* together with *quicksand*. "They discovered the Ocean of the Courts to be all over full of flats, shelves, shallows, quicksands, cargs, rocks, gulfs, whirlpools, sirts," wrote Traiano Boccalini in *New-found Politicke* in 1626 of two neophytes loose in political intrigue.[7]

Only authors certain of learned audiences used *sirt* or *syrt* alone. The King James translators of the Bible rendered Acts 27:17 as "fearing lest they should fall into the quicksands," not as Paul described one harrowing moment in his passage to Italy, "fearing lest they should fall into the Syrtis." But Milton, for example, noted how one "fury stay'd, / Quencht in a Boggie *Syrtis*, neither Sea, / Nor good dry land," making one character in *Paradise Lost* founder onward, "treading the crude consistence."[8] Of course, Milton presumed an audience steeped in classical tradition, and salted with a little knowledge of ocean-borne trade.

Syrtis designates either of two inlets on the North Africa coast—the Gulfs of Gabes and of Sidra—known to Strabo and other classical authors as the location of treacherous quicksands, vast reaches of crude consistence occasionally recalled as the place where, at the beginning of the Christian era, a voice called out that Great Pan was dead. Essentially an immense marshland protected from the sea by a low, sometimes broken barrier beach of sand, and in some places rock, the region presented real difficulties to anyone trying to move anywhere near or in its fifteen-hundred-square-mile area. Strabo suggests that in classical times at least one inlet between sea and marsh proved a navigable refuge from the storms that whipped the shallows and altered tides, but most travel behind the beach appears to have been by horseback or on foot, except when the rainy season made the whole place impassable to small boat and animal alike. Since the marsh contained almost only salty or brackish water, extended adventuring demanded that water be carried, which deflected locals and sorely taxed early nineteenth-century British explorers searching for classical ruins.[9]

British antiquaries saw in the Syrtis region a spectacular hunting ground for artifacts. Lightly inhabited and rarely traversed by local tribespeople, the place might have contained ruins, fragments, and inscriptions from the classical era, and so Frederick William Beechey and others plodded through it. Beechey's opus of 1828, *Proceedings of the Expedition to Explore the Northern Coast of Algeria from Tripoly Eastward*, meticulously describes the quicksands along with the general topography of the region, and concludes that the great barrier beach had risen. Not only had the channel designated by Strabo as the Gulf of Zuca disappeared, but the ports visited by the ancients had crumbled into dust—or sunk into dry or wet sand. As Beechey picked his way from one quicksand to another—at times finding a thin crust of dry sand floating on twelve-foot-deep reaches of water that terrified any horse that stumbled into them—he dismissed the classical horrors as hyperbole, then understood that the entire marsh might be drying as the barrier beach rose against the sea. Just offshore, a British warship poked and probed near the coastal shallows, now and then dispatching a boat ashore to leave heartening notes for the explorers splashing parallel a few miles inland. The *Adventure* poked not for ruins but for cartographic reasons, its captain charged with exploring an especially dangerous, piratical coast charted only two years earlier by the French warship *Chevrette*. Superpower politics necessitated the charting, of course, since both nations had frequently fought each other just offshore, and trade, too, might benefit from good charts. But the legends of Syrtis had frightened off mariners for millennia, as Beechey realized. While searching for ruins, he recorded the remnants of channels and the vanished Gulf of Zuca as matters immediately useful to his king. No mariner, not even Royal Navy men charged with a coast survey, would venture far into the marshes beyond the beach.[10]

Military men detest quicksand, hating it worse than mud and mire. As Beechey discovered, it stops a galloping horse almost instantly, and panics the calmest of horses stepping into it. Neither water nor land, it mocks the efforts of sailors, soldiers, and even marines, turns organized charges and retreats into chaos, and often requires entrapped men to cast off weapons and other equipment in order to save themselves. The Battle of Solway, waged near the border of England and Scotland in the reign of Henry VIII, lingers in military memory as the worst example of quicksand-induced horror. An entire troop of cavalry galloped into a solid-seeming morass, "which instantly closed upon them." William Gilpin, the indefatigable British landscape aesthetician whose determination made him study every sort of landform, ugly or not, rode to view Solway Moss late in the eighteenth century and learned that "the tale, which was traditional, was generally believed, but is now authenticated. A man and horse in complete armor were lately found by the peat-diggers, in the place."[11] Yet quicksand offers powerful protection, and wise defenders seek it.

"But where you find quicksands, quags, and such like, there must you not work much of the foundation at once," cautioned Paul Ive in 1589 in *The Practice of Fortification,* "lest the quags waister you." The wise military engineer obtains "great chalk stone" and carefully paves a platform sunk below grade, in the heart of the quicksand. By mortaring the stones together, it becomes possible to build "the foundation of graveling wrought upon a quicksand," and then to build a standard foundation atop it.[12] Ive knew that such a fortress was almost impregnable, for attackers mired in the quicksands about it, struggled, and died of exhaustion or drowned if the defenders did not shoot them first. Beechey found one such tiny fortress, which he termed a castle, of indeterminate but ancient years, still inhabited, and still almost inaccessible except by twisting paths connecting solid ground to solid ground.

Such a castle just inland from the barrier beach, perhaps adjacent to a tiny estuary opening on the sea, proved a perfect headquarters for the corsairs that preyed on the merchant shipping of the newly independent United States. The Syrtis corsairs intrigued late eighteenth- and early nineteenth-century American men and women of education and affairs, for few missed the symbolic significance of sending the frigate *Constitution,* named for the classics-inspired document at the heart of the new national government, to destroy an enemy lurking in the backwaters of the old Roman Empire. Only the canniest of mariners and military engineers at first questioned the dispatch of deep-draft frigates against shallow-water pirates protected by quicksand and sandbars. But when a shallow-draft corsair lured the frigate *Philadelphia* onto a reef and captured it and its crew, Congress felt the sting of humiliation and appropriated funds for naval officers to purchase trabaccolos locally to carry marines inshore, into Barbary quicksand.[13]

But although African-coast piracy focused the attention of the educated elite on classical quicksand and naval blunders, the typical inland American probably knew nothing more of quicksand than local experience and hearsay, perhaps enlivened by such occasional fictional treatments as Sir Walter Scott's romance *The Bride of Lammermoor* (1818), in which a Scottish horseman rides into coastal quicksands called Kelpie's Flow. Only the appearance of Wilkie Collins's best-selling novel *The Moonstone* in 1868 changed the popular understanding—or misunderstanding—of coastal quicksand and began the still-powerful popular-culture image of quicksand as mysterious and malevolent.

"The sand-hills here run down to the sea, and end in two spits of rock jutting out opposite each other, till you lose sight of them in the water. One is called the North Spit, and one the South." So begins Collins's delicate effort to create a natural place symbolizing the depths of human psyche into which anything, moonstones included, may vanish. From the fir-plantation-covered cliffs above, his characters gaze upon natural horror. Bracketed by two spits of rock that thrust seaward, "shifting backward and forward at certain seasons of the year, lies the most horrible quicksand on the shores of Yorkshire. At the turn of the tide something goes on in the unknown deeps below, which sets the whole face of the quicksand quivering and trembling in a manner most remarkable to see, and which has given to it, among the people in our parts, the name of The Shivering Sand." Avoided by children, local fishermen, even shorebirds, the flats sheltered by sandbanks half a mile at sea are covered not so much by waves but by heaving, silent swells. At every turn of the tide "the broad brown face of the quicksand began to dimple and quiver—the only moving thing in all the horrid place," making one character think of bodies writhing beneath the surface, another—a hardened detective—start with fear, and then a bold young man nearly lose his nerve.[14] The sands claim a young woman, and become a leitmotiv for hiding, secreting, swallowing, stealing.

Collins's phenomenally successful novel—often called the first detective novel—slammed quicksand into the popular imagination, especially into the popular inland imagination, first in England, then immediately after in the United States. Since *The Moonstone* popular writers have used quicksand as though some sinister intelligence governs its motions. Hollywood has only sharpened the image, refining it to include dry-sand quicksand, a geological impossibility, and placing it within a larger genre of sea-monster films. The camel-boy who flounders to his death in David Lean's film *Lawrence of Arabia* is perhaps the best-known example of cinematic stupidity, but a recent popular movie, *Blood Beach*, demonstrates beyond doubt the continuing fascination with quicksand and the visual force of deliberate distortions.[15] *Blood Beach* especially emphasizes the horror of women dying in a place they suppose to be free of danger.[16]

Blood Beach and other films, and pulp-fiction, too, subtly shape the mind-set of many first-time beachgoers, who find in the shining sand flats and bars exposed at low tide, and in guzzles, visual evidence that wet sand might be quicksand, waiting to swallow their children, if not themselves. Anywhere away from the exposed sand stretching seaward from recognized, well-populated bathing beaches, first-time beachgoers move tentatively, approaching tide pools, sandbars, and guzzles with probing feet, often linking hands against the slobbering swallowing that might lurk ahead. About quicksand no visitor says much, but occasionally local boys, aged twelve or so, warn of it just to see the summer folk react. After all, a guzzle is wet, and might be quick.

Nowadays *guzzle* prospers as a verb, but a verb scarcely scented with salt air. Certainly the college kids home for the summer guzzle beer at night, around the driftwood fire hidden among the dunes, and the weekend fishermen guzzle beer grabbed up from ice-filled coolers secured in bilges. As they guzzle perhaps they ponder the odd, recent connection between beer-drinking—and beer advertising—and old coastal terms. Perhaps they wonder why a tall glass of beer is now called a *schooner*, or why noisy drinking is *guzzling*. But probably

they merely guzzle the beer, and think about nothing maritime, not even the guzzles along the beach. Most one-time beachgoers, most inlanders who frequent beaches in the company of other strangers to seashore, indeed, many regular beachgoers who jog or walk with nothing more than a nod or smile at the locals fishing or beachcombing, never learn the old vocabulary of which *guzzle* is a prime component. If the Inuit have a dozen words for different forms of snow, the beach locals retain a few for sand in different configurations.

Along the stretch of coast that shapes this book, *guzzle* is an old term, as old as those of hazards just beyond the outermost breakers, and a very precise one. This sliver of Massachusetts shore retains in that localism an echo of its Kentish settlement in the early 1630s, for *guzzle* is a dialect word, and not even "Kentish," but peculiar to the coastal realm of Kent, a realm of sandbars, dunes, sandy beaches, and shifting inlets. Thoreau understood its meaning, for he uses it correctly in *Cape Cod,* wondering about a whale "dragging in over the bars and guzzles."[17] Along with so many other American topographical localisms, it is long gone even from the finest unabridged dictionaries, and only local, usually oral, usage keeps it vigorous.[18] People who love beaches, especially those who fish the surf, amateur naturalists, runners and hikers, shallow-water small boaters, and the silent men who begin their mornings nursing cups of coffee as they stare out into the nothingness beyond their pickup windshields all may know and use the term. But they use it either among themselves, among people who know the immediate usefulness of the rich vocabulary of coastal topography, or they use it tentatively when addressing an inlander, usually in giving directions.

Any careful reader of nineteenth- and even early twentieth-century topographical description immediately encounters *guzzle,* but without period dictionaries the reader is likely to stumble. Thoreau knew guzzles as linked with inshore sandbars, for example, and George S. Wasson, in *Home from Sea* (1908), knew the term as identifying a safe but shallow harbor: one of his characters tells how he "followed a fisherman clean up into this 'ere little guzzle-hole."[19] Wasson, along with his local-colorist friend Sarah Orne Jewett, strove to record not only the sound of Maine-coast English but its peculiar terminology, say a "flirt" of snow, more than a skift, more than a flurry. Weather terminology meant much to local fishermen, lobstermen, and their wives, who understood the ramifications of the slightest changes in local weather conditions, and nowadays the ship-to-shore radio channels still sound with a meteorological jargon as old and precise as that of beach topography, a jargon Brocklesby would recognize instantly. Yet without some firsthand understanding of the vocabulary, the words echo as oddly pronounced mysteries, perhaps even vague threats. After all, might not a guzzle swallow up the family vacationing from Kansas?

A guzzle is simply a low spot, usually on a barrier beach, over which the sea now and then flows into the salt marshes inland from the dunes. Usually a guzzle floods only during a severe onshore storm, and then often only if the tide is high, but often, too, its sand remains wet even on a bright July afternoon. Sometimes tiny rivulets of water course along its middle, and now and then a shallow pool forms, attracting toddlers into its inches-deep warmth. But a guzzle can be other things, albeit similar ones. It may designate the very shallow channel through a sandbar exposed at low tide, a channel usually navigable only by rowboats or other very small craft—or by a dying whale thrust landward by the surf, as Thoreau implies. And it may be a piece of land well inland, but in reach of rare great storms, those expected once each century or so.

Such a guzzle lies at the foot of a cornfield, well inland—and downhill—from great cliffs that face the sea. Half a mile inland from the cliff, past expensive summer and year-round homes, then (slightly lower) past some new condominiums, the walker heading away from the sea encounters a golf course running down to a tiny drainage ditch, some scruffy trees, and a poorly paved road that crosses a diminutive culvert. Then the walker starts uphill, perhaps wondering how a cornfield first planted in the 1630s endures surrounded by upscale resort architecture. Nothing in the scene hints at the sea, for the sea is obscured by rising land and by mature juniper, beech, maple, and pine trees, and what water lies stagnant in the ditch is not even brackish. To be sure, the air often smells of the sea, and during great gales the sound of the surf striking the outer ledges, and the base of the cliff, booms everywhere, but nothing maritime marks the place. And all that is odd is the cornfield, lingering ghostlike from a colonial agricultural past.

It lingers because the ditch, properly called a gutter, opens into secrecy, as all gutters did, long ago. Once *gutter* connoted more than a narrow, hand-dug ditch across a field or a trough at the edge of pavement or roof. Once *gutter* meant a secret passage, a trace of which meaning

Now and then locals make a guzzle, usually to flush some inland salt pond. (Photograph by Barbara Norfleet)

survives in contemporary cracks about people whose minds are stuck in gutters. Fresh, stagnant, or foul, gutter water often flowed in secret, and secrecy makes foulness worse.[20] Following the gutter that separates cornfield from golf course proves difficult, although the straight course of the ditch hints its built origin as it helps the trespassing historian, for the gutter leads straight into dense undergrowth, largely poison ivy. In either direction the gutter reaches inland, toward salt marshes in from barrier beaches. The gutter, ordinarily a drain for spring rains, is the very spine of a guzzle, a guzzle so old, so rarely flooded, that only local agricultural memory recalls the November Gale of 1786 or the subsequent gale of October 1829.

In 1786 the sea roared through the gutter. Connecting Scituate Harbor and the salt marshes to the north with the immense salt marshes to the south, the surge carried an entire gristmill across the cornfields. The memory of that storm endures in oral tradition better than the later storm, perhaps because farmers remembered the loss of the gristmill, perhaps because the November Gale came with blinding snow. The November Gale reinforced vague notions of digging a ship canal across the great guzzle, thus connecting the estuary salt marshes with the harbor, and in June 1829, Congress appropriated $15,000 for a survey.[21] Despite vivid evidence three months later that the sea might flow through the guzzle, federal authorities decided against the canal. Subsequent great storms did not flood the guzzle, apparently because wind and tide did not synchronize perfectly, and even the hurricane of 1938 left it in peace. But tales of where the sea had once reached kept even condominium-builders away from the cornfield.

And the night before Halloween, in 1991, the tales brought a handful of locals to the gutter, locals already awed by the immense surf churned up by an odd gale somehow stuck far offshore, locals wondering if abnormally high tides might combine with the odd storm to make the gutter flow. In the gathering dusk, raised arms shielding eyes from the first stinging drops of wind-driven rain, the locals watched the black water rise in the gutter, spread outward toward cornfield and golf course, and begin to surge higher and higher. From the catbrier, from old pines and oaks and chokecherry came the sea, crashing from inland, from landward, following the salt marshes to the guzzle only the locals, and only the knowing locals, knew as a guzzle, crashing and roaring, covering the culvert, reaching toward upland. Without talking, the men backed their trucks further from the salt water that surged near the feet of the women, then walked back to the guzzle. As darkness closed over the water, rain started to mix with the sand already whipping inland from the dunes disappearing before the gale. Dry sand had become quick, too.

It had before, of course. The first colonists mistakenly grazed their few cattle on the meager grasses holding the dunes in place, and watched in mounting amazement as the dunes began to move, to fill old guzzles with dry sand, and to open new. Before the near-constant onshore wind, the sand drifted across the new barrens made by colonists who felled trees just inland from the dunes, and sometimes the wind scoured the sandy soil from among the stumps, driving it further inland. Though they wrote little about the spectacle itself, the settlers quickly forbade grazing and tried to stop the loosened sand, sometimes by heaping up seaweed, less often by planting grass. In the desperate larger effort to clear forest and build houses, however, staying blowing sand received scant attention, for aside from the grass, sandy beaches at first glance offered nothing much else of use. Not until well into the eighteenth century did coastal towns now and then regulate the use of beach materials. One town meeting voted in 1765, for example, "to choose four persons to take care of the town's beach, in order to prevent other towns' people carrying off sand, rockweed, and clams, and trash, or other rubbish off said beach."[22] By then locals knew the many uses for beach resources, and needed several. Sand ballasted ships, mixed with crushed clamshells to make mortar, and lightened clayey soils, while rockweed and seaweed made excellent fertilizer and, when dried, crushed, and mixed with broken clamshells, an effective house-wall insulation. Clams supplemented the diet, of course, and "trash, or other rubbish," refers chiefly to driftwood, which by the late eighteenth century helped many owners of half-exhausted woodlots eke out their house-hold fuel requirements, but trash included stranded whales and blackfish, too, whose rare appearance enabled townspeople to obtain oil and fertilizer without charge.[23] Such conservationist thinking came late, however—after seventeenth-century calamity.

By the middle of the eighteenth century, blowing sand had blocked shallow channels behind barrier beaches and begun engulfing woodlots, cropland, and even villages. The early nineteenth-century geographer Justin Winsor noted the effects of colonial timber exploitation on a sandy island in Duxbury Bay. "Clark's Island in some parts possesses a mould, which if equalled, is scarcely surpassed in the county," he wrote of an eighty-six-acre "garden" in 1849, but "of its original forest of red cedars, only three decayed trunks now remain."[24] A great barrier beach sheltered Clark's Island from the violent wave-generated erosion that destroyed Webb's Island, Slut's Bush Island, and other small islands off Cape Cod, but sand *blowing* from the dunes to the east endangered it nonetheless. A neighboring island called Saquish had become a peninsula of Duxbury Beach some time in the late eighteenth century—having first become connected by a guzzle covered by ordinary high tides—as blowing sand filled in what had been a deep channel. "The people of Duxbury, of the last generation, used to go there in small lighters, enter the creek, take off wood, pass thru' the creek between the Gurnet and Saquish, and proceed to Boston from that outlet, then sufficiently open for that purpose," recorded a local jurist in 1829 of a channel too insignificant for eighteenth-century chart makers. "Time has blown in sand banks and choked it and the very men who used to pass thru' it, in their old age could easily describe the precise place."[25] A judge had the best access to period description of the havoc wrought by careless woodcutting, for deeds and other seventeenth- and eighteenth-century land records offer the best insight into the gradual—and sometimes sudden—blowing away of dunes and the burying of inland acreage. But the deeds make little sense to the contemporary reader, for often the sea claimed lots first smothered by blowing sand, and in the absence of landmarks, surveyors' descriptions, and the land itself, ordinary methods of landscape reconstitution fail.[26] Deeds become as false, as misleading as any of the waste-paper charts John Smith followed along the coast of New England.

Blowing sand intrigued thoughtful observers, especially the educated elite fascinated by Barbary pirates, weather patterns, geology, and local history. As early as 1668, an English writer described the blowing sand of Suffolk for readers of *Philosophical Transactions*, the journal of the Royal Society. Thomas Wright blamed injudicious plowing of sandy hills for the extraordinary dunes that marched across the fields near his home, in time burying the

houses of the nearby village. The great dune "met with this advantage, that till it came into this Town, all the ground it passed over, was almost of as mutinous a nature as itself, and wanted nothing but such a companion to set it free, and to solicit it to this new invasion." Clearly the scouring of the sand killed field crops, exposing the sandy soil to the "impetuous blasts" of prevailing winds, and so increased the amount of blowing sand. Not only did the weight of sand crush buildings, the volume actually choked three miles of the Thetford River. Given wealth and leisure, Wright experimented with stopping the sand, finding that planting fir trees worked but raised sandbanks twenty feet high around his house, that putting hundreds of loads of muck atop the sand stabilized it in small areas, and that eight- or nine-foot-high walls protected his inmost acreage.[27] Wright worried as much about the exact process of the phenomenon as he did about stopping it, however, and his letter typifies the inquiries made by educated Britons through the subsequent century.

Sand-study once complemented aesthetic inquiry. Just as Gilpin traveled to examine quicksand because he felt obligated to place quicksand in a larger frame of landscape aesthetics, so other observers integrated geological musings with appreciation of scenery. In an essay in the Royal Irish Academy *Transactions* in 1797, for example, William Hamilton combined antiquarian insight with geological surmise, probing at the conjunction of subjects that intrigued Beechey in the Syrtis region and puzzled so many other eighteenth-century explorers of seashore. Hamilton found "vestiges of ruins, traced with difficulty amid heaps of barren sand," near the mouth of the Bannow, in the barony of Forth, and tried to account for the destruction of a village recorded as flourishing in 1626. Further along, he encountered a recently abandoned manor house, half engulfed by the blowing sand that had already buried twelve hundred acres. "At present every object in this place presents to view peculiar charac-

Postcard views of the 1930s frequently depict moving dunes as romantic or almost gothic, never as the product of failed ecosystems. (author's collection)

ters of desolation," he wrote of an eerie, empty place disfigured by abandoned formal gardens "totally denuded of trees and shrubs," by garden walls knocked down by giant dunes, by sand that filled the first-floor rooms of the great house. Everything artificial had failed, giving "free passage to this restless enemy of all fertility," and as far as his eye ranged, an entire agricultural landscape had been "reduced to one undistinguishable scene of sterile uniformity." Hamilton had walked into something for which classical education had scarcely prepared him, modern ruins. He understood the correct response if "embattled walls or marble piers should start up amid the sands, suggesting ideas of ancient elegance and festivity," but could only grope toward an appropriate response to new things newly buried.[28] A decade earlier, while hiking along the west coast of Donegal, he had encountered a man failing where Wright had won, a man about to move his house further inland, away from acres upon acres of moving beach sand, and perhaps that encounter made him wary of jumping to aesthetic conclusions about the Bannow scene. Not only had the sand conquered a vast area, but it had done so incredibly quickly. Nothing of antiquity, scarcely anything of age gentled the scene before him, and new things jutting from sand jarred his antiquarian aesthetic, forcing him back to the old agricultural attitude that sterile sand is ugly because infertile.

Eighteenth-century educated observers knew, moreover, that greater historical and even theological issues might be implicit in any examination of areas of moving dunes. Even as the wind swept sand ahead to bury fields and houses, it scoured sand from below grade, revealing all sorts of troubling things. European observers quickly discovered revelation upon revelation beneath the sand, all suggesting immensities of time. Scholarly journals and literary periodicals began publishing descriptions of ancient forests uncovered by blown-away coastal sand, descriptions that helped confirm geologists' theories that the earth had a deep, prebiblical history. Some geologists studied safe topics like the interactions of water-eroded guzzles and wind-blown dune sand, but others moved from studying the burial of coastal villages by moving dunes to the cosmological implications of fossils in blown sand.[29] More than any other geologist, Charles Lyell synthesized reports of European coast walkers and African desert explorers, not only clarifying something of the extent and process of blowing sand but suggesting its historical and prehistorical significance. As his three-volume *Principles of Geology* appeared between 1830 and 1833, educated Britons and Americans learned that although some geological processes occur over immensities of time, others are quick indeed. And some of the quickest, Lyell pointed out, could be found in the sands of any coastal realm, the realm that enabled Thoreau, and other well-read Americans, to see history through the prism of European geology. When Thoreau crafted his "The Sea and the Desert" chapter of *Cape Cod*, he wrote under the spell of Lyell's synthesis, as brief comments like those about "the moving sand-hills of England" and swallowed-up villages "on the shore of the Bay of Biscay" reveal.[30] For Dwight, Thoreau, and their lesser-known successors, geology moved as quickly and as disturbingly as any Cape Cod Bay dune.

However well-read in the nature of quicksand and North African geography, antiquities and sand dunes, guzzles and geology, the educated Americans rode or trudged through sand familiar to locals unschooled in higher philosophies. Locals persisted in discounting sandy beaches, in maintaining the old agrarian prejudice against sandy soil. An editorial in the *Farmer's Cabinet* "On the Nature of Soils" insisted in 1836 that sand, "from the looseness of its texture, admits heats too freely, and is not capable of retaining a sufficient degree of moisture for the purposes of vegetation." Essentially, the editor classified soils according to the medieval four-element philosophy favored by the first colonists, concluding that "all sands are hot and dry" and that "no plants will come to maturity on mere sand, except such as extend their roots very deep, and attract nourishment from a stratum below it."[31] Two centuries earlier, a newly arrived settler wrote home about New England soil in almost the same terms, dismissing much of it as sandy. "The soil, it is, for the nature of it, mixed; the upland rather participates of sand than clay, yet our rye likes it not, an argument it is both cold and barren."[32] Hot and dry or cold and barren amounted to the same verdict, "useless for crops," and the settlers moved inland rapidly in search of British-like soils, particularly alluvial loams. Those who stayed behind on the sandy soils just in from the dunes turned quickly to livelihoods other than farming, but they never lost the ancient soil-classification attitudes or vocabulary, as Thoreau learned.

In a cryptic chapter of *Cape Cod*, Thoreau recounts his overnight visit in the home of a Wellfleet oysterman, an aged man no longer oystering but still keeping up his fragile agricultural operation. "When I asked what they did with all that barren looking land, where I saw so few cultivated fields," Thoreau writes, the oysterman answered, " 'Nothing.' " Thoreau then asked why he bothered to fence his fields, and the man replied, " 'To keep the sand from blowing and covering up the whole.' " Moreover, the oysterman explained, " 'The yellow sand has some life in it, but the white little or none.' " Thoreau encapsulates the typical nineteenth-century collision between well-educated beach visitors and unlearned locals. The oysterman patiently explains old-time surveying techniques useful on land smothered in dunes, tells of horses drowned in a guzzle and a ship breaking up on a sandbar just beyond the dunes, and proudly shows off his half-acre garden planted from seeds washed ashore from a shipwreck. And he explains what he calls " 'the rut,' " the "peculiar roar of the sea before the wind changes," something important to a man gardening in a coastal hollow. In the end, as "he loomed so strangely, that I mistook him for a scarecrow" in the middle of a cornfield, Thoreau calls the oysterman a "wizard," and unknowingly reaches the limits of book-learned seashore visiting. In remarking that "the desolate hills" between the oysterman's house and the shore "are worthy to have been the birthplace of Ossian," Thoreau stumbles badly, citing the great fraud of nineteenth-century English poetry as a sort of window on the plain-speaking oysterman snug in his sandy hollow, a hollow the oysterman describes in antiquated terminology, in words so old they might have been salted.[33]

What of the wizard's remark about fencing fields, for example? Surely he set up no walls like the British country-gentleman Thomas Wright, for after all, Cape Cod lacks rocks. The oysterman's fences could only have been made of driftwood, and likely did little to confine his single cow. But as any local knows, fences make wind-borne sand behave in the oddest pos-

sible ways. Nowadays concerned townspeople erect wooden snowfences to contain wayward dunes—the old custom of embedding Christmas trees withered when ornithologists discovered that the branches impede the nesting of piping plovers and other birds—and puzzle summer visitors, who wonder why the sand does not escape the inchwide openings. But like rural children who know that the swimming hole forms always *below* the dam, beach people know that a few horizontal rails or vertical slats may be enough to distort the path of windblown sand, to send it scooting upward or down, or actually stop it. Even a rope stretched low to the sand between two posts can work wonders, usually unpredicted, against the sand, and a bit of fence combined with poverty grass and other indigenous grasses can stabilize large areas of sand. In the nineteenth century few geologists, and fewer generally educated walkers like Thoreau, had even begun to guess at the forces channeled by a simple fence, and nowadays the forces remain as scarcely understood as looming in Jefferson's time, in part because experimentation is little practiced in such fragile ecosystems, in part because natural vegetation—beach pea, goldenrod, bayberry, beach rose, and even pitch pine—can hold the sand if left unmolested, in part because few geologists care to recreate the sand-channeling fences of Thoreau's era.

THE BEACH, DENNISPORT, MASS.

Farmers still care, of course, although few "saltwater farms" endure in a coastal region of spectacularly high land values. But farmers and gardeners remember the accidental lessons first learned in the 1830s, when Cape Cod farmers, determined to do something with sand-covered, nearly sterile acreage, first planted pitch-pine seeds "to protect the lands within, and to prevent the sand from blowing and forming extensive dunes, like that in the center of Wellfleet, and other places upon the shores of Cape Cod," in the words of one Massachusetts farmer remembering in 1864 three decades of hard work. "Pitch pine is now considered as certain a crop as can be planted," continued S. B. Phinney, addressing farmers assembled from everywhere in Massachusetts. "The days of *experiment* have passed."[34] Between 1832 and 1864, coastal farmers experimented not only with pitch-pine seeds but with seeds brought from Norway and France, and they determined that local seeds sprouted better and made stronger, tougher trees. Nowadays their efforts are everywhere along the sandy coast, although sometimes nearly buried if a particularly nasty gale has forced inland vast mounds of sand. Pitch pines, that species so wonderfully adapted to salted wind and sandy soil, stand as essentially the only enduring designed feature of the sandy realm. Everything else alongshore people have tried gets buried sooner or later.

To be sure, many continue to experiment. Anyone who lies on a sandy beach can experiment if the wind blows hard, and many do, in desperation. People curl up together, children hunch down under blankets, grown men seek shelter behind coolers. And sometimes the shelterers notice the patterns made around them by deflected sand, although often only children start to admire the designs, then try to alter them. The oldest trick brings to mind the swimming-hole dam, for anything solid and vertical, say a full soft-drink can, creates sculpture about it, especially if placed away from other wind-shaping objects. Almost invariably, always if the wind holds true, the downwind side of the object appears to sink, perhaps because sand builds up on the windward side, perhaps because a sort of Venturi effect scoops out the downwind sand. Taller objects make concentric rings, or arcs, downwind, the ridges of sand distinct but hard to explain. Coolers, cans, oars, even pocketknives shoved blade-down in the sand create wind deflections usually evident in sand patterns, making the sand somehow surround the object as the wilderness surrounds the jar Wallace Stevens sets down.

*In this **Life** drawing of 1901, Charles Dana Gibson depicted his "Gibson Girl" and friend oblivious of the incoming tide that floods the beach. (Harvard College Library)*

I placed a jar in Tennessee,
And round it was, upon a hill.
It made the slovenly wilderness
Surround that hill.[35]

A jar, not a Grecian urn or other classical object favored by eighteenth-century beach-walkers like Hamilton, shapes the sand, for anyone who chooses to experiment, to wonder at sand alive.

But few wonder. Instead they swipe at sand, abrading their sunscreen-covered skin, irritating the soft white flesh beneath bathing suits, knocking sand from hair to shoulder to waist. Only a few summer visitors pause to wonder where the sand goes when they knock it free, and in an era of dune-protection signs, fences, and behavioral codes, few walk through the dunes and pitch pines to the salt marshes that separate the outer barrier beaches from the upland to the west. Beaches attract the tourists and rare photographers who follow Dwight and Thoreau, swipe at the sand, and hippety-hop across guzzles-that-might-be-quicksand, but marshes attract almost no one. Marshes lie low in late twentieth-century landscape thinking, for they rise not from sand but from mud, mud that makes sand, even whirling, North African sand pillars, seem gentle, clean, almost friendly. And mud opens not on guzzles but on salt creeks, narrow, sinuous watercourses that might mask quicksand, that flow over thigh-deep mud that terrifies first-time steppers, that flow through immense seas of light-green grass standing shoulder-high against the horizon. Nothing in American culture, not even tales of great explorations in the tall- and short-grass prairies of the Mississippi basin, prepares summer visitors for a ramble in the marshes—or a thoughtless paddle into the maze of salt creeks.

4 SALT MARSHES

Nowadays only locals distinguish the usefulness of different sorts of marsh grasses. The fine grass makes far more nutritious hay than the coarse, taller grass that surrounds it. (JRS)

Chartreuse photographs badly. Even under ideal conditions, the most experienced photographer risks bizarre results when using contemporary color film. Film chemistry simply misses chartreuse, as it does other colors. In Reference Sheet E-73, the Eastman Kodak Company warned photographers in 1985 that some colors, "such as shades of chartreuse, lime, pink, and orange, may not reproduce so well. It would be possible to design a film that would reproduce these colors better, but then some of the more important colors would suffer."[1] Lime, pink, and orange, perhaps unimportant inland, blossom everywhere all alongshore. Bikinis alone sometimes constitute the only artificial color along a whole stretch of beach, and many navigational aids—and the long slash on every Coast Guard bow—depend on orange for visibility in bright sunlight and heavy rain. Bright "neon-lime" bikinis and "day-glo-orange" buoys photograph badly, but not so badly as chartreuse bikinis, or salt marshes.

Spring turns the salt marsh into acres upon acres of chartreuse grass, or more properly chartreuse-yellow grass. Through the 1930s, Americans distinguished among several related colors, all themselves related to a now almost lost range of liqueurs loosely called chartreuse. Chartreuse-yellow, named for the sweetest of the liqueurs first made by monks at La Grande Chartreuse, indeed properly designates the dominant springtime color of the marshes, but as the season advances, the grass moves through its own spectrum, from chartreuse-yellow to chartreuse to chartreuse-green. According to *Webster's Second Unabridged Dictionary* (1934), chartreuse is properly "a color, yellow in hue, of medium saturation and high brilliance. It is less red and of higher saturation and brilliance than canary yellow." Kodachrome, the color transparency film introduced a year or so after the definition appeared, perhaps had the long-term effect of eroding the color vocabulary of educated marsh-walkers, and perhaps caused the truncation of dictionary definitions of color, too.[2] While Thoreau delighted in the soft autumn colors of Cape Cod plants, "the incredibly bright red of the Huckleberry, and the reddish brown of the Bayberry, mingled with the bright and living green of small Pitch-Pines, and also the duller green of the Bayberry, Boxberry, and Plum, the yellowish green of the Shrub-oaks, and the various golden and yellow and fawn-colored tints of the Birch and Maple and Aspen," let alone the colors of the sea itself, subsequent visitors tend to make photographs, and eschew verbal description of colors.[3] Of course, what strikes contemporary readers is perhaps Thoreau's easygoing identification of species, not his keen eye for color and his graceful determination to record hues in words. Yet in a salt marsh, or on its edge, say when looking down from a hilltop or even a dune, the contemporary observer can scarcely rely on color film, for except in winter, salt marsh is a remarkable palette of green, and the chief green is chartreuse. And chartreuse photographs wildly, as any film-processing laboratory knows.

Yet even photographers enamored of black-and-white imagery rarely approach salt marshes, at least with landscape imagery in mind. Marshes may attract wildlife photographers, especially photographers fascinated by nesting birds, but they prove awkward landscape subjects for even the most experienced, determined photographers. One turn-of-the-century geologist dismissed them in a sentence that perhaps explains why so few photographers, and so few summer visitors, examine them for long. "As scenic features they are monotonous and uninteresting in the extreme because of their lack of relief and uniformity of appearance."[4] But *when* salt marshes became so monotonous and uninteresting, at least as visual subjects, raises troubling questions about the impact of black-and-white photography on alongshore scenery values. Until the era of photography, poets and painters routinely sought out marshes for their visual richness, more or less agreeing with James Russell Lowell that marshes reward close scrutiny:

> In Spring they lie one broad expanse of green,
> O'er which the light winds run with glimmering feet:
> Here, yellower stripes track out the creek unseen,
> There, darker growths o'er hidden ditches meet.

In four lines, Lowell demonstrates his understanding of green mixed with yellow, of the topography of marshes, and even of the curious regional terminology affixed to them, but his precision now means as little to most readers as does his meter. Only locals now grasp the meanings in his "Indian-Summer Reverie," the importance of glimmering, the Yankee use of *creek*. His lines are not at fault. Something has changed.

Lowell recognized the change. Earlier in the poem he pities the observer who cannot see beauty in the marshes, "who sees in them but levels brown and bare." But his condemnation, "Vain to him the gift of sight/Who cannot in their various incomes share," rings wrong, somehow falsely condemning the one who sees no beauty.[5] In winter, marshes are indeed levels brown and bare, and anyone remotely familiar with them knows it, for all that the bare brownness may be splendidly beautiful, too. Perhaps Lowell knew the difficulty of photographing summer marshes. Even in his black-and-white-film-only era, marshes photographed poorly, exactly as the geologist said. In an age devoted to the stereoscopic view, the textured, three-dimensional landscape image, salt marshes lacked punctuated foreground and middle distance, and often appeared deadly dull no matter what season enticed the stereographer.[6]

On a foggy day, the edge of a marsh at high tide still promises the mystery and magic Sarah Orne Jewett describes in **A Marsh Island**. *(JRS)*

Lowell's lines make clear his super-photographic interest in sustained reverie, in scrutinizing vastness as the locals still scrutinize the glim, in looking further into vistas than photography penetrates.

Only two or three decades after Martin Johnson Heade created a sensation with his Luminist paintings of salt marshes, most educated observers simply ignored the very places he and other painters so prized. What entranced Lowell and his contemporaries, what brought Heade and other painters again and again to the marshes just in from barrier beaches, had simply lost value, perhaps because photography made so little of it, perhaps because farmers had made so little of the marshes themselves. Photography did nothing, absolutely nothing, to rescue salt marshes from artistic oblivion, let alone to make them vicariously accessible to people afraid of quicksand and even more frightened of the seductive voluptuousness of chartreuse green. "As the sunset deepens, the salt-meadows are clad in a golden green moss, each dry blade and bramble on the wind-swept hill gleams like a javelin, the red flowers burn in crimson flames, the cranberry swamp, too, is on fire, and the bridge in the distance looks as if it led to paradise," Sarah Clark told *Harper's Monthly* readers of the salt-marsh realm at sundown in 1882, but the cognoscenti no longer cared.[7] Farming had not made marshes modern, safe, and prosperous, and indeed this failure made them almost frightening, as frightening as failure.

By 1885, when Sarah Orne Jewett explored the salt-marsh world in a novel, *A Marsh Island,* both painters and farmers had shifted perspective, although so subtle did Jewett find the shift that it forms the spine of the novel. From the first lines of her story, when the local folks dismiss "with some contempt the bit of scenery" chosen by a wandering painter as his subject, the declining status of the salt marshes as artistic material grows ever clearer. The "foreground of pasture, broken here and there by gray rocks" opens only on an estuary and marshes that seem "to stretch away to the end of the world." Incapacitated by a twisted ankle, then captivated by the wide skies and flat grass, the painter devotes a summer to exploring and painting the place that "looked as if the land had been raveled out into the sea, for the tide creeks and inlets were brimful of water." Hilltops afforded views of distant barrier beaches crowned with sand dunes, and tiny harbors marked by haze-shrouded fishing-schooner masts, but the painter finds such long-distance views bewildering. The soft browns of the haystacks in the marshes reassure him, and those he paints, deciding to remain into the autumn not only to pursue the daughter of his farmer-host but to embrace the changing colors of the marsh, especially the bright greens snaking in along the creeks lacing expanses of browns and grays "embroidered here and there with samphire."

Jewett's painter fails in love, in his bid to paint the marshes, even in his attempt to walk about the marshes. Though his paintings enjoy a successful New York show, he is clearly a dilettante, a painter of means but little energy, a painter content with salt marshes. Two-thirds of the novel pass before Jewett makes clear his utterly second-rate nature, in a brief passage recounting his drifting about the tide creeks in a borrowed dory, his falling asleep on a bit of high land, and his discovery by the farm girl, who has walked to a place he thinks inaccessible by land. The painter admits he cannot find his way about the marshes on foot, that the incoming tide imprisons him on any high ground, and that he is not even sure how to find his way back in the dory. " 'There is something mysterious about the marshes to me,' " he concludes before attempting again to beguile the woman. But she only points out that the tide has risen enough to float the boat, and the painter, feeling "a faint sense of mistake and disappointment," understands his place after she "only smiled when he demanded the oars which she had taken." Jewett describes how she "pulled with strong, steady stroke, as if it were a relief and welcome defense against threatened discomfort," and talked of the marshes in winter, when they "look so dead and desolate, with great black cracks in the ice, like scars." Rowed home by a woman, a woman who knows the marshes in winter, the painter blurts out that he could not stand the long winter, that "town is the place when the snow comes." Jewett deftly makes her point. "He did not like to have Doris row the boat, and a great insecurity and indecision took possession of him." As they run ashore on dry land, at the edge of the marsh realm, the painter calls himself a coward.

In the remainder of *A Marsh Island*, Jewett underscores the tough competency of her otherwise gracious, modest heroine. The farm girl finally ventures across the marshes before dawn, illuminated only by the moonlight reflected from frost, following the faint footpath from one hummock to another, then taking out an ancient boat kept as the salt hayers' ferry. "The boat leaked and went heavily; the oars that she pulled from their familiar hiding-place were short and heavy, and splintering at their handles. But Doris rowed as if this were a race, and looked over her shoulder, until at last she heard the dry sedges of the farther shore rustle and bend, and she could step on dry land and be on her way again." She leaves the marsh and runs along the dunes just visible in the dawn, dunes that "appeared to recede as she advanced, mocking her like a mirage, and at last coming close when she thought they were still far away." Beyond is an apple orchard smothered in sand drifted in from the beach, "a desert waste of sand, white as bone, deep and bewildering," then a fox keeping her company "over the white desert."[8] Rowed across the last inlet by two boys going hunting, she arrives at the harbor village to astound her local lover, someone who knows the marshes well enough to know that her passage through them evidences true love.

Unlike the painter, whose knowledge of the marshes is merely dilettantish, Doris's knowledge is deep and tested, valuable even at night, when the painter sees nothing. She not only knows the secrets of walking about the marshes at any stage of the tide and the ways of glimmerings and mirages, she knows where the oars lie hidden and how to use them. Doris Owen is one of the most active, competent heroines in nineteenth-century American writing, perhaps because she inhabits a farm almost entirely surrounded by marsh, by estuaries, by barrier beaches, by wilderness. But as Jewett perhaps expected some late nineteenth-century readers to understand, Doris lives surrounded by men scarcely more competent than the summer-visitor painter.

A Marsh Island remains one of the best fictional treatments of saltwater farming in its long, gray afternoon. Heade and other Luminist painters ventured into salt marshes at the heyday of saltwater farming, when the marshes struck everyone as prosperous, well-managed appendages to thrifty Yankee farms. By 1885, however, the rising expectations of agricultural engineers and other experts seemed unlikely to be met by the marsh owners, and an air of poverty and failed promise hung over the marshes like autumn mist. Jewett knew the situation, and weaves it everywhere in her novel, beginning by mentioning that the farmer's only son, killed in the Civil War, lies buried on high ground overlooking the marshes, and ending by observing that local people thought the old farmer had begun to lose his touch, that the farm had begun to decay. Age suffuses the novel, and antiquity rules Jewett's description of haying.

Old Israel Owen ventures to mow salt grass only after launching his hay boat, a massive, scowlike craft Jewett describes as "stupid looking" and "square," as slightly narrower than the salt creeks are wide. The old craft, flung inland by a ten-year-high tide the previous spring and requiring horses and rollers to move, leaks as Owen and his crew head into the estuary, towing a dory filled with scythes and rakes, and while old farmer defends the aged vessel to his youngest hired hand, the whole hay-making effort seems more ancient than the boat. The men work all day with scythes, not horse-drawn mowing machines, and they lug the cut grass to the boat to be floated home with the incoming tide, off-loaded, and dried. They take their nooning under a tent made from the small sail of the dory, eating lunches they brought with them, not by the farm women, and go back to scything grass.[9] They work as men worked two centuries earlier, when colonists found few meadows to support their livestock and invented the long-lived enterprise called "marshin' " or "marsh-haying," a farming based on salt meadows and hay boats called *gundalows*.

Marshin' dates to the first years of the Plymouth settlement, when half-desperate colonists gathered salt-marsh grasses for livestock fodder and thatch for roofing. Within a few years, certainly by 1640, the great salt marshes in from the harbors at Plymouth, Duxbury, Scituate, and Marshfield, and along the estuaries running inland from Scituate, had been apportioned as very valuable real estate indeed. Families hard-pressed to clear forest and plant crops knew that nutritious salt hay could be had only for the harvesting, that the vast marshes provided an extraordinary agricultural resource, and an extraordinary nuisance.

Boundary conflicts came early among the owners of the marsh lots, most so far into the marshes that men visited them only in the summer haying season. Unlike inland meadows and other fields, no stone walls marked the edges of marsh holdings. Not anywhere in the entire marsh lay a rock useful as a marker, and from the middle of the seventeenth century onward, owners marked marsh-lot boundaries with corner stakes. Rare high tides, especially in spring, and sheet-ice movement all winter often disturbed the stakes, sometimes obliterating whole groups of them, and making hay-season locating difficult, and hot-weather tempers even hotter. Plymouth Colony court records suggest that real trouble had struck the marshes by the 1660s, for trespass case after trespass case required decision. Plaintiffs and defendants were always men of good standing, town responsibilities, and property, something that suggests honest mistakes made over very valuable grass. Humphrey Turner's complaint against Abraham Sutliffe in 1667 "for trespassing upon said Turner, by mowing his meadow, and carrying away his grass," is typically worded. Often litigants settled just before trial, as when John Palmer dropped his charges against John Silvester in 1673 for cutting hay "between the meadow of Joseph Barstow, the said Barstow's meadow going on both sides of the said meadow from whence the said Silvester carried the said hay or grass." Out-of-court settlements sometimes resulted when everyone determined either honest confusion—Palmer's wedge-shaped meadow made boundary mistakes easy—or the impossibility of proving anything. As early as October 1651, however, trespass became assault, when Edward Jenkins testified that "Joseph Tilden hired him to mow grass for him at the marsh before the island, called Hatch's Island, and that he had not been long there, but Morris Truant came to him and did forbid him to mow there; and afterwards came with a pitchfork, and bade him leave, of which if he did not, he would break his scythe." [10] Disputes along this stretch of coast never deepened to the severity of those in Essex County marshes inshore from the wreckage of Thacher's pinnace—marshes in which the men of Rowley went armed, for example—but the unfrequented salt marshes witnessed intermittent confrontation. [11]

Throughout the eighteenth century, confrontation lessened and cooperative harvesting increased, not so much because Yankee tempers mellowed but because marsh hay grew ever less valuable. Schooners brought cheap upland hay south from Nova Scotia and other Canadian provinces, and long-accomplished forest clearing gave many farmers the option of keeping fields in Timothy or other "English Grass" even more wholesome than salt hay. By the beginning of the nineteenth century, just before the railroad nearly buried coastal farmers in cheaply delivered, high-quality hay cut far inland, marshin' had become a tradition-bound, modestly profitable enterprise of much value to part-time farmers who supplemented a little crop-raising with fishing, boat-building, or other occupations. Marsh hay fed a span of horses and a cow or two, and thatch proved useful in banking houses against winter gales, in bedding livestock, even in mulching gardens with a purely seedless top-dressing. After all, locals argued when cornered by agricultural experts, marshes required no manuring, no weeding, no maintenance at all to produce tons of high-quality hay. Jewett well understood the timelessness of saltwater farming, and crafted *A Marsh Island* to depict a farm lost in time, an island in progressive agriculture.

Marshin' depended first on the tides, especially on understanding cycles of low high tides and high high tides, the last created by the conjunction of sun and moon. With the moon

at apogee—something every farmer knew from the *Old Farmer's Almanac*—mowing might begin three days before the lowest low tide, but with the moon in perigee, mowing began three days before highest high tide, though with the risk that a storm might float off much of the cut hay. Depending on the cycle of lunar and calendar months, therefore, a salt-marsh haying season lasted from eight to fourteen days, something of incredible importance to farmers dependent on salt hay for fodder, and seasons occurred at different patterns in the months from June to October. Men entered the marshes at dawn and worked steadily all day, pausing only for lunch, knowing that the combination of sunny weather and low high tides would last only a short while. Two or three days after mowing began, the men would begin to stack the cut grass, having first carried it to particularly dry spots on stretcherlike hay poles, two men to a pair. Salt-marsh haystacks stood on "staddles," twelve-foot-diameter clusters of poles sticking about two feet above the marsh, just enough to keep the quick-drying grass above tide level. Typically, a staddle held from one to three tons of hay—a sixteen-acre lot might produce twelve tons of hay, stacked on six staddles arranged to spare the men long walks with laden hay poles—and farmers tried to stack hay so it might shed water. Some farmers roped down their stacks, often with ties made from twisted thatch, and others covered theirs with old sails or even crude wooden roofs, for many expected the stacks to stand untouched until late autumn, when the last hay came ashore.[12]

Moving hay from staddles to shore required the ugly vessels Jewett so accurately describes, the infrequently used, built-forever, grossly heavy gundalows. Essentially long, rectangular

Coastal realm farmers learned to stack hay on staddles, hoping that no late-summer gales carried off their stacks. (author's collection)

At high tide farmers rowed dories over the marshes, pausing at each staddle to inspect their stacked and staked hay. (Ipswich Historical Society)

scows, the salt-marsh gundalows perhaps take their name from Venetian gondolas, although nothing graceful characterized them—or their operation. While similar vessels moved along the Piscataqua, sometimes under sail, and through the South Carolina rice plantations, salt-marsh gundalows existed only for calm-weather, protected-water voyaging only, usually with the tidal current, since their great sweeps and poles barely moved them, even when empty.[13] Everywhere around the marshes linger tales about gundalow operation, tales told and retold not so much as "good stories" like those from rum-running days but as potentially useful instruction. Just as old men instruct their grown sons in the mysteries of giant block-and-tackle techniques for moving boats usually shoved about by tractors and Travelifts, they pass on the fundamentals of gundalow gunkholing.

Launching a gundalow demanded the effort Jewett recounts, especially if abnormal tides had driven it further ashore than oxen or horses had dragged it at the end of haying season. Usually men tried to shove the gundalow into the water, perhaps at high tide. Only rarely could they harness draft animals to the channel-end of the gundalow, and often they wound up digging under the thirty-five-foot-long craft to insert rollers and pry poles. Aiming the slow-moving craft down the narrow creek leading into some broader one opening on a larger estuary was critical, for a gundalow wedged catty-cornered in a creek often stayed wedged for hours. Some June launchings went well, but others lasted hours, and every shove, every heave with pry poles might puncture the bottom planks, necessitating jacking up the gundalow for repairs at the start of the short hay season.

Conning a gundalow required as much skill, as much brute force. In constricted creeks, men moved a gundalow by poling it along, trying to keep the poles from sticking in the muddy bottom, and fending the boat off at each bend. Once in estuaries broad enough, men rowed at giant sweeps, oars sometimes twenty feet long, mounted in the bow, while one held a steering oar in the stern. Arrived at the loading place, the men bailed out whatever water had worked its way in and began stowing hay, piling it roughly everywhere in the gundalow except at each end. If all went well, they drifted homeward with the incoming tide, the men at the sweeps bellowing rudder directions to the man at the stern, whose view of everything ahead stood utterly blocked by the vast heap of hay. Nautical terms rarely echoed over the marshes, for many harvesters at the sweeps and steering oars knew only farming or carpentry or other dry-land callings, and "port" and "starboard" led to running aground or collisions with other gundalows. Old timers still counsel listeners to yell "over to Norwell" or "a little toward Marshfield" when guiding laden gundalows—and in extreme conditions, say in rising

Farmers infrequently went "marshin'," but when they did, they used flat-bottomed gundalows to carry home their hay. Some specialized in hay transport and owned sail-carrying gundalows like this one moored in a salt creek. (Society for the Preservation of New England Antiquities)

Small gundalows floated to the edge of creeks, where farmers used them as platforms to transfer hay to wagons, as in this scene from Wallace Nutting's Connecticut Beautiful. (Harvard College Library)

wind or at night, to put a "cunner" atop the hay pile and follow his conning directions, especially when approaching low bridges.[14]

Well into the nineteenth century, marshin' endured as it had for at least a century before, although some innovations provided improvement or amusement. Here and there farmers tried floating mowing machines and teams of horses into the marsh, using gundalows as ferries. They quickly learned that horses sank into the marsh, and so they invented wooden shoes that locked around the horses' hooves, distributing their weight over more ground. Horses had to be taught to walk comfortably in the shoes, and in time they learned, but the cast-iron mowing machines proved extremely awkward in the marsh, and devilishly tricky to move on and off gundalows. And even the calmest draft horses might bolt when bitten by greenhead and other marsh flies, and in galloping break a leg in a ditch or grass-covered hole. Even after farmers no longer needed salt hay for fodder, they harvested it for sale to nurseries, for gardeners had learned that unlike upland hay or straw, salt hay sheds no seeds to sprout into weeds, making it the perfect winter cover for delicate ornamental perennials.[15] So the gundalows moved into the 1910s, sometimes later, until the last wore out, their ribs still noticeable in the upper marshes.

Tales of marshin' are of real value late in the twentieth century, however, for they keep alive understanding of public landings at the edge of the salt marshes. Places where farmers long ago landed gundalows piled high with cut grass or cured hay may have never seen any

other sort of commerce, not even a fishing boat landing fish, let alone a schooner off-loading cordwood. But they endure in tradition as public landings, places that now look like simple gashes in salt marsh, gashes opening on estuaries and reached by dead-end dirt roads called "lanes." Oral tradition, carefully recounted at town meetings, keeps the landings public, keeps the access for canoeists and kayakers free of private encumbrance. The town meeting knows that the state attorney general will ask if the landing place served the needs of public commerce, and knows the answer derived from back-of-the-barn tales of conning gundalows, of making the marshes pay.

Early in the nineteenth century, the first visionaries began reshaping attitudes toward marshin' and urging changes in traditional coastal farming that would make marshes pay as they never had paid. As early as 1820, William Tudor condemned salt marshes as "a reproach to our agricultural management." His *Letters on the Eastern States* argue that pioneer farming belongs nowhere along the seacoast, that farmers ought to learn from coastal examples in Flanders, the Netherlands, and especially England, that early New England polder-making experiments failed for want of determined maintenance following the initial diking.[16] Tudor advised coastal farmers to form diking cooperatives, wall out the sea, and plant vegetables and grain on the wonderfully fertile soil just under the marsh grasses, thinking perhaps of the great successes of drainage companies in the Fens of England—and forgetting the decades of effort and disappointment that preceded British success.[17] For three or four decades, few followed the advice of Tudor and other reformers, and the marshes attracted painters, especially Heade and other Luminists, and other educated visitors entranced by their beauty and modest prosperity. Until about 1860, salt marshes remained as something very rare in the Republic: expansive, agrarian vistas unchanged for centuries, vistas speaking if not of antiquity, then at least of age.[18] Not until the agricultural reformers reshaped public opinion did marshes become second-rate views, views attracting dilettantes like Jewett's painter.

The reshaping began early along this stretch of coast, at least as early as 1789, when a group of Cohassett farmers diked some salt marsh well inland—so inland, in fact, that they might have called it "sea green," meaning meadowland washed by the tide only once every twenty years or so—and in time planted it to upland grass, having watched their dike collapse and need renewing only three years into the experiment.[19] Away from places distant from storm-induced waves, currents, and high tides and so almost perfect for diking—if not naturally becoming freshwater marsh anyway—reclaiming salt marsh proved much more problematical. Agricultural newspapers like the *New England Farmer* and the *Cultivator* make clear the progression of reclamation thinking that followed the appearance of Tudor's book. In the 1820s, letter-writing farmers argued that the price of salt hay, combined with the nonexistent costs of fencing and manuring, made diking a poor investment. By the 1840s, as the price of harvested salt hay dropped to one-half that of upland hay, editors began to suggest diking experiments in all sorts of salt marshes, but not until the 1850s did reports arrive from farmers.[20] The reports suggest that rather than dike the salt marshes, many farmers struggled to improve their holdings by ditching. Not only did ditches make the marshes drain better after high tides, but the mud made wonderful fertilizer, especially after mixing with manure.[21] Moreover, the ditches provided permanent, or nearly permanent, boundary lines, and so lessened the chances of trespass. But even as the reports of improvement arrived, other science-minded farmers began to extol the salt content of salt-marsh hay, which offset the cost of salt bought to supplement upland hay fed to cattle.[22] No one knew exactly what to do, and tentative improvement efforts aside, few did much.

The Civil War, and an 1861 federal report published as part of the *Report* of the Commissioner of Patents, nudged coastal farmers toward change. Wartime scarcities drove up the price of hay, and the arguments of William Clift, a Stonington, Connecticut, farmer who successfully diked salt marshes and visited other modified marshes, perhaps provided the catalyst for change. Clift counseled improvement for moderate gain—his essay opens with "Whoever expects sudden riches by reclaiming these meadows may as well pass this article for a more congenial theme"—and emphasized the simplicity of his approach. Although no agricultural society or government agency evaluated the long-term effect of his advice, many marshes still provide evidence that some farmers tried to follow it.

Clift started with two premises that indicated his practical experience. First, he suggested that the inmost salt marshes be diked and drained first, since they stood far from the reach of ocean storms, and second, he urged farmers to use railroad embankments as dikes. By 1860, railroads linked many coastal villages, snaking their tracks as closely parallel with the shore as possible and building causeways across salt marshes. Clift knew the impact of the Boston–New York railroad line, and many other farmers, say those farming the coast between Boston and Portland and Boston and upper Cape Cod, knew the effects, too. Marshes were divided by well-footed railroad embankments, usually gravel and rock with riprap on the ocean side and protected by deep, wide ditches, sometimes fourteen-foot-wide drainage canals on the inland side, and while the drawbridges irritated the crews of heavily laden gundalows, the railroad rights-of-way seemed like free, first-class dikes. Inland of the railroads, he argued, farmers ought to convert salt-marsh ground into meadowland.

All they needed to do Clift explained, using his own experiments as examples. After building a dike, or using a convenient railroad embankment, the innovative farmer ditched his marsh holdings, allowing salt water—and later rainfall—to drain from the land quickly. In the embankment he built a tide gate, a simple floating gate set in a sluiceway; outgoing water forced open the hinged or pivoted gate and so drained into the ebb-tide estuary, but incoming water forced shut the gate and kept the marshes dry and fresh. "The essential things in the gate are, that it should move rapidly upon its hinges, opening with the current from the drainage-water as the tide falls, and shutting with rising tide; that it should shut close, and that it should be made of durable material," Clift advised. "It is also desirable that the gate should be lined with yellow metal, or with copper sheathing, to prevent the gnawing of muskrats, which are one of the greatest pests about these reclaimed marshes." Gate built and muskrats stymied, the farmer needed only to cross-ditch his holding, using the sod and mud removed from the ditches to build a raised wagon-way into the marsh to afford access to horses and hay wagons. Clift sold his surplus mud for a dollar a cord and set his income against improvement costs, and though he could offer nothing specific about how deep and how long to make ditches—he told farmers to trust their instinct and re-ditch if rushes and sour grasses appeared in subsequent summers—he urged farmers not to plow the dense sod and thick mud but to sow English hayseed "within six months of the time when the tides are shut off" and clover in the first fresh spring. Done right, marsh reclamation had to work.

Done wrong, Clift admitted after observing failed efforts, reclamation proved useless and expensive. Much of his detailed essay recounts failures from Massachusetts to New Jersey, failures caused by poorly fitting tide-gates or collapsing ditches. He advised farmers to hire carpenters "who can make a close-fitting joint" and favored farmers who built embankments from upland soil, not marsh mud and sod. Shallow or widely spaced side ditches turned some salt marshes into freshwater marshes covered with bluetop, dock, dandelion, and four-o'clock, places still inaccessible to hay carts, let alone fully loaded hay wagons, and failure to manure led to poor yields within a few years of freshening the marsh.[23] And taking on too much, especially trying to reclaim marshland adjacent to marsh still open to the tides, led to disaster, too. In the end, cooperative ventures to reclaim entire marshes seemed safer to Clift than rugged individualism.

Marsh reclamation or "marsh subduing" quickly enticed other reformers, especially those anxious to find New England farmers something new to do in the face of competition from Western prairies. Charles A. Goessmann of the Massachusetts Agricultural College provided a lengthy analysis of reclamation efforts in the 1875 issue of *Agriculture of Massachusetts*, an annual, state-supported report aimed at improving farming and advertising tested experiments. Goessmann, like other nonfarming experts, devoted much attention to the stupendous successes of Dutch, Belgian, and Danish farmers, who reclaimed marshland for haying, pasture, and tillage crops, although he neglected to say much about actual reclamation processes or government support. The meat of his report focuses on this stretch of coast, especially on the Green Harbor reclamation project that epitomized late nineteenth-century efforts to make salt marshes into something better. Along with other well-traveled experts, Goessmann found at Green Harbor a miniature Netherlands, a Yankee polder.

In July 1872, a private company began building a dike eighteen hundred feet in length and nearly seven feet high, enclosing 1,412 acres of marsh, some 200 of which lay so far inland that everyone considered them essentially fresh. So vast an undertaking required state mandate, and the legislature mandated a tax to be paid by every landowner who benefited from the diking, a wise and prudent move indeed, since the dike cost $30,000. Sluices emptied water into Green Harbor River, two large tributaries, and an intricate skein of creeks, and in the first months all went well, encouraging everyone involved with the project. But within a year the secondary drainage ditches proved too shallow and too far apart, and heavy rains began to pool on low spots, killing the upland meadow grass planted to replace the salt-marsh grasses. Better drainage, including underground tile drains, struck Goessmann as the best immediate improvement necessary, and by 1874 the land could definitely sustain vegetables, at least in test spots.[24] Twelve years later, when another expert visited the site, the meadows had become exceptionally productive. Nathaniel Southgate Shaler, perhaps the preeminent coastal geologist in the nation, learned that "with a slight plowing" the fields produced oats and rye, onions and potatoes, but very little corn, since maize is exceptionally intolerant of salt. "The land has proved fertile and tillable," Shaler determined. "The ordinary crops of the country do well upon it, and their growth promises to prove more and more satisfactory as the ground is longer tilled." Although he noted several of the same problems Goessmann discovered at the start of the experiment—Shaler argued that the dam cost twice what it should have and that many ditches were still too shallow—his sole concern was the perceived subsidence of the reclaimed marshes. As near as he could make out, the surface of the reclaimed areas had

settled nearly eight inches, and he guessed it might settle eight more or so. In Shaler's expert view, and in the view of the farmers who owned the land, the experiment was a success.[25]

Unlike his predecessors, Shaler worried about the long-term complexities of the Green Harbor experiment. Everyone, he admitted, knew far too little about the natural forces twisted by reclamation efforts. "The remote and picturesque coral reefs have long proved fascinating subjects to the geological student, while these near-at-home structures, which are in their way almost as interesting as the work of the polyps, have never been adequately studied," he remarked of the salt marshes. "In the northern part of Europe they have been the subject of much economic consideration, but there seems to be no scientific literature whatever on the subject."[26] However much he was devoted to agricultural experimentation, Shaler suspected that the Green Harbor enterprise might be a mistake. Others, more locally involved, knew it was a mistake, right from the start.

Gundalow-poling farmers irritated fishermen when their hay-stacked scows wandered across the bows of seagoing vessels, but the Green Harbor farmers ignited a storm of protest when they built their dike. Green Harbor had long been a gigantic sea of salt marsh and salt

ponds behind a low barrier beach, a beach of such low and shifting dunes that one night in 1810 a party of fishermen dug through it, creating a channel that quickly scoured the sand and made a harbor of a shallow saltwater pond. While the fishermen did nothing more than reopen a passage closed since settlement days, they reaped an immediate good—a small, shallow, exceptionally useful harbor, one that quickly became modestly prosperous, and mightily disadvantageous to the owners of salt-marsh lots suddenly swept by salt water. When the farmers organized to dike the marshes, they understood their effort as a return to status quo.

But the dike dramatically decreased the flow of water through the channel dug by the fishermen, which began to silt up immediately. The fishermen went to court, and lost.

After hearing the court decision, and fearing for their livelihoods, the fishermen determined on more direct action. On two occasions they blew up the dike with dynamite, once with spectacular results to the timber-lined sluiceway. When Shaler arrived in 1886, a third attempt had just been frustrated and a man was under arrest.[27]

Salt-marsh reclamation badly upset tidal currents, something no one considered when musing about lush meadows and rockless fields of potatoes perfect for mechanized harvest.

So narrow-minded had agricultural reformers become by the late 1860s that the interests of fishermen, waterfowlers, and even alongshore merchants simply did not occur to them. By the 1880s, however, hydrologists had begun to learn something of the movement of water, and forward-looking government officials began to employ their arguments in rejecting massive building efforts. As early as 1882, for example, the Massachusetts Board of Harbor and Land Commissioners argued against allowing railroad companies to fill beneath several trestles that jutted into Boston Harbor, concluding that "to do so would clearly be not only to violate all the traditions and teaching of scientific research," but "possibly, and even probably, to conflict with or defeat other schemes of harbor conservation and improvement." Simply put, "there are unknown elements, particularly in the phenomena of tidal action, the results of which it is difficult to predict." [28] Boston Harbor was too important to experiment upon, the board decided, not wholly in a spirit of scientific open-mindedness. As it well knew, the legislature held the board responsible for fixing mistakes, and the board was fast learning how difficult and expensive such fixing could be.

By 1898 the board had been charged with repairing Green Harbor, a still-shallow place used by fishermen and increasingly favored by pleasure boaters, pleasure boaters who always voted and often knew influential people. After erecting two cut-granite jetties, one on either side of the harbor mouth, the board found that it had also to erect a "timber training wall" to deflect what little remained of the Cut River current straight through the harbor "so that it should pass out directly through the Narrows, instead of running across the harbor and obstructing the flow in the channel of the main river." A year later, however, the board had still not determined when to dredge the harbor, and although it grudgingly admitted that the ebb tide had begun to scour out some sand, more work clearly lay ahead. [29] The westerly stone jetty, built on soft sand, settled in the winter of 1899–1900, and although some harbor dredging had deepened the main anchorage, the board worried that sand from the sea drifted over the depressed westerly jetty and commissioned a "timber bulkhead" to be built inshore of the jetty to catch the sand. Unless it dredged the entire harbor, however, it could do nothing about the sandbars that materialized everywhere around the main anchorage. [30] Within a year the board found the main anchorage shoaling rapidly at its inner end, "where the current on the ebb tide washed down quantities of sand from farther up the harbor." The board admitted that "these changes were not unexpected," but it argued that "their extent could not be anticipated." Having spent some $88,000 over three years, the board determined that the channel "has substantially as much water as there was on the bar previous to the construction of the dike, and is giving great satisfaction," but it well understood its permanent commitment to dredging. The fishermen had been joined by pleasure boaters, and the fishermen had begun using "naphtha launches," the first powerboats out of the harbor.[31] Reclaiming the salt marsh by interrupting the tidal currents had committed the Commonwealth of Massachusetts to an eternity of dredging.

Green Harbor River drained a watershed of roughly seven-and-a-half square miles, not half of which lay in salt marsh at the start of the reclamation effort. The dike not only proved hideously expensive to build and maintain—even discounting damage done by dynamite-wielding fishermen—but only Shaler continued to study its long-term larger effects. In Massachusetts Board of Agriculture report of 1891, "The Inundated Lands of Massachusetts," he directed far more attention at reclaiming freshwater marshes, and while arguing that the reclaimed Green Harbor marshes did produce valuable crops, admitted that the slow decay of the underlying peat and the continual presence of salt decreased their value. Calling the Green Harbor experiment "imperfect," while citing the larger environmental consequences faced by the Board of Harbor and Land Commissioners, Shaler sounded the death knell for

further large-scale salt-marsh draining. Salt marshes, he concluded, are part of the agricultural "reserves" of the commonwealth, something good for something someday.[32]

Between 1860 and 1910 salt marshes spoke only of failure, laziness, and failed promise, striking many educated visitors as did the abandoned, grown-over farms of northern New York and New England.[33] Everywhere the marshes demonstrated the incapacities of Yankee farmers, men who could not master the sea as well as Dutch and Danish farmers, men who started things and gave up, or men like the farmer in Jewett's *Marsh Island*, men who never tried improvement. Farmers knew differently, of course. Not only did many choose to invest in other farm improvements, but many deliberate farmers watched neighbors invest much capital and time in reclamation efforts only to spend more and more in maintenance. A strong onshore wind combined with abnormally high tides, particularly late-summer or early-autumn tides, devastated marshes planted to upland hay or vegetables. Even a balky tide gate, or a sluice-way undermined by muskrats, could be disastrous, and always the reclamation effort produced intricate subsequent effects, especially on tidal currents. Ditches, more than one innovative farmer learned, need current to stay clean, and without the current, marshes became mudflats after heavy downpours. And as the automobile spelled the end of the horse-and-buggy era, farmers watched the market for upland hay—and salt hay—shrink every year.

In the 1910s began a subtle shift, one retrospect makes distinct. While the federal Department of Agriculture published *Tidal Marshes and Their Reclamation* in 1911 and summarized the best of experiment and hope, a year later Charles Downing Lay published "Tidal Marshes" in *Landscape Architecture*.[34] Lay understood that marshes might be reclaimed for agriculture, but knew, too, the problems and costs involved. He argued for something different, something essentially visual, something he called "picturesque ditching." If the homeowner could afford some marsh reshaping but lacked the funds or desire to drain or fill the entire marsh for the purposes of making a lawn, he or she might excavate part of the marsh "to make ponds or creeks, or lagoons with islands." Homeowners would add to the beauty of the view beyond their high land, and make some part of their marsh accessible. Lay wrote as the rediscoverer of salt-marsh beauty, emphasizing the "changing color with the advance of the seasons" and the wonderful way marsh beauty directs attention to the sky. "Nowhere," he claimed, "is there greater beauty of line than in their curving creeks and irregular pools."[35] Moreover, he wrote in the deeper tradition of small-scale, private-lot landscape architecture, a tradition that faded in the late nineteenth-century public parks era but began to flourish again in the early twentieth century. Before the Civil War many educated, well-to-do Ameri-

Only fools race
speedboats along salt-
marsh creeks. The
smallest of mistakes, the
merest misjudging of
water depth, even
momentary failure of
nerve results in disaster
in locations inaccessible
to salvage vessels. (JRS)

cans cherished a love of half-wild, half-cultivated scenery, the "borderlands" that lay beyond cities and new suburbs. They loved scenery and animals farmers loathed or tolerated. Musk-rats, for example, epitomized the rich nature of swamps and marshes, and aesthetes who delighted in finding them haunting boathouses and wharves little considered the plight of farmers trying to keep muskrats from salt-marsh sluices.[36] By 1912, however, the earlier love of nature had reasserted itself, and Lay knew that upper-class families had turned from urban parks because the parks had become too tame, too cultivated.[37] He knew that the families, his clients, had long discounted the old fears of salt marshes as unhealthful and had redis-covered them as beautiful, as abutting estuaries and creeks that afforded perfect access for small powerboats.[38] And he knew—or suspected—that the failed reclamation ditches and earthworks made salt marshes even more attractive to people eager to enjoy summer or year-round homes without getting their feet wet, eager to own tiny docks for small boats, and eager for private, saltwater swimming holes.

Well into the nineteenth century, indeed perhaps until the 1950s and 1960s, salt marshes endured as male preserves. Jewett's Doris Owen is not the only salt-marsh-walking, salt-marsh-confident heroine of turn-of-the-century fiction. Kathleen Norris created another, one even more active, in 1910 in her *Atlantic Monthly* short story, "The Tide-Marsh." Mary Bell plunges into a storm-tossed, high-tide-swept marsh to rescue a group of young boys caught by the storm, tide, and nightfall, and urges them landward along a tottering fence, some relic of failed reclamation and pasturage.[39] Her story, part of the turn-of-the-century competent-women-in-wilderness genre, pushes its heroine further than Jewett pushed hers, for Mary Bell actually gets wet, almost drowns. Jewett, Norris, and other female writers tested the broadmindedness of their readers by presenting heroines who moved confidently in a half-land, half-water near-wilderness that many men found daunting, that most women knew as places where women wet their walking boots, then their skirts and petticoats, but where men might wear very little, and even swim.

Marshin' often involved noonday swims and frequently meant unexpected plunges into waist-deep water masked by drooping grasses. Jewett only touches on the issue, but male writers well understood the peculiar pleasure of salt-marsh haying. Unlike upland haying, perhaps the hottest, driest, dustiest work of the farm year, marshin' meant the continuous ability to get wet. And women, even women who knew how to get to hay grounds dry shod, understood enough to know that the haymakers took their lunches with them in part so that no woman arriving at noon with hot food and cold drinks might find the whole crew naked, cavorting in the nearest creek. Men and boys accepted salt-marsh nakedness as a right, a right they valued more and more as they lost the right to swim naked on beaches. In the nineteen-teens, old men remembered when boys swam naked on beaches and from harbor wharves, their lack of any attire wholly acceptable. "No restriction of its free and boundless practice was imposed upon us," remembered Joseph E. C. Farnham in 1915 of swimming fifty years earlier. "Bathing costumes were unknown, at least not required, no fear of interference by the 'cop' disturbed us, and every situation was perfect." Summer visitors, especially women, brought changed attitudes, however, and from the 1880s to the 1960s, salt-marshes endured as places for boys and men to swim naked and for young boys to learn to swim in the "pots" or "holes" in the creeks.[40] Moreover, the shallow water of the creeks and small estuaries was often so much warmer than the ocean surf that boys and men preferred it. Few women ventured far into the marshes, therefore, and those who strode from hummock to hummock or pulled their own small boats understood the unspoken code, a code contemporary old timers remember lasting at least into the 1920s, often much later, that women disconcerted by male nakedness had only to look away.

Now almost no one except locals ventures far into the marsh. Salt marshes turn back all

Low tide strands the salt-marsh navigator, forcing the decision to sit still for several hours or else tramp about the gutters. (JRS)

but experienced explorers, and the barefoot historian examining behavior in the frontier zone where beach and marsh mingle notes over and over the cowardice, the confusion of first-time visitors to the chartreuse realm. Almost always, he classifies behavior into one of two categories, each slightly dependent on the peculiar topography of particular frontier zones.

Where sand ends at marsh, where windborne sand has actually spread onto the marsh grass in from the dunes, bold summer visitors step almost without thinking from one ecosystem to another. For such visitors the short-grass marsh seems almost as accessible as a rank lawn or uncut hayfield, and they step into it casually, curious about something, perhaps a stand of five-foot-tall sea-oats, a few yards beyond. Suddenly they stop, almost at once if they walk barefoot, perhaps a few steps further in if sandals or sneakers protect their feet. Marsh grass cuts at toes, not as badly as dune-binding razor grass slices the soft skin between toes, but quite effectively all the same, and bits and pieces of dead marsh grass, rigid in brokenness, stab unprotected skin. And the marsh somehow squishes, even at low tide when the soft ground at the base of the grass seems almost dry, when the bits of dead stalk are dry enough to hurt. Barefoot and shod visitors stop to think, to look down and around, to decide how to proceed.

At that moment, while they stand wondering about this novel environment, the greenhead fly bites. Big, lumbering, and really quite iridescent in its chartreuse greenness, the greenhead bites with abandon, literally removing small pieces of flesh, hurting its quarry instantly. Easily killed when it is biting, the fly is hard to hit in the air and nearly impossible to ward off with flailing arms. Usually it attacks the back, or the backs of shoulders and legs, and though the bite hurts slightly less than a hornet sting, after a few minutes each bitten place swells and begins to itch. Through the 1930s, when horses still worked near and in the salt marshes, swarms of greenheads drove draft animals and wandering cattle to distraction, usually attacking their stomachs, and even now they will bite a dog on its unprotected underside. Late in the 1960s, as the DDT-spraying era ended in a spasm of horror at dead ospreys and other birds—and a creeping suspicion of human illness—some unsung local inventor perfected a cheap greenhead trap.[41] The devices, each a box roughly two by three feet by three feet deep, stand atop four thin posts and contain a V-shaped wire funnel that channels attacking flies upward and contains them until they die. Scattered here and there along the edge of the marshes by locals ecstatic with their nonchemical efficiency, the homemade cow-impersonating traps have largely eliminated the greenhead population, except near beaches crowded with summer visitors. Along the sand an occasional fly still attacks, then returns to its native habitat to encounter the nearly nude explorer of salt marsh. Within two or three minutes, rarely more, that explorer abandons all thought of venturing into the grasses and runs for the sand, perhaps figuring out that a quick swim might relieve the pain and itch.

Few casual visitors discover the other way into the marshland, although any high vantage point reveals it. But locals know the way, honor and love it, and conspire to keep it secret, not only because they respect its ecological fragility but because they value its emptiness, even on a hot July Fourth, when the marsh opens onto wonders of solitude, even on secret beaches far distant from upland. In speaking among themselves they use one of the rarest of Yankee-isms, the word Lowell catches in his musings about salt-marsh-seeing, the word Jewett uses in unthinking Yankee snobbery. They know the word *creek*.

All alongshore, New Englanders use *creek* only when speaking of the meandering tidal inlets that lace the salt marshes. All other watercourses smaller than rivers and larger than

rills are *brooks*, and no other term more distinctly identifies a local speaker of local language. Southerners use *creek* or *branch* when speaking of brooks, and Westerners use *stream*, but coastal New Englanders persist in their specific use of *creek*.[42] In the salt marshes along this stretch of coast, creeks meander from ancient drainage ditches and landing places, opening in time on estuaries leading always to harbors, albeit ones usually far more tiny and shallow than Green Harbor. Creeks rise and fall with the tide, and offer rare high tides routes far inland, just as they offer children rowing leaky skiffs or paddling inner tubes the easiest routes seaward. "My mother's house stood half-way down the lane that ran past Warren Turner's shipyard and a hundred yards back from the bay," recalled Frank Tooker in 1920 of a house by a creek. "It was a long, low story-and-attic house, rising white from a clump of cherry trees, with an apple-orchard crowding in on the south and a little creek winding through salt meadows up to the back door." Once, just once, a great spring tide sent water coursing above the creek, above the marsh, above his kitchen floor. But most of the time, the creek struck Tooker as a sort of path. "The harbor always seemed the natural road to the house, and whenever I think of it, it is of the rear, with the creek and the salt meadow and the white beach

that lay like a barrier against the bay."[43] A creek is a natural road, a twisting way through an almost exotic place, a place of vast vistas and, at low tide, extraordinary unevenness. And, too, a creek is a secret road, known now mostly to locals who remember the gundalow landings, who push off into a sea of head-tall grass, watch it open before the bow of their skiff or kayak, and suddenly burst into protected space.

All the old reclamation structures—and the railroad embankments—protect salt marshes from incursions by large boats, by crowds, by summer people. Late in the nineteenth century, railroad companies realized the slow decay of marshin' and began petitioning state governments to replace drawbridges with fixed spans just high enough to allow loaded gundalows to pass inland at low tide.[44] The Board of Harbor and Land Commissioners granted the petitions, and so closed estuaries to the rare schooner, catboat, or other masted craft. By 1910 or so, pleasure boaters found highway bridges—also drawless—paralleling the railroad trestles, and learned that even at low tide many boats simply did not fit. And so began the long, chartreuse sleep of marshes inland from railroad trestles and highway bridges, a sleep scarcely noticed by anyone, except perhaps by Charles Wendell Townsend, whose *Beach Grass* appeared in bookstores in 1923, bound in splendid chartreuse cloth.

5 SKIFFS

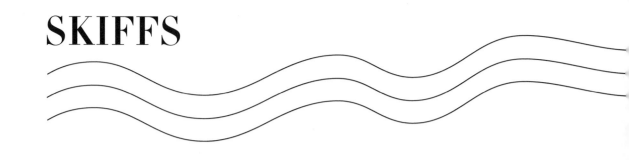

S alt creeks remain ways to harbors, ways inland, ways to nowhere in particular in the great green expanse of the marsh. At low tide the ways grow secret, and small boaters, kayakers, and especially swimmers move along them below the level of the marsh, invisible to anyone looking across what seems an unbroken surface of waving grass. Creeks and estuaries meander marvelously, sometimes winding nearly twenty miles to reach inland a straight five, sometimes branching and branching again, each creek growing brackish by the mile until it terminates in some hidden freshwater swamp or some colonial-era dam or nineteenth-century salt-marsh-reclamation dike, usually long decrepit. Most visitors to the coastal realm ignore the marshes, and so ignore the creeks, and even long-distance yachtsmen devoted to the detailed instructions of coast-pilot books like *A Cruising Guide to the New England Coast* can only ponder what lies inland from deep sea, through the treacherous, guzzle-like openings in cliffs and barrier beaches.

"A bell buoy and five nuns are guides through the entrance, where the tidal current can be strong. The entrance to this harbor can be very dangerous to a small boat and a large one cannot enter," warns the *Cruising Guide*. "The tide runs out hard over a shallow bar, estimated vaguely as having 'less than six feet,' and meets the swells of the open ocean. A really perilous breaking sea can develop here very quickly. Every year lives are lost on this bar." So the North River bar keeps out large boats, and endangers small ones trying to reach the sea or scurry back into the safety of the estuaries, especially when an onshore wind slams into the ebb-tide current pouring from square miles of salt marshes just behind cliffs and great beaches. But inland from the bar and its dangers, explorers find views "pleasing rather than spectacular," and if the boater owns a craft small enough to venture under the derelict drawbridge he or she "can continue into quite wild country." In the marshes so inpenetrable from the land, so difficult to decipher from the densely forested hilly ground, the small boater finds "an interesting place to explore."[1] Only penetration leads to discovery, and penetration from seaward proves difficult and dangerous, the worst sort of gunkholing.

Navigational aids, warns the *Cruising Guide*, are maintained not by the federal government but by town governments and boatyard owners, and they are sometimes missing or damaged. Federal authority not only seconds the *Cruising Guide* but offers a caution even more brusque. "Uncharted private navigational aids exist in the Scituate North River. These aids are frequently shifted with changing conditions. Use only with local knowledge," warns Chart 1233.[2] The federal authorities intermittently attempt to chart the channels and give low-tide depths, but the sands shift so rapidly that no chart is wholly accurate. Locals get around satisfactorily enough, but far too many strangers wreck while probing the opening from seaward, or trying to pass it toward the ocean beyond. So deceptively dangerous is the sandbar that the Coast Guard recently abandoned any hope of marking it. In 1991 it announced its intention to remove the two buoys marking the sandbar's northern end, and only determined petitioners led by an up-estuary marina-owner convinced it to keep the buoys and to place three more at the southern end. But the Coast Guard has the last word. At the boat-launching ramps inland from the sandbar, signs announce the danger of the bar, cautioning tourists duped by the smoothness of the salt-marsh-bordered creeks into launching small boats.[3]

And always the boater is the "small boater," that curious creature of legal language, advertising, local lingo, casual colloquialism. Male or female, young or old, the small boater is not, definitely not, a yachtsman. Sometimes a legal definition applies, say the Coast Guard one, which distinguishes boats under sixteen feet from those longer, but a seventeen-foot boat is no yacht. Sometimes a boat retailer mutters something about yachts being boats with inboard engines and enclosed toilets—"heads" in nautical terminology—or boats too large to trailer. The small-boater designation sounds slightly ridiculous, suggesting that all small boaters are scrawny or stunted, that they are small-minded or poor, or that they make only puny voyages, missing out on the adventures accruing to proper yachtsmen, those who make long passages like that attempted by the crew of *Tyehee*. Of course, even the smallest pram or skiff, kayak or round-bottomed pulling boat floats as a legal yacht if its owner keeps it for pleasure, not for commercial fishing or clamming, not as a ferry, not as a vessel for hire, say one chartered for the pleasure of clients. A yacht, Joseph C. Hart explained patiently in 1848 in his apology, *The Romance of Yachting*, is "a pleasure vessel." Hart probed deeply into nautical termi-

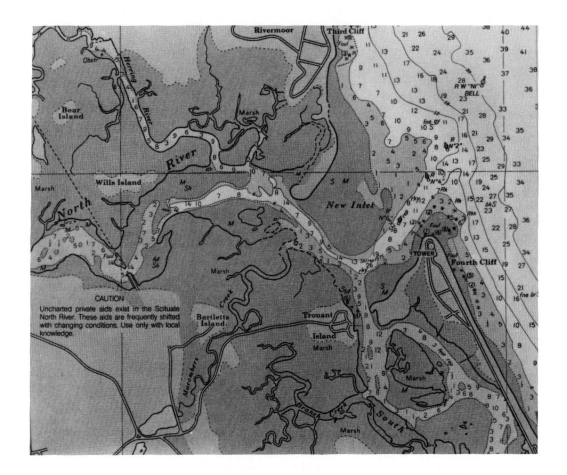

Federal chartmakers warn mariners to seek local knowledge before exploring the place still called New Inlet for want of a better name.

nology most useful to yachtsmen, providing an entire "Vocabulary of Sea Terms" as part of his book and a spirited defense of an eighteenth-century reference work, "the old Dictionary of Bailey—the best technical and general Dictionary extant, now nearly but undeservedly out of use." About the chief word, the very keel or mainmast of his book, however, Hart said little.[4] Of course, Bailey, writing in 1814, said little, too, remarking in his *Dictionary* merely that a yacht is "a small ship or pleasure-boat," more or less setting both general and marine lexicographers on a course far removed from precise definition.[5] The term still drifts through English, a veritable Flying Dutchman of a word.

Yacht is a Dutch term, afloat in English since 1557 at least, its meanings as vague as its orthography. In the sixteenth century, the term defined a fast, small sailing vessel built in Norway or the Netherlands, but by the middle of the seventeenth, the English used it to designate sailing vessels kept by nobility for pleasure.[6] "I sailed this morning with His Majesty in one of his yachts (or pleasure boats)," wrote John Evelyn on October 1, 1661, "vessels not known among us until the Dutch East India Company presented that curious piece to the King."[7] At the beginning of the eighteenth, the term defined a small warship, mounting perhaps four to twelve light cannon and crewed by no more than forty men, but within half a century *yacht* again identified noble pleasure vessels, usually rigged as ketches, except that of the king, which had three masts, making it a miniature full-rigged ship.[8] *Yacht* never lost its mid-eighteenth-century connotation of nobility, even royalty, and early nineteenth-century American lexicographers puzzled about its meaning in a republic, a republic with few pleasure boats anyway.

Noah Webster brusquely defined *yacht* as a vessel meant for pleasure or for conveying heads of state, but Joseph Worcester went much further.[9] A yacht might be "a small ship or vessel of state, usually employed to convey princes, ambassadors, or other great personages, from one kingdom to another." But—and here Worcester voyaged far beyond his competitor—*yacht* is "a name also given to a private pleasure vessel when sufficiently large for a sea-voyage. A first-class yacht is one above thirty tons burden."[10] In defining a yacht as a seagoing pleasure vessel, and by actually specifying the displacement of one rated first-class, Worcester went further than many marine lexicographers, most of whom dismissed pleasure craft as unimportant in a world of warships, merchantmen, and fishing boats. In his *Dictionary of Sea Terms* (1851), Richard Henry Dana, the author of *Two Years Before the Mast*, defined *yacht* as "a vessel of pleasure or state," one of the crudest of his entries and scarcely less useful than one offered by John G. Rogers in his *Origins of Sea Terms* (1985): "A privately owned pleasure boat."[11] Occasional fine unabridged dictionaries echoed Worcester's insight—in 1896 the *American Encyclopedic Dictionary* insisted that a yacht must be decked, in 1914 the *Century Dictionary* emphasized that small steam launches might be classed as yachts, and in 1928 the *New Standard Dictionary* noted that yachts are seagoing pleasure vessels equipped with "luxurious furnishings"—but most dismissed the term while lavishing attention, often on the same page, on other marine terms like *yaw* and *yawl*.[12] Only Gershom Bradford's *Mariner's Dictionary* of 1952 advanced Worcester's argument into the twentieth century and into a changed world of pleasure boating.

"A sail, steam or motor vessel used for pleasure, and usually a fast, fine-lined craft," begins the Bradford definition. But it soon raises an issue ignored in earlier dictionaries. "The question may arise when a boat becomes a yacht? An answer might be: When the navigating and maneuvering of the craft becomes a task greater than the owner's capacity and the services of friends or hired hands are employed."[13] Anyone who stands long on the shore, who walks observantly along the great barrier beaches, who follows the salt-marsh creeks and estuaries into harbors and small bays understands the nuances underlying the Bradford definition. Boats cruise everywhere inshore, sometimes approaching close alongshore, while yachts move in deeper water and need all sorts of help.

Help, servants, crew, all traditionally at least partially distinguish a yacht from a boat, and have since colonial times. Typically a fishing vessel of some kind, even a sloop or small schooner, became a yacht-for-a-day, often on a summer Sunday, when its owner invited family and friends aboard for a short cruise or a passage to some picnic grounds. Southern plantation owners first explored the possibilities of keeping full-time yachts, using them chiefly to visit other planters along Chesapeake Bay. One Maryland planter attracted attention as early as 1689 for owning *Susanna*, "a small yacht or pleasure boate" he kept only for visiting, for waterfowling, and for sailing. Nearly a century later, one observer remarked that on Sunday morning the bay came "alive with boats," many carrying families to church, others pleasure fishing, others "sporting," the last simply sailing or rowing around. In 1772, Virginia exempted pleasure boats from toll-paying in the proposed Archer's Hope Creek to Queen's Creek Canal, and two years later the planters enjoyed regular racing off Tappahannock. Almost every vessel carried some sort of crew, even the boats rowing four or more oars, so that girls and women might venture without great effort and male owners and guests might devote themselves to sporting or to the women, not work.[14]

Planters, merchants, in time mill owners and eventually financiers had the means to build, maintain, and crew vessels that proved hideously expensive to own. While a fisherman might have a larger craft than the small sailing yachts about him, his smelled of fish, "stank of fish" some sensitive would-be guests might say. Keeping a boat sweet, using it only intermittently, and storing it in winter challenged the financial resources of many would-be yachtsmen, but the chief difficulty lay in finding a crew. Planters might use slaves, but other yachtsmen had to employ competent men who usually had regular employment, or else wanted Sundays off. Not surprisingly, merchants and other businessmen intimately involved with seagoing enterprises found yachting somewhat easier, since they had ready access to wharves, boatyards, and—above all—crewmen. By the end of the nineteenth century, yachting had become a rich man's sport, one boasting everything from *America's* Cup races to yacht clubs, one making clear to any telescope-equipped watcher that far offshore, beyond the casual glance of most Americans, existed a sporting world known only to the very rich, one dependent on servants, the crew many offshore yachtsmen still call "boat niggers" or "BNs" from their tanned skin and gross exploitation. Only rarely—very rarely—does anyone mark the extraordinary social and economic gulf manifest between the tourist on the wharf and the immense sailing yachts anchored in the outer harbor.

"In a sort of effortful casualness, he said to her, 'There's a boat out there to die for,'" reports Richard Todd of a pale-legged, Bermuda-shorted man speaking to his wife of the *Sumurun*, "a ninety-three-foot antique ketch with a teak deck, brightwork that glowed like furniture, huge raked masts, a cockpit big enough for dinner for twelve." On the wharf in a Nantucket Island harbor visited by the New York Yacht Club annual cruise of 1989, Todd glimpsed something ordinarily out of sight, something that loomed for an instant, only an instant, over the social or economic horizon. "For some reason this ordinary remark sent a little dagger into me. It seemed inexpressibly sad. As if he could actually imagine himself owning such a boat. As if the boat could be bought for something as relatively inexpensive as a life." *Sumurun* makes something crack and split just by appearing before ordinary beachgoers, before utterly ordinary Americans peering harborward from the edge of a pier. "It is a charm—perhaps a strength—of our society that we deny the vastness that lies between the middle class and the very rich, but this self-willed ignorance can be depressing."[15] Most tourists never encounter *Sumurun* and like yachts, the giant, antique sailing yachts that normally cruise far offshore, usually calling at select, otherwise unfrequented harbors, harbors with gated and walled yacht clubs. Yachts like *Sumurun* are real yachts, not imitations, and jerk any viewer toward the early twentieth-century economic gulf fitfully bridged by Depression, New Deal, and income tax. But the income tax taxes only income, not wealth, and anyway, *Sumurun* sailed before the income tax amendment, in the halcyon days of private railroad cars, immense estates, whole sections of coast held in fiefdoms like those along the western shore of Long Island or some reaches of Narragansett Bay or North Haven and other coves north of Bar Harbor. And yachts like *Sumurun* remain aloof, unapproachable, isolated by protective boat-nigger crews, stern officers, and far-out, deep-water private moorings reached only by yacht-tenders.

Yachtsmen require tenders to approach the shore—and to hint gently at the magnificence of the yacht anchored beyond easy public gaze. Fair and slim, built of white oak and cedar, usually varnished to a glistening, high-maintenance gleam that stuns observers unused to the finish of boats from times before income tax laws, yacht-tenders subtly announce seaworthiness and other-worldly luxury. Rugged yet sybaritic, their twelve to sixteen feet of varnished length often costing more than outboard-engine-equipped fiberglass boats far bigger,

they whisper as softly as Chopin's whispering sea. Tenders whisper secrets of the tall-masted yachts anchored in the distance, whisper of genuine canvas sails, fine food and wine, women long free of bikini tops, and men who own executives to execute orders in far-off offices. They whisper above all of near limitless time for long, long cruises, voyages to Labrador or Frenchman's Bay. Yacht-tenders cruise whisper-quiet across social gulfs, moving effortlessly in the margins of American social reality. Yacht-tenders move under oars, under white-ash or spruce oars almost perfectly silent.

As they carry owners and crews inshore from anchored, deep-draft yachts too massive for shallow water, dally at some public pier or beach, then return laden with groceries and chandleries, they momentarily connect the realm of the very rich with the everyday zone of ordinary Americans. Among the last small craft made to move superbly under oars, to move in silent, wakeless grace, yacht-tenders do more than exemplify the tending—what Bailey in his *Dictionary* calls the "nursing"—all yachts need, they accentuate the gulf that separates yachts from shore, yachts from boats. As a type of small boat, they have an odd, unremarked history that differentiates them from run-of-the-mill ship's boats, as William Falconer makes clear in his *Universal Dictionary of the Marine* in 1769: "a small vessel employed in the King's service, on various occasion; as to receive volunteers and impressed men, and convey them to a distant place; to attend on ships of war or squadrons, and to carry intelligence or orders from one place to another." Handsome, fast, darting hither and thither among the anchored fleet or squadron, the eighteenth-century tender carried impressed men and admirals as easily as the late twentieth-century tender carries boat niggers and old-money millionaires. And just as their Royal Navy counterparts did, contemporary tenders voyage long distances, only aboard yachts, hoisted on davits over the stern or, more properly, hoisted inboard, then spun upside down and secured atop deckhouse or deck.[16] A yacht is a pleasure vessel big enough to carry, not tow a tender, a real tender, not an inflatable excuse for one, and usually it carries the tender inboard, secured for offshore cruising. And always unlike a boat, a yacht cannot be beached, run up on the sand among run-of-the-mill Americans, those who own no yachts, or else own only boats, being small boaters.

All alongshore, boating for pleasure evolved much, much differently than yachting, and while pleasure boaters sometimes trick themselves into believing themselves yachtsmen— say in belonging to a yacht club that hires one or two summer-only wharf hands to help make fast boats or a cook and waiters to prepare dinners in token of full-time crews and servants attached to individual yachts—boaters understand that something separates them from yachtsmen, something as vast as the gulf Todd notes between the middle class and the very rich. Boaters stay close inshore, enjoy shallow harbors and bays, and rarely make long passages, say from Scituate to Bermuda, Duxbury to Cannes, Hingham to Honolulu. But their boats deserve sustained attention, for small boats—and the small boaters aboard them—furnish the seaward view, enliven the glim.

Boats are tricky furniture, however, and are often motionless, either moored just away from land or drawn up on the sand or shingle, sometimes upside down. Moreover, they prove devilishly tricky to photograph, in fact so difficult that many experienced professionals and thousands of amateurs refuse to photograph anything but big boats. "Some boats are more difficult to photograph in an interesting way than others," warns Bruce C. Brown in *Watershots: How to Take Better Photos on and around the Water*, a pithy, high-tech-oriented guide by an experienced alongshore photographer. "Unless your audience is made up entirely of avid dinghy sailors, small boats don't have the interest of large boats. They are also harder to shoot, because you have to get closer to fill your frame." While it is "relatively easy" to fill a frame with a forty-foot sailboat, especially when it carries all its sails, Brown continues, "it is quite another matter to fill the frame with a 14-foot International Finn." [17] By "interest," Brown apparently means complexities of detail, and small boats, even very small sailboats, lack the intricacies of rigging, steering gear, and other equipment that dignify—or clutter— much larger vessels. Small size and stark simplicity conspire to make boats difficult photographic subjects, therefore, and most nautical coffeetable-book and calendar-art imagery seconds Brown's advice on choosing large, complex craft or focusing on details, even tiny details, of small boats. Despite the proliferation of full-color nautical imagery, especially in

calendar illustration, the small boat is remarkably absent, or present only in some detail, say a thwart or oarlock or fragment of lapstrake planking.

Then, too, even stranded boats present extraordinary difficulties of parallax, difficulties beyond the level of Brown's book and known chiefly to painters. As early as the beginning of the nineteenth century, professional painters struggled to explain boat-rendering to struggling amateurs, amateurs far closer to their subject than Thackeray's cliff-top painter of seascapes. Samuel Prout struggled longer than most, beginning in 1816 with *Studies of Boats and Coast Scenery for Landscape and Marine Painters*, a guide-by-example to painting coastal realm subjects, especially beaches decorated with boats pulled up from the sea. Capturing the curving sheer of boats frustrated amateurs, and four years later, in *Easy Lessons in Landscape Drawing*, Prout offered more examples, routinely implying that boats might be approached most easily from the side.[18] Prout's books remained in print for decades, offering focused, pointed advice to painters trying to relieve the emptiness of water by depicting boats, or determined to paint stranded boats realistically. "The vessels are arranged on an ellipsis, and it will be

Depicting stranded boats as J. M. Turner does here in an illustration from his **Picturesque Views of the Southern Coast** *demonstrated an artist's ability to render very difficult subjects. (Houghton Library, Harvard University)*

seen, in both examples, that unless this line of objects be continued from the horizon to the fore-ground, the picture will be incomplete, and that, without the near boats, the distance would press too much on the eye, and hang over the fore-ground," he cautioned students of his *Hints on Light and Shadow.* Always he warned student painters and amateurs that "the lines of the sea and distance are diverted by vessels," and that without the vessels the viewer sees only so much blankness, an emptiness that destroys any foreground rendered on canvas.[19]

Contemporary manuals of advanced-level drawing and painting approach boat illustration with precision and trepidation, counseling the determined to conceive of the boat as a grid and to perform a subtle calculus of curves within it. In *Painting Seascapes in Oil,* for example, José M. Parramon advises students to follow the grid-the-canvas-first "boxing method" of Cézanne, especially when drawing a boat bow- or stern-on, and warns readers that small boats made fast to wharves present difficult but surmountable issues of parallax and double vanishing points.[20] Round-bottom boats, the sort Prout and other accomplished artists preferred for their delicacy and sophistication of line, rarely sit plumb on the sand, and artists discover that an imaginary grid must cant slightly in order that symmetrical gunwales appear so in final illustration.

For as long as the alongshore American merchant marine consisted of small wooden sailing vessels, particularly coasting schooners and brigs, artists could scarcely avoid the challenges of boat-rendering, for almost every harbor, every estuary sheltered some schooner or
other vessel with its boats ashore. Away from larger harbors, professional and amateur artists encountered fishermen's boats drawn up everywhere on the sand, usually slightly listing.
Although they might choose to ignore the boats, they knew also that a well-drawn or finely
painted boat demonstrated their mastery of their art, and fetched praise and high prices.
First-rate painters like salt-marsh devotee Martin Johnson Heade and luminist Fitz Hugh
Lane delighted in painting boats out of water, often making them the focal point of their can-
vases, and later painters, especially Winslow Homer and William Partridge Burpee, proved
their competence by accepting similar challenges. More than zeal for realism inspired the
delicate but accurate rendering of small boats in Heade's *Stranded Boat* or Homer's *Green
Dory* or Burpee's *Green Dory*. Throughout the nineteenth century, and well into twentieth,
painters understood that each section of the coast had evolved its own particular types of
small craft and that observant summer visitors delighted in noting the subtle details that distinguished one type from another—and in buying paintings that recorded such subtleties. In

In **The Stranded Boat**
(1863), Martin Johnson
Heade demonstrates a
fidelity to small-craft
line and detail that
delights any experienced
small boater. (Museum
of Fine Arts, Boston, Gift
of Maxim Karolik for the
Karolik Collection of
American Paintings,
1815–1865)

In **The Green Dory,**
Winslow Homer *demons-*
trates his understanding
that dories are indeed
local boats, varying
slightly from harbor to
harbor, and often
serving purposes other
than for which their
builders designed them.
(Museum of Fine Arts,
Boston, Bequest of Tracy
Cabot, 1912)

distinguishing between a general-purpose dory, the subject of his *Green Dory*, of 1891, and a Swampscott dory, the subject of a painting of that name a year later, Burpee demonstrated his acute understanding of nuances of boat design—and his awesome ability to capture them in paint.[21] Unlike Jewett's *Marsh Island* dilettante, Burpee perceived the uniqueness of a place, a uniqueness manifest in dories, banks, and Swampscott.

All along the coast, painters knew particular types of small craft as the key indicators of local color, local landscape, local seascape. "The type of craft used in any locality is invariably the outcome of local conditions," argued one writer in 1902. "The first and most serious of these forces is the depth of water in the approaches to the harbor in which craft find refuge either for shelter or traffic; the second, the width of the fairways; the third, the average strength of the winds; and fourth, the business in which they are employed."[22] An accomplished painter who produced an accurate rendering of a specific variety of small craft simultaneously revealed to educated viewers the context of the rendering, its alongshore setting.[23] A contemporary marine historian, even a knowing small-craft enthusiast, scrutinizes period paintings by Heade, Homer, and others not so much to examine the boats but to understand the local conditions, the peculiarities of locale that shape rig and hull.

Along this stretch of shore, nineteenth-century painters knew the catboat as the signature craft. Contemporary painters still do, as paintings hanging in waterfront art galleries demonstrate. And locals—and visitors with an eye for coastal-realm detail—still know the catboat as a "local" boat, something "from around here," good for local conditions, good for gunkholing under sail.

A catboat, even a fiberglass one sporting an aluminum mast and Dacron sail, glows with folk-art luminosity. Unlike other New England firsts from schooners to clipper ships, catboats originated in the simplest of early nineteenth-century Cape Cod and Narragansett Bay boatshops. Conceived by boatwrights who worked by rule of thumb and whittled half-models rather than from the elaborate plans so worshiped by marine historians today, the exact origins of the catboat lie beyond the horizon, always on the verge of looming clearly, for an instant. Small-craft historians have argued since the 1920s that Yankee builders modified a colonial New Netherlands hull like the shoal-draft ones depicted in so many seventeenth-century Dutch oil paintings, or tinkered with centerboard-sloop craft. But more likely, trial and error ruled the experimenters along Cape Cod Bay, southern Massachusetts Bay, and Narragansett Bay.[24] By 1850 everyone from Salem south to Cape May, New Jersey, knew the catboat as a distinct type, and—away from Barnegat Bay—identified it as a southern New

England craft. No one, however, knew about its name. "Turns quick as a cat" may explain it, or "Goes to windward like a cat," although a catboat, most unlike a cat, moves into the wind reasonably well. What really mattered to owners and observers, and to occasional passengers, was how the catboat performed as a fishing boat.

Yankee thrift shaped catboat form. The single sail and wide beam—catboats are often almost half as wide as long—meant ease of handling and generous work and stowage space. A man and a boy could work a cat as efficiently as two men and a boy could work a sloop, especially since its single big sail stayed mostly inboard, a real advantage in winter seas. But stability and roominess mattered less than the shallowness of draft.

The centerboard, the great wooden board that increased stability and kept the boat from drifting to leeward, moved up and down at the will of the crew. When crossing sandbars and other shallows, and when approaching the beach, the boy raised the board, letting the catboat, all thirty feet of it sometimes, float in just six inches of water. Men living along the shallowest of harbors found the catboat perfect for commercial fishing, and the boats went after mackerel and bluefish and in winter often towed scallop drags. Though a few made three-day

trips to Georges Bank and other offshore fishing grounds, most hugged the land, for the great cockpits, once flooded, meant almost certain sinking and so kept prudent mariners away from ocean rollers. For half a century after 1860, catboats patrolled the easternmost reach of New England identity, the outermost hazards and banks.[25] Westward-bound seamen and passengers knew Yankee waters, knew the shallow waters of inshore fishing, after they passed through the offshore fleet of American, Canadian, and Portuguese fishing schooners and encountered the first catboat, its man and boy working efficiently.

In the 1880s, perhaps even ten years earlier, catboat fishermen began chartering their little craft, taking parties of summer visitors to islands, down estuaries to picnic grounds, or simply for sails around shallow-water bays. Fishermen spent a week or so in May cleaning and repainting their catboats in anticipation of the summer trade. Day-trippers, hotel- and summer-cottage residents, and especially picnic parties delighted in the catboats that whisked them from railroad station wharves, town piers, and even beaches to distant salt-marsh islands or to fishing grounds. Many catboats worked regular runs, wedging themselves into summertime memory. One catboat, *On Time*, carried twenty-five YWCA girls at a time around Martha's

Vineyard, and the thirty-eight-foot *Lillian*, a massive black L sewn on its sail, ran between Nantucket Village and Wauwinet, one splendid day with a hundred and five passengers on board, a record for the broad-beam, wonderfully roomy fishing-boat-turned-taxi. In *Rugged Water*, *Mr. Pratt*, and the rest of Joseph C. Lincoln's countless best-selling Cape Cod novels, catboats performed similar routine, and sometimes heroic, service.[26] Stable, spacious, and always arriving gracefully in the shallows so inaccessible to keel-bottom sailboats, the big charter-cats set summer people to thinking.

As early as 1896, the *Rudder* had begun publishing in-depth articles about two sorts of catboats built only for pleasure, the ones built for wealthy owners determined to race, to push the

At the Mystic Seaport
Museum, a catboat
ghosts through the still
water, its giant sail
catching a faint breeze.
(JRS)

A catboat boasts an immense cockpit that comfortably seats a dozen and a movable centerboard that lets the boat probe the shallowest of waters. (JRS)

All along Cape Cod Bay, the catboat became a piece of harbor furniture. (author's collection)

hull and rig "to its limits," and the ones for families eager to explore shallows by themselves. Catboat racing was a short-lived phenomenon, one laced with innovation, rule-making, and near-disasters as owners built ever narrower catboats topped with taller and taller masts, and ballasted with movable iron bars or sand-filled bags. Shifting ballast to balance the immense single sails proved awkward and often dangerous, and by the turn of the century some writers, including Thomas Fleming Day, the editor of the prestigious yachting magazine, had begun to complain that some owners had bought catboats and lead mines together, the one for sport, the other for ballast. Moreover, a few very wealthy owners had begun to hire paid captains,

fishermen experienced in the ways of catboat behavior, and when other owners complained, the rich moved on, to other classes of boat-racing.[27] By 1908, when the *Rudder* published a two-part article on catboats in Massachusetts Bay, racing had become distinctly secondary in importance and the class a bit old-fashioned, certainly no longer attractive to the rich. "Some could afford to go into a more expensive class, others could not; but all are in the cat class from choice, and therein lies its strength," remarked Winfield M. Thompson. "They fit out their own boats, and are never happier than when burning off or scraping paint, putting on new coats of white or bottom green or red, and touching up spars and bright-work with varnish." Traveling salesmen, dentists, middle-level managers in the fruit-importing business and other industries, and one owner of a shoe factory struck Thompson as typical owners.[28] In the quiet that followed the turn-of-the-century racing flurry, catboat builders had found a new market for the slow, roomy, and stable catboats fishermen preferred.

Thompson chronicled the change in catboat-owning, and his understanding of the middle-class appeal of the boats intrigued Day, perhaps because the editor of the *Rudder* glimpsed the possibility of broadening his subscriber list. In "A Catboat Sailor's Yarns," a *Rudder* article of 1906 defending his choice of boat, Thompson explained that he would not be a deep-water amateur sailor if he had the chance, that he lacks the muscles, time, and money for long passages, that he has "a fondness for a snug harbor and a stout mooring at night," for comfort aboard. "Let the other fellow shiver in the lee of the foremast as he ploughs the raging main on a dark night, but give me a cushioned transom, a blanket and pillow, and a clear

This photograph from the August 1908 **Rudder** *is not at all unusual. It merely records a typical catboat outing. (Harvard College Library)*

conscience, and I'll pass the hours in a pleasanter way." After financial reverses, wondering about buying a yawl or knockabout, and a long spell of looking at overpriced secondhand catboats, Thompson bought a new twenty-five-foot catboat and immediately rebuilt the cabin interior. His article focuses more on the "grub box" for vegetables and canned goods, the twenty-gallon water tank, and the "convenience station" chamber pot than on the rig or hull, clearly indicating his preference for a shallow-water sailing boat plentifully supplied with cushioned bunks, good food, and a tiny stove for warding off the evening chill. For *Twister*, fully equipped, and a new rowboat to serve as tender, Thompson paid $860, of which $6 went for dishes and $30 for cushions.[29] Thompson expected to win no races but to cruise comfortably in shallow bays and harbors.

Others quickly grasped the implications of comfortable gunkholing in shallow, uncrowded places. As early as 1902, Cape Cod catboat builders began furnishing their boats with gasoline engines, making entering Green Harbor and other silted-up salt-marsh-backed harbors far easier, and soon all sorts of conveniences beyond chamber pots became common.[30] *Sea Wolf*, a twenty-eight-foot cruising catboat Thompson described in a *Rudder* piece of 1907, easily carried ten or twelve people on its cockpit seats, and slept four in its spacious cabin. More than an electric-start engine graced *Sea Wolf*. The catboat carried an ice chest, a stove fitted with a kettle for cooking "a peck of clams or a dozen lobsters" at a time, a full-length clothes locker, and a variety of other conveniences from shaving mirror to chart cabinet, all purchased for just under $1500.[31] Exactly as Thompson had guessed a few years earlier, the cruising catboat had come into its own, not for its speed, not for its deep-sea cruising ability, not even for what it purported to be, a shallow-water cruising vessel, but simply because it provided enough roominess to store all the comforts of home—and the shallow draft to let it approach any beach well within wading-ashore distance.

For thirty years after the turn of the century, summer people and year-round residents bought catboats from local builders only too pleased to spend winters framing up eighteen- to twenty-four-footers complete with roomy cockpits, small but comfortable cabins, and, as always, the movable centerboards and sternward-jutting "barn-door" rudders that let catboats roam the shallows. Easily handled, and so immediately accessible as a rental boat to tourists eager to get out on the water, the small catboats sailed matter-of-factly up to any sandy beach or stayed perfectly upright when miscalculation of tides set them aground at low tide. Inexpensive, especially in the twelve-foot size that nevertheless provided much room, rugged, and nimbly handled by children, catboats were safe in squalls and wonderfully able to reach lonely beaches. People sailed catboats not only for the pure joy of sailing but to get *to* alongshore places, to the special beaches and islands away from crowds arriving by motorcar, to the clam flats too far for rowing, to the inlets beyond the reach of deep-draft sloops and other sailboats, to inlets too shallow for engine-boats decried by the early twentieth-century gentry as unpardonably noisy and smelly.

Yet no matter how roomy, no matter how comfortable, catboats remained boats, not yachts. Thompson and Day missed the widening gulf between yachting and boating, between the "salt" and the yet-unnamed inshore boater, the gunkholing, comfort-seeking, shoal-draft mariner anxious to keep near the beach, to ferry self, family, and friends from one picnic spot to another. And Thompson and the editor of the *Rudder* missed, too, the fact that catboats had grown beyond the means of many boaters, those boaters whose craft lacked rudders, those boaters who rowed, who "pulled." No amount of careful editing nor cautious mixing of articles and advertisements could make the *Rudder* equally useful to post-1930s deep-water yachtsmen, inshore boaters, and summer-vacation rowers. In time the great generalist magazine failed, and a variety of specialized magazines filled its place, one or two of some use even to rowboaters, the smallest of small boaters.

Fishing seduced many landlubbers into rowboats, especially the rowboats available at turn-of-the-century boat liveries, usually ramshackle sheds and wharves sheltering five or twenty rowboats. Duck-hunting ventures helped rent rowboats in autumn, and the prospect of a cool breeze on a torrid summer day attracted customers, too, but fish lured men and boys—and some women and girls—into the small craft that served ordinary Americans eager to venture onto salt water.

Until late into the nineteenth century, parties of sport fishermen routinely chartered small fishing boats to take them "out to the fish," usually a mile or so offshore. The scrubbed-clean fishing boat often carried girlfriends or wives and children, too, and the fishing excursion doubled as a sort of fresh-air outing, a one-day, fair-weather cruise. Ordinarily the voyages began shortly after dawn, and frequently attracted people "who had never been on the water before," like those young men and women who accompanied the anonymous author of a *Knickerbocker* article of 1842, "My First and Last Sea-Fishing." Such excursions made intriguing visual adventures at least, as "first the tall spars of the multitude of shipping at the wharves melted and mingled in one confused, interminable mass of light tracery-work; then long blocks of substantial warehouses of brick and granite gradually lost their individuality," then the entire city vanished into the haze, and the party-goers imagined themselves on some transatlantic voyage. Interrupted by the usual foolishness—one member attempted to hail a passing schooner by shouting technical questions and received "Go to hell!" for an answer—fishing involved mostly the baiting of hooks with raw clams, the attaching of sinkers, and the dropping of lines, all under the supervision of the master and one-man crew, and the catching of one or two fish. A stop at an island for refreshments and courting, a brief thundershower, and then a slow, rather tiring sail back to land completed the adventure. "Among other things, she had learnt this," remarked the author of one young woman, "never to go a fishing in a new challey dress and rose-colored stockings."[32] And the author had learned, among other things, that chartering a fishing boat did not guarantee catching any fish.

Half a century earlier, Timothy Dwight had learned to expect no fish, indeed to plan on purchasing fish from fishermen encountered during the fishing trip. "The day was very fine, and the company agreeable," he wrote of what became in time the typical outing of the educated elite. "Our business professedly was fishing, in which we were perfectly successful. It is true, we neither caught, nor attempted to catch, any fish; but we made an excellent dinner of very fine ones."[33] But in 1797 enough of the old work ethic lingered to make men want some reason for spending a day on the water, and fishing provided it. So long as the men brought home dinner, and perhaps tomorrow's dinner, too, the day had been spent profitably, not in wanton idleness. Within fifty years, however, the fear of idleness—and the consequent desire to return with fish—had almost wholly vanished from the unmarried young men and women who chartered the boat described in the *Knickerbocker*, although the men hoped to catch fish to demonstrate their manliness and skill to their sweethearts. By the 1890s, parties no longer expected charter captains to cruise to fishing grounds at all, as many catboat skippers discovered to their delight. A sail alone justified the expense, or a sail punctuated with a visit to a sandbar exposed at low tide. But by then, finding a small fishing boat had become considerably harder.

As George Herbert Bartlett makes clear in his masterpiece, *Water Tramps; Or, The Cruise of the Sea Bird* (1895), the fishing industry had begun to change dramatically in the 1880s, so much so that a group of college men could actually pay for their summer vacation by fishing from their rented sailing boat—and retailing their fish to summer residents along the New England coast, or now and then selling it wholesale. Crammed into *Sea Bird*, a sloop so small that it came with twelve-foot oars to move it in calms and so deep-draft that it came with a tender, the young men are adept not only at catching all sorts of fish but at selling them door-to-door, extending their sailing vacation week by week and their cruise ever further northward from New York. *Water Tramps* at first appears improbable, but Bartlett clearly understands that small harbors and inlets no longer sent forth full-time fishermen, that the fishing industry had moved to larger and larger schooners and other craft, to specific ports like Gloucester and New Bedford, and to engine-powered vessels that delivered cargoes to wholesale markets in Boston and New York.[34] As immigrant Irishmen and Italians took over the larger-scale inshore fishery off much of New England, often by embracing innovations like engine-driven dories

scorned by tradition-bound Yankees, summer visitors began expecting to buy fish delivered by train.[35] Bartlett's book makes clear the root cause of the transformation that made fishermen anxious to charter catboats to picnicking families and that made fishing boats scarcer and scarcer in shallow-water harbors and inlets.

A variety of factors combined at the turn of the century to make sport fishing increasingly difficult. Pollution from municipal and industrial sources drove fish first from many rivers, then from estuaries, then from small bays. Over-fishing, especially taking fish using staked nets and weirs, decreased breeding stocks dramatically, as did trawling from steam-powered vessels. And the increasing popularity of sport fishing thrust more and more amateur fishers into competition with professionals. Moreover, cyclical downturns in fish populations may

Wharves, bridge approaches, and skiffs provided access to channels filled with large fish, but only the sure-footed ventured along the stringers that topped the spiles. (Mystic Seaport Museum)

have coincided with human impacts to make inshore fish scarce.[36] Some full-time fishermen turned to oystering and lobstering, and many began to fish only part-time, working on land during the newly named "off season."[37] Others spent their summer days charter fishing, taking out "sports" after bluefish or whatever species was "running," or simply ferrying families from one deserted beach to another, then spent the winter building catboats or rowboats to sell to the summer crowd.

Full-time and part-time boatbuilders faced technological challenges by 1915, however, as the outboard engine delighted and infuriated experienced and novice boaters alike. Within a matter of decades, certainly by the late 1930s, the outboard engine had transformed pleasure boating and permanently refurnished seacoast scenery. In 1935, when it closed its doors in the face of economic depression and impossible, insurmountable change, the Crosby Catboat Company had built some 3,500 catboats since its first in 1850, but even its superb shallow-water sailboats could not compete against outboard-equipped boats. To be sure, the hurricane of 1938 devastated boat shops all along the New England coast, driving many builders permanently out of business, but it destroyed thousands of small boats, too, including scores of Martha's Vineyard catboats.[38] After the storm, however, many saddened owners decided on a "modern" boat, not some local type peculiar to—and especially suited to—local waters. Boatbuilders who rebuilt their shops were faced first with customers demanding novelties, then with the difficulties of a prolonged war. And peace brought only bigger outboards and more technological changes, changes that quickly manifested themselves everywhere in the coastal realm.

Outboard engines properly belong in ponds and lakes, maybe in rivers, in any body of water that lacks tides. In spite of their much-advertised "tilt-up" feature, outboard engines do very badly indeed when run into rocks or sandbanks, as inexperienced, teenaged boys learn when beckoned ashore by teenaged girls. Who are these micro-bikinied girls, waving at the boys speeding past, just beyond the low surf, if not the sirens of the late twentieth century? Secure and smug in wisps of nylon, confident in sunscreen and blowing hair, the girls wave and laugh as the fiberglass speedboat parallels the low breakers—then they turn and skip along the wavelets. Beyond the surf occurs conference, then decision, and the fast boat noses beachward, crew standing, holding the windshield frame, watching as the girls skip away, somehow deaf to shouted greetings. The outboard engine touches bottom once, then again. The boat pitches forward, slews sideways, beam on to the last roller, the first breaker. One boy jumps overboard to hold the rolling hull, finds himself afloat in water too deep to stand, finds himself bucked up and down by the boat. Another boy drags him aboard, someone else shoves forward the throttle and turns the steering wheel toward the sea, toward deeper water. The engine roars, tilts, throws up the "rooster tail" that delights the locals, the longtime summer visitors reading racy novels, even the historian distracted from guzzle-pondering. Another rooster tail, another breaker, and the engine chokes on sand, coughs, and dies. The heavy fire-engine-red hull swings broadside again, rolls ponderously twice while its panicked crew probes with boathook and six-foot oar, then struggles to remember where someone stowed an anchor. The low surf wins, and the boat grounds in the shallows, attracting the five-year-old set fresh from making sand castles, then the mothers of the five-year-old set, then the tourists eager to glimpse a boat up close, a boat not awash, not wrecked, but surely and definitely aground, most decidedly stranded. A boy jumps out, tilts up the engine, then locks it in place. One or two locals, old-timers walking back from surf-fishing maybe, or from beach-combing, or from walking for no other reason than to watch minor boating disasters, saunter over and suggest pushing the craft to sea before the tide falls further. The teenaged boys, reduced to awkward little boys, begin to push, struggling to hold bow to breakers, feet to sand. Beachgoers join in the fun, in launching the giant soap-dish-shaped fiberglass thing wallowing now in the surf, almost consciously refusing to reach water deep enough to drop the outboard engine. Far enough off to be unrecognizable, but not too far off to see everything clearly, the girls stand and watch, then turn and caper along the wet sand of the ebb, perfectly pleased with themselves.

Outboard engines belong away from the coastal realm, away from rocky and sandy hazards like those that frightened Champlain and the *Mayflower* crew and passengers. They belong in the Great Lakes, in the waters for which Ole Evinrude and other inventors originally intended them, in places more manageable than the surf, however gentle the surf may be. Experienced outboard-boaters fear the shallows, fear the damage inherent in any collision, fear ripping off the whole lower half of their engines, fear "shearing a pin" and being powerless while they make temporary repairs. Nowadays, only the brave or foolhardy bring an outboard-powered boat to shore, unless they have a very small boat indeed, perhaps one of the last wooden flat-bottomed skiffs, or unless they know the location—and tide tables—especially well.

For generations, shallow-water boatmen used flat-bottomed skiffs as reliable transportation not only across the shallows that covered sandbars and clam flats but from moored catboats and other vessels to sandy beaches a few yards away. The traditionally built "flat-bottom" skiff rows well, its bottom being flat only from side to side; looked at from the side, a flat-bottom skiff drawn up on the sand shows its pronounced "rocker," its bottom rising from the sand at its bow and its stern, much as a rocking-chair rests with its ends above the floor. Combined with a slightly narrowed stern, the rocker makes the skiff row easily enough, not as easily as a round-bottomed yacht-tender, of course, but well enough, especially over short distances. Most important, the rocker permits the skiff to glide up on sandy beaches and to sit upright when aground at low tide, both useful traits for fishers and others transporting cargoes—and passengers—ashore.[39] Skiffs sacrifice ease and speed under oars for efficient beaching and for great initial stability.

Unlike the turn-of-the-century yacht-tender, the turn-of-the-century skiff shifted only slightly when an adult stepped into it from a larger craft or wharf. The most clumsy of first-time passengers, even inland folks renting a rowboat at some salt-creek boat livery, found the stability reassuring. Novices sometimes stepped onto the seats of skiffs, and still the little boats stayed upright. Rowboats built for speed, say narrow punts, lacked such initial stability, and savants like William Cooper warned against them. "I have witnessed some very ugly accidents owing to the incautious handling of small, cranky punts," he wrote in 1888 in *The Yacht Sailor*. "In fact, I would strongly recommend you to avoid those very small, handy-looking cockle-shells; they are very neat looking and pretty to the eye, but are nothing better than mantraps."[40] Experienced boaters—and Ratty and Mole in *The Wind in the Willows*—favored such vessels for calm waters, but few boat liveries rented them, since they only appeared stable, unlike canoes, for example, which appeared tippy at the start and so warned even novices. Yacht-tenders tipped—and still tip—perilously when novices clambered into them, and nearly capsized when someone stepped onto a seat rather than into the center of the bottom, but unlike skiffs, they boasted a splendid degree of subsequent stability, being able to cruise through waves, wind, and swells, to do everything better after the beginning, except land on a beach. Yacht-tenders and other round-bottomed boats eased ashore badly and at low tide tipped to one side, requiring attention to cargo. Not surprisingly, then, turn-of-the-century builders favored the easy-to-build skiff as the vessel best suited to landlubbers who demanded an inexpensive, fairly stable, more-or-less easy-to-row, shallow-water craft.

These skiffs delighted pioneer outboard-engine manufacturers and users, since their wide, flat sterns sank very little under the weight of the engines hung over them. Even at rest, floating next to some landing stage or small wharf, the outboard-fitted skiff looked seaworthy, even comfortable. And under way it did not bury its stern in the sea, or swamp during turns. Round-bottom rowing boats, including the finest of yacht-tenders, behaved incredibly badly when fitted with outboard engines. Out-of-trim from the moment the outboard dropped onto the transom, round-bottom boats demonstrated an extraordinary, almost malevolent propensity to capsize during routine turns, say the turn made by a boater avoiding a half-submerged log or attempting to come about in a seaway.[41] By 1920 the flat-bottom skiff had triumphed over all but the finest of yacht-tenders belonging to die-hards who loved to pull a boat under oars or hired an oarsman or two to pull. Subtly reshaped for the heavier and more powerful outboards that appeared by the year, the flat-bottom skiff soon boasted less and less rocker and an ever wider stern, the better to support the heavier engine. While sometimes dismissed as "floating flatirons" or "isosceles triangles," the wedge-shaped skiffs that no longer rowed well—indeed scarcely rowed at all—delighted more and more seashore visitors.[42] Even catboat owners knew by the late 1930s that novice boaters preferred to buy an outboard-equipped boat, that sailing had become almost as obsolescent as rowing a skiff onto a sandy beach and unloading a picnic basket.

Powerboating had entranced wealthy families since the mid-1880s, when many purchased steam-propelled launches and learned to endure heat, smoke, and cinders for the sake of chugging past becalmed sailboats. But coal-fired boats were tricky and tiresome to operate as pleasure craft, and by the mid-1890s, when the *Rudder* established its "power department" column, naphtha-, electric-, and gasoline-powered boats had begun to appear as the novelties of well-to-do boating families.[43] "Runabouts" designed around gasoline engines mounted amidships quickly supplanted earlier naptha-fueled launches and dispatched storage-battery boats to lakes that lacked currents and tides, soon earning the emnity of pleasure boaters devoted to rowing and sailing. As automotive technology improved, inboard-mounted engines became more reliable and more powerful, and although traditionalists condemned them as "stinkpots" and worse, by 1910 the engines had begun to appear as auxiliary power in larger sailboats, even in catboats.[44] Runabouts could not approach shallow water; their downward-jutting propeller shafts proved too fragile for collisions even with sand, and what shaft and propeller did not hit, rudders did. So for a brief moment, perhaps from 1895 to 1905 if boating magazines are a fair indication, die-hard rowboat enthusiasts, canoeists, and swimmers enjoyed shallow water free of motorboats, and confirmed sailing enthusiasts cruised far enough offshore that they rarely encountered the vessels equipped with unreliable, often balky engines.

But as powerful, reliable engines and ever-smaller powerboats ended the interlude of segregation, sailing enthusiasts found themselves besieged by a new generation of boaters and began arguing that powerboat operators ought to be regulated, licensed, and even banned from inshore waters. By 1908, in response to complaints from sailboat owners, and apparently from masters of tugboats and other commercial vessels, the federal government had determined to regulate the use of runabouts according to the steam-navigation rules in place since 1852.[45] The *Rudder* and other magazines, understanding that traditionalist boaters still comprised a majority of their readers, began to feature both pro-regulation articles and articles condemning the behavior loathed by pro-regulation writers. "The manner in which the hundreds of newcomers handle their ships is a matter of amazement to the old hand," opined Walter M. Bieling in 1908 in one article bluntly titled "Mishandling Boats." "A favorite pastime to the owner of a little 15 by 4 tub, with a firecracker engine, is to dash madly through crowded anchorages, cutting as close as his unlimited nerve and limited skill will permit to the sterns of anchored yachts, the consequent pitching and rolling of the boats and the ill-concealed annoyance of their owners causing the marine pest to chortle with joy." Bieling insisted that some regulation had to occur. "I can today, figuratively speaking, buy the largest gasoline pleasure craft in the world, and I can start her up the Hudson River or through the East River in

charge of a ten-year-old boy from the plains of the Dakotas." No matter how young, no matter how ignorant, anyone could take charge of a gasoline-powered runabout, whatever its horse-power and size. But not so a steam-engined craft: "Were this same ship equipped with steam, or were she only a six-foot steamer, she would have to be handled by experienced, licensed men."[46] The family in the comfortable catboat cabin, snuggled down for the evening accord-ing to the dictates of Winfield Thompson, soon learned to despise the wake left behind every runabout, had learned to fear capsize and ramming, sometimes at the hands of ten-year-olds.

Young boys, even ten-year-olds, learned about motorboating not only by watching *Viper* and *Flash* and other similarly named boats slicing through the water and slamming over rollers but by reading a best-selling series of juvenile novels by H. Irving Hancock. The "Motor Boat Boys" swung the imagination of thousands of adolescents away from pirates and sailboats to the thrills of engine-boating. "Up to within nearly two hundred yards of the dock the *Rest-less* dashed in at full speed," Hancock writes of a typical landing in *The Motor Boat Club and the Wireless*, one of the volumes written in 1909. "Then signalling for half speed, next for the stop, and finally for the reverse, Captain Tom swung the yacht in almost a semi-circle, run-ning up with bare headway so that the boat lay in gently against the string-piece."[47] Unlike the sailboat-equipped teenagers in Arthur M. Winfield's *Rover Boys on the Ocean* (1899), Hancock's novels present boys thoroughly up-to-date, as totally confident with inboard en-gines as they are with wireless telegraphy, wholly able to instruct grown men in the fine points of engine-boat operation.[48] Whether in *The Motor Boat Club at Nantucket* or *The Motor Boat Club off Long Island* or in any of the other four novels, however, Hancock's young heroes are paradigms of recklessness under power, of engine boats mishandled.

As late as 1907, safety concerns meant the concerns of runabouters, concerns like drift-ing helplessly when an unreliable engine chose to quit, or burning to death when a leaky gasoline tank blew up. Boating magazines ran articles with such titles as "Superstition on the Decline" and "The Safe and the Dangerous Gasoline Tank," all educating motorboaters in the dangers of careless use of new technology.[49] Even in the summer of 1907, *Country Life* and other general-interest magazines featured articles like Payne Martyn's "Era of the Motor Boat," which emphasized that most motorboats are "family boats," safe, comfortable craft under thirty feet supplanting "naphtha launches, sail boats, and canoes," with only a minority built for speed.[50] But after that summer, the public dangers of the speedboat, as runabouts and engine launches came to be called, began to inform magazine articles, especially those emphasizing "courtesy afloat" and the "rules of the road." Hindsight makes discernible the

cause-and-effect relation of something early twentieth-century writers failed to notice: motorboating became wilder and more dangerous not only as engines became more powerful but as state after state passed restrictions governing the safe operation of automobiles.[51] The sea became the refuge of the "speed demon," the show-off ruled off city streets and state highways.

Outboard-engine-equipped boats only added to the safety issue, not only because they moved in water too shallow for inboard-fitted motorboats but because they actually moved faster. Although yacht clubs and the boating industry—and boating magazines—lobbied intensively to prevent the rules of steam navigation from applying to motorboats and struggled to educate powerboaters in navigational rules and simple courtesy, the outboard engine was so cheap that it enticed thousands more families onto the water, families whose votes swayed Congress from regulating engines. Throughout the 1930s, as firms like Evinrude and Johnson emerged as national manufacturers of reliable, portable power, outboard-boating became the most common sort of boating, distinct from yachting and sailing, utterly distinct from rowing and canoeing, and increasingly relaxed, refined, and safe, as outboarders exercised subtle but growing pressure on speed demons, giving them a bad name. By 1939, when Congress authorized the U.S. Coast Guard Auxiliary "to promote safety on the water" and otherwise aid

EVINRUDE DETACHABLE ROWBOAT AND CANOE MOTORS

In planning your vacation this year, don't fail to include an EVINRUDE in your list of "THINGS NEEDED."

An Evinrude really is needed if you want to enjoy every minute of the time. Think of what it means to be able to take with you to lake, river or seashore a powerful little marine motor that can be clamped in two minutes to any kind of craft —rowboat, sailboat, houseboat or canoe —and will run four hours on less than a gallon of gasoline.

The Evinrude drives an ordinary rowboat at the rate of 7 to 8 miles an hour canoes 10 to 12 miles an hour. Women and children can operate it the first time they try. It starts by giving the flywheel a quarter-turn and is stopped by pressing a push-button. May be carried anywhere as easily as a valise, or shipped in an Evinrude trunk as baggage.

Let us send you (free, of course) a handsomely illustrated booklet describing this wonderful little marine motor that numbers among its users: Peary, the discoverer of the North Pole; Stefansson, the explorer; Theodore Roosevelt, who purchased two Evinrudes for his South American Expedition; the Governments of 22 countries, including the United States; and thousands of sportsmen and pleasure-seekers in all parts of the civilized world.

There is an Evinrude dealer in your town—we'll send you his name on request. Just say:—"Mail me a copy of your booklet and tell me where I can see an Evinrude."

EVINRUDE MOTOR COMPANY, 100 Evinrude Block, MILWAUKEE, WIS., U. S. A.

Distributing Branches: 69 Cortlandt Street, New York, N. Y.—218 State Street, Boston, Mass. 436 Market Street, San Francisco, Cal.—182 Morrison Street, Portland, Ore.

Very early, as this Vanity Fair advertisement of May 1915 implies, manufacturers of outboard engines focused on women as an important market. (Harvard College Library)

In 1939, Evinrude advertised its "Mate" engine, a lightweight outboard engine that made every woman a captain and made every strong-backed, strong-armed man unnecessary to the success of a small-boat cruise. (Antique Boat Museum)

the Coast Guard, many older safety fears had already disappeared, flat-bottomed, triangular skiffs had been replaced by V-bottomed boats designed especially for outboard engines, and families had discovered the joys of light, comfortable, reasonably fast boats.[52]

More important, however, women had discovered the outboard as something especially delicious.

Early in the Depression, outboard-engine manufacturers had begun targeting female would-be users, partly to broaden their markets, partly to convince wives that husbands should have outboards. Far more suffuses their advertising illustrations than athletic women in swimsuits, however, for in many advertisements the woman is not scantily clad at all, or even scarcely visible. Often she is in an automobile, gesturing with delight at the engine lashed to a fender or running board, the boat conspicuously absent and presumably waiting in all its skifflike flatness at a boat livery down the road. Sometimes she—or she with another woman—bends over the outboard, preparing it for sea, perfectly competent and perfectly away from men. And by 1940, especially in Evinrude ads for the half-horsepower, ten-pound "Evinrude Mate," she stands proudly erect, holding the mate that has set her free not only from rowing but from having a man row. Magazine illustrations, even magazine covers, quickly seconded the advertising, often by depicting groups of women voyaging far away from men.[53]

Whereas turn-of-century women pulled yacht-tenders and other made-for-rowing boats as easily as the farm-woman heroine in *A Marsh Island*, by the late 1930s many had become prisoners of men almost literally prying made-for-outboard skiffs through still harbors and along salt creeks. Once boat-livery owners understood how many renters arrived with outboard engine in hand, they began to rent skiffs designed for outboards—triangular skiffs that rowed badly. The hapless powerful man, let alone the hapless woman, athletic or otherwise, who rented one of the outboard-less skiffs quickly determined to give up rowing and rent a boat fitted with an engine—all to the profit of the livery owner. And the outboard-fitted skiff, renters quickly learned, could approach the beach scarcely closer than any inboard-equipped boat. Women delighted in their freedom from male rowers but found their freedom to approach the shore markedly proscribed, even when they hauled up the outboard engine and proceeded across shallow water under oars. Skiffs big and beamy enough for outboards were hard to move across the last few hundred feet of shallow water and, once aground, difficult to move off beaches and mudflats.

In 1912, at the moment engine-boats and even outboards began their three-decade tri-

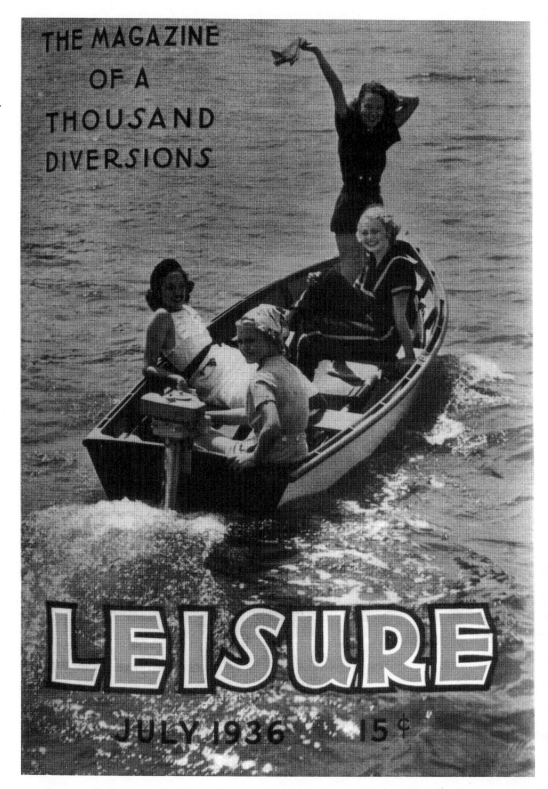

THE MAGAZINE OF A THOUSAND DIVERSIONS

LEISURE

JULY 1936   15¢

umph over local oar- and sail-propelled craft, best-selling writer Joseph C. Lincoln glimpsed the emerging small-boating problems of summer visitors to Cape Cod, and warned that "a lap-streaked, keel-bottomed dingy, good enough as a yacht's tender or in deep water," made "the worst boat in the world to row about Denboro bay at low tide." *The Rise of Roscoe Paine* advertises the common sense of locals by describing their boats, especially their flat-bottom, made-for-rowing skiffs. "The pull was a long one, but I enjoyed every stroke of it," the narrator relates. "The tide was almost full, just beginning to ebb, so there was scarcely any current and I could make a straight cut across, instead of following the tortuous channel. My skiff was a flat-bottomed affair, drawing very little, but in Denboro bay, at low tide, even a flat-bottomed skiff has to beware of sand and eel-grass." In his skiff he rescues two summer visitors, first towing them back toward land around one mudflat after another, all the time

All alongshore, locals viewed powerboats with disdain and envy, but above all with suspicion, worrying always that engine failure might lead to shipwreck. Here an illustration from Joseph C. Lincoln's Rise of Roscoe Paine *shows how the local boy gets the girl away from the rich tourist. (author's collection)*

arguing about the channels and guzzles too shallow for their yacht-tender, then shifting one into his skiff while still towing the other. Roscoe Paine's rise in the world begins on a falling tide, when he and his local boat, a well-rockered, flat-bottom skiff, rescue a lovely summer woman from the clutches of a summer dandy aground in a yacht-tender, a dandy so ignorant of boating that he loses an oar. The woman, so beautiful that the local man keeps his eyes on the bottom boards of his skiff rather than look at her, understands her earlier danger only as a squall strikes the bay. "The channel beyond the flat," an opening so shallow as to be a mere guzzle, "was whipped to whitecaps in a moment and miniature breakers were beating against the mud bank where the dinghy had grounded." Safe from shipwreck—or boatwreck—and a good wetting, perhaps even drowning, the woman only vaguely realizes that the skiff, and its owner, are special, both local, both fitted for the place. "The tide was too low to make use of the little wharf, so I beached the skiff and drew the towed boat in by the line," the hero concludes, and the heroine steps dry-shod from the local boat so daintily touching the beach.[54] As Lincoln guessed, within ten years stepping dry-shod had become as anachronistic as pulling a local boat. By the early 1950s such landings were almost unknown.

Engine-powered boats, especially outboard-powered boats, fared worse in shoal-water bays and inlets than the dandy's one-oared, lapstrake yacht-tender. Unlike catboats, made-for-pulling skiffs, even light dories and other local boats, post–World War II motor boats could scarcely close the land, nestle against sandy and rocky shores, discharge anyone, heroines or otherwise. Instead they zoomed faster and faster, pulling fishing lures, then aquaplanes, then waterskis, roaring from one deep-water anchorage to another, becoming specks in from the horizon, beyond scrutiny of swimmers and beach walkers. After the late 1950s, when fiberglass boats began to replace wooden ones, even backyard builders despaired of owning a "picnic boat" or a "beach boat," something made for the seashore, for low tide. Coast Guard crews began their long change of course from search-and-rescue efforts to policing speeding, to enforcing the rules of the road, to inspecting boats too large, too deep-draft to nudge into beaches. Yachts remained offshore, over the horizon, only now and then visiting anchorages useful to tenders. Engine-powered boats and deep-keel fiberglass sailboats stayed inshore from yachts, but always away from beaches, from rocks, from salt-marsh creeks, from all the shallow-water obstructions so devastating to propellers and keels. And as the small boats stayed offshore and only the smallest of small boats poked their lonely way along the beach and salt-marsh shallows, all alongshore everything changed.

Everywhere, especially around harbors already nearly bereft of small commercial fishing vessels, the alongshore concatenation of space and structure dependent on wooden boats vanished not into thin air but into air-conditioned restaurants and condominiums. "Boating is quickly becoming 'yachting' again, in the Morgan tradition," complained one writer in 1986 in an issue of *Offshore: New England's Boating Magazine*. Utterly unaware of the colossal irony implicit in the title of the magazine in which his article appeared, David J. Damkoehler argues in "The Vanishing Marina" that as boating changed so did the whole boating-support enterprise, but so subtly that no one noticed.[55] Yet in noticing what had left the coastal realm, Damkoehler unwittingly points readers at the taken-for-granted importance of marinas, marine railways, ship chandleries, and other businesses dependent on traditional small boats, on local boats, on everyday close-inshore maritime activity.

When the Coast Guard determined to remove the buoys that marked the shifting channel into the North River and its salt-marsh creeks, a marina owner marshaled opposition. To be sure, self-interest must have been some motivation, but then again, a marina owner is both on the water and onshore, on a spot where any small boater in catboat, yacht-tender, skiff, or fiberglass speedboat can draw breath, relax, and mutter "safe," but find no railway, no shop, and little help.

6 HARBORS

P lastic boats last forever. No maintenance. Strong enough to endure groundings, even hitting rocks. Perfect for powerful outboard engines. Utterly modern. Such was the song the siren salesmen sang.

Like the girls who lured ashore the boys in the fiberglass speedboat, mid-1960s boat-show sales forces lured small boaters away from wooden craft. Plastic boats, soon routinely designated fiberglass in a rapid-fire effort to distinguish hull material from styrene and other likely-to-shatter plastics then appearing in toys and housewares, slipped perfectly into the futuristic, science-oriented, what-chemistry-does-for-you 1960s. First a polymer-resin-based patching material, quick-fix stuff to slop over gouged planks or an annoying patch of rot, then a near-miracle synthetic to be painted skinlike over every exterior inch of a new-made wooden boat, fiberglass soon was molded, formed into whole boat hulls.

Plastic boats first pleased no one, not even devotees of mass-produced, instantly identifiable boats like those made by Chris-Craft, Elco, Lyman, and other great firms that built hundreds or thousands of standard or "stock" boats each year. Plastic boats sounded odd, especially when propelled by powerful outboards across steep, choppy waves. While not the nerve-wracking "thrang, bang" made by post–World War II aluminum boats, boats so light that fluky gusts of wind sometimes flipped them over as they planed above waves, the "thuckety-thuckety" sound of early plastic boats implied imminent cracking, splitting, or whatever these boats did when the plastic fatigued and gave way. Then, too, the sound derived from the density of the stuff, as well as its thickness, and plastic boaters soon learned how much their maintenance-free boats weighed. Many touched sandbars and beaches to become almost immovable objects, and smaller ones defied lifting, even dragging up beaches to boathouses. And, of course, they sank.

Local boats once comprised the foreground of almost every harbor view, especially at low tide. (Peabody Museum of Salem)

Most unlike wooden boats, plastic boats sank when suddenly filled with water. Certainly engine-weight pulled many wooden boats under immediately after they swamped, but astute small boaters knew enough to detach the outboard and let it sink to keep the hull afloat. A wooden hull floating awash kept its former occupants above water, and with luck it might be bailed out. But a plastic hull, even one bereft of its outboard engine, dived or glided under the waves, leaving many small boaters reminiscing about wooden boats like Lymans.

Throughout the 1950s, Lymans ruled the inshore waters of this stretch of coast in a wholly autocratic fashion, one nearly impossible to imagine today. "Lymans were popular? I'd say so," writes Peter H. Spectre. In a recent article in *WoodenBoat*, he recalls the Cape Cod Bay to Massachusetts Bay dominance of the boats. "A few miles from where I lived there was a small inlet west of Harwichport called Allens Harbor. In the 1950s there were no sailboats in that harbor, only powerboats, and every one of them—and I mean all of them—were Lymans." And most Lymans in that and other coastal harbors were one model, one special model. "The preponderant model, the boat nearly every adolescent Cape Cod boy desired more than Marilyn Monroe reclining on a red satin sheet, was the Lyman Islander." What was it, that Islander? Spectre describes it lovingly but brusquely. "An honest-looking 18-footer with plenty of freeboard, it had a flared bow, a tumblehome stern, a low windshield, a walk-around engine box amidships, a fold-down canvas top, and best of all, side steering." An Islander went fishing, voyaging, adventuring. "No sissified automobile-style steering wheel on this baby. You could stand there next to the engine box with a faraway look in your eyes and steer your way to Chatham or Hyannis or Martha's Vineyard or Paradise." After this paean to his boyhood dreamboat, Spectre identifies characteristics that made Islanders the only boat in Allens Harbor, the typical boat in most nearby harbors. "Lymans were hardworking, hard-driving utilitarian watercraft that were at home on the sea and were worth every penny of their modest price."[1] Mass-production brought down the price of the excellent wooden boats, brought it down so much that alongshore boatbuilders struggled to sell catboats and other local boats to men—and boys—smitten with Lymans made inland, in Sandusky, Ohio.

In 1874, Bernard Lyman, a German immigrant cabinetmaker living in Cleveland, built himself a little lapstrake boat for fishing and rowing-for-fun. Within a year he had quit cabinet-making and begun to build copies for individuals, boat liveries, and amusement parks, and by the 1890s he had begun to build sailboats, too. At the turn of the century, Lyman had been building powerboats for three or four years, and in the first years of the 1900s he re-designed his hulls to accommodate the first outboard engines. By 1924, Lyman Boat Works had acquired a national reputation as the builder of eleven-foot boats that sped along between eighteen and twenty-two miles per hour when equipped with simple two-cylinder outboards. William Lyman moved the firm to Sandusky when his father retired in 1928, buying a disused tractor factory on the Lake Erie waterfront and devising assembly-line mass-production methods for wooden boatbuilding. Demand for inexpensive outboard-equipped boats slackened in the Depression, but the firm survived by building inboard-engine runabouts for the rich, and during World War II the company built special-purpose small craft for the military. Boom times returned with peace, and by 1949 the firm had six models for sale: one rowboat, one boat for rowing and for outboard engine use, three outboards, and one eighteen-foot inboard. In 1958, perhaps the peak year of production, Lyman Boat Works produced some five thousand wooden boats, planking them from the new wonder material, marine-grade plywood, while using traditional materials—oak for steam-bent ribs, bronze for hardware—elsewhere in every boat. Yet thirteen or fourteen years later the assembly line stood idle, and the great firm, its work force shrunk to a fraction of its late 1950s numbers, specialized in repairing damaged Lymans.[2] Fiberglass competition had nearly destroyed the company.

Coincidence made Lymans perfect for the inshore waters near Allens Harbor, Scituate Harbor, and other inlets along this stretch of coast. Quality craftworking and competitive price undoubtedly prompted many postwar buyers to invest in Lymans, but the lapstrake hull design developed for the shallow western basin of Lake Erie, water notorious for its short, choppy waves, worked ideally in certain alongshore locations. Although Lyman had dealers in every state in 1951 and advertised its boats as perfect for any body of water, the lapstrake hull and deep-V-configured, generously flared bow proved magnificently seaworthy in the vast shallow areas of Cape Cod Bay, where sudden weather changes make reliability absolutely necessary.[3] Moreover, the big lapstrake hulls easily handled pairs of outboard engines and large gasoline tanks, making the boats fast enough to pull waterskiiers and to outrun oncoming storms, and able to cruise long distances without refueling. Lyman hulls also gave a dry ride at speed, since the overlapping strakes deflected spray away from the boat. The foredeck was a perfect place to sunbathe and read, and passengers could sit with backs against towel-covered windshields while others trolled for bluefish and striped bass. Almost as seaworthy as the eighteen-foot inboard Islander model, the fifteen-foot outboards enabled many families to invest in a boat that could be upgraded every three or four years with a more powerful outboard engine or pair of engines. The same families understood, too, the trade-in advantages of the Lyman reputation, or learned of them while reading almost any boating-oriented publi-

*Lyman boats appeared everywhere in maga- zines during the 1950s, as in this **Popular Science** article of March 1955 on high-speed performance. (author's collection)*

cation. Small Lymans could be sold easily when families determined to buy an Islander or one of the twenty-four-, twenty-six, or thirty-foot-models. Introductory books like Ernest Venk's *Complete Outboard Boating Manual* of 1958 either explicitly recommended Lyman boats to neophytes or implied Lyman supremacy by illustrating argument after argument with spectacular photographs of men, couples, and families enjoying Lymans—all furnished by Lyman Boat Works.[4] Throughout the 1950s and 1960s, sporting and outdoor magazines repeated the boating-magazine message—a Lyman, or a Thompson or other Lyman look-alike, meant a safe investment, an investment as safe as the eighteen-foot Lyman *Popular Science* featured in March 1955. Overloaded with thirteen adults, powered by two Mercury forty-horsepower outboard engines, the Lyman lifted its bow from the water and planed along at twenty-four to twenty-six miles an hour.[5] Could one find a better boat to buy in any dealer's showroom, or to buy used, at the end of the season perhaps, as September rain swept some salt-marsh creek, some small harbor?

Until late in the 1960s, a Lyman in good condition sold almost immediately, and one in poor condition, say one that had collided with rocks or been neglected in a backyard, could be repaired easily and cheaply in any of the alongshore boatshops or boatyards. Boatbuilders familiar with Lyman hulls knew how to estimate repair costs accurately, knew that accessories and metal fittings, once ordered, would arrive immediately from Sandusky, and knew the devotion of Lyman owners. They knew, moreover, that Lymans, like all wooden boats,

Trimotor boats? It's being tried—
with beefed-up transoms

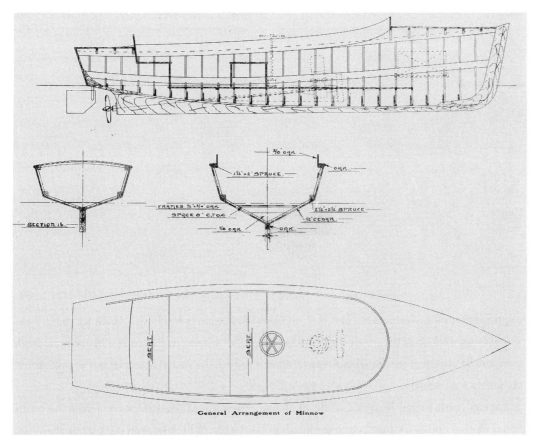

SECTION 16.

3/6 OAK

1½"×2" SPRUCE

OAK

FRAMES ½"×3/4" OAK
SPACE 6" C.T.O.C.

2½"×2½" SPRUCE

½" CEDAR

3/6 OAK

OAK

General Arrangement of Minnow

In "How to Build a Launch for $100," Rudder explained the building of the powerboat Minnow, beginning with an analysis of its plans. (Harvard College Library)

Perhaps to reassure readers that its project boat could be built, in June 1908 Rudder published a photograph of Minnow framed up. (Harvard College Library)

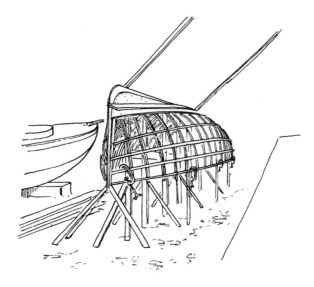

Small boat set up, bottom upwards.

required annual maintenance, and that many owners expected to have it done by the professionals who worked in the cedar-shingled boatshops all winter, who emerged in spring to work outdoors in adjacent yards, whose appearance out-of-doors meant the start not of spring but of "fitting out" season.

Factory boats meant boatyard work, of course, but a sort of boatyard work unlike the traditional skilled craftwork that began to vanish in the 1930s. Where once every harbor sheltered full- or part-time boatbuilders "working up" one or more local boats each year, by the 1950s a new generation of men worked in the shops and yards. While still craftsmen, still routinely called boatbuilders, the new men had become expert not only at repairing Lymans and other factory-built boats but in working with the wonder material of World War II, plywood—and its partner, the resin-based, fighter-plane glue called Weldwood.[6] Plywood and Weldwood married to change not only the building of boats in factories and boatshops but the whole coastal realm, too.

Since the first years of the twentieth century, readers of boating magazines had known the abbreviation "KD," which stood for "knocked-down." Knocked-down or kit boats came in sizes from eleven to fifty-three feet and attracted competent home handyworkers too poor—or too determined—to buy a finished boat but not competent enough to build a boat from scratch. Depression-era would-be boaters found kits an inexpensive way to obtain boats ranging in quality from seaworthy and fair to dangerous and ugly, and after World War II, a new generation of builders embraced them enthusiastically.[7] As late as 1966, James P. Kenealy explained in *Boating from Bow to Stern* that "build-it-yourself boats" cost "approximately 60% less than finished ones," but warned that "only competent craftsmen should attempt the completely disassembled kits that call for complex angles and rounded chine lines. Handymen should purchase kits with the hull completely assembled or nearly so."[8] After 1970 or so, however, authors of how-to-get-started-boating books like Kenealy's shifted their emphasis wholly away from wooden boats and rarely mentioned kit-building as an inexpensive alternative to buying one ready-to-float. Amateur boatbuilding had declined in popularity, and that decline had radically changed coastal landscapes.

No longer did the road to every harbor, the meandering, sand-covered road paralleling every salt marsh, the dirt lane dropping down to some muddy landing pass one boat-under-construction or one boat-under-renovation after another.

Amateur boatbuilding, routinely defined as the building of boats by people anxious for ownership rather than sales, blossomed among teenage boys and grown men late in the nineteenth century and attracted an extraordinary amount of attention. Unlike most other do-it-yourself projects, boatbuilding prospered as a very outdoor, very noticeable activity. And its very publicity enticed other boys and men, until by 1915 it had become a modest but widespread fad.

In "Yacht Building," Charles G. Davis took adult Rudder readers through the building process step-by-step, explaining the process of "lofting" the plans onto the shop floor and explaining what tools to use along the keel. (Harvard College Library)

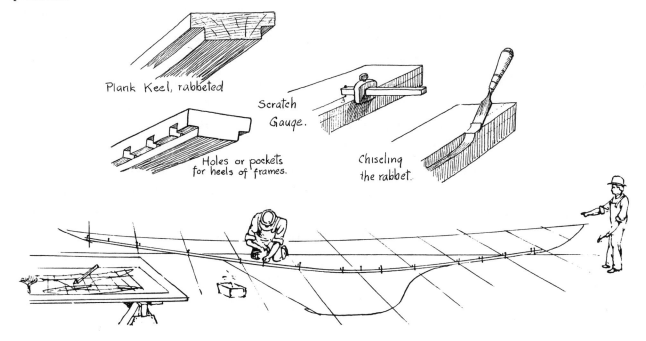

Plank Keel, rabbeted

Holes or pockets for heels of frames.

Scratch Gauge.

Chiseling the rabbet.

Alongshore boys had always built their own boats, learning from boatbuilders how best to proceed, then using their skiffs and other small craft for inshore fishing, lobstering, and other profitable enterprises preparatory to full-time on-water work. But in the 1870s, perhaps earlier, some boys began to build small boats for pleasure use only, often in partnership with fathers or friends, devoting a summer or two to building a skiff or larger vessel in some back or side yard, imitating the fishermen whose on-land activity had begun to make alongshore landscape seem "salty" or "quaint" to tourists.

What boys accomplished after school and on Saturdays, without power tools of any kind, and often with only the sketchiest of plans, seems stunning in an age when boatbuilding has become almost a mystery, a magic art. But in a time when boys habitually made toys and sporting equipment from wood, and when professional builders still worked in shops open to boys who knew when to appear, when to look on in silence, and when to ask questions, a boatbuilding project often was the last boyhood project, something subsequent to building a double-bobsled, something preliminary to high school graduation.[9] *Harper's Boating Book for Boys*, a compendium written in 1912 by Charles G. Davis, a *Motor Boat Magazine* staff writer who had earlier contributed regularly to the *Rudder*, best summarizes turn-of-the-century effort, although it slights the capacities of alongshore boys raised from infancy in the world of for-profit boatbuilding and fishing.

Davis emphasizes boats useful "on small enclosed salt waterways where the wind and tide are moderate" and wholly ignores what he calls "heavily constructed" oceangoing small craft. Yet his plans and detailed, step-by-step instructions clearly reveal his understanding that boys would row or sail homemade boats into risky situations. "A boy can make a dory from twelve to sixteen feet long, but a fourteen-foot dory will be quite large enough to hold from four to six boys comfortably and safely," he advises. "A dory of this description makes an ideal fishing boat where the water is rough, since it can be rowed either forward or backward." In the same way, he warns builders of his twelve-foot sailing skiff that "care should be exercised in handling the boat; and be sure to reef the sail in case of a strong breeze." His designs and directions focus on rugged construction before simplicity, therefore, and clearly assume not only competent, risk-taking boys but time and space to build.

In an age before the motorcar, and so before the garage, boys who followed Davis's plans or those published in so many magazines often worked outside, beginning construction in late spring. Given the still-limited ability of many young builders and the difficult outdoor con-

Turn-of-the-century amateur boatbuilding often began in the shadow of boatyards, where children built and sailed model boats. (author's collection)

Harper's Boating Book for Boys *detailed the process of building several boats, including, of course, the simple skiff. (author's collection)*

ditions in which they worked, Davis and other writers sometimes advised hiring expert help for tricky operations. "It would be better to let a boat-builder or carpenter make this board the proper size and shape to fit the trunk, for it is the most difficult thing to construct about a boat and somewhat beyond the ability of many boys," writes Davis of centerboard building, for example.[10] Nearby boatbuilders might or might not be helpful, but local lumber dealers almost invariably helped, since amateur boatbuilding sold wood. Until the lumber shortages of World War II, boys could easily and cheaply obtain pine boards sixteen feet long and often fourteen inches wide—and wholly free of knots.[11] Two such boards made the sides of a simple skiff, as boys who read Jack Bechdolt's *Handy Book for Boys* discovered in 1923. "It can be done by any boy who will take the pains to measure and cut accurately and who will be careful to follow directions," Bechdolt advised, before explaining how to fasten one end of board to a transom board, then draw together the two free ends. When a rope is looped over the two free ends, then the loop is twisted over and over with a stick, the two boards meet in a perfect

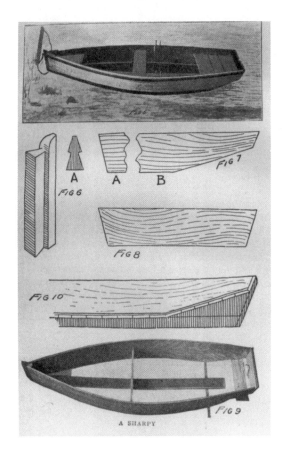

bow, ready for the stem piece.[12] About all that remained a boy could do in another Saturday—installing a seat and planking the bottom. Painting on a warm, dry day completed the project.

Book after book appeared, each aimed at competent boys who were intrigued by boat-owning and understood that boatbuilding must come before boating unless parents bought a boat ready-made. *The Boy Mechanic*, a frequently updated tome published by *Popular Mechanics*, first appeared a year after *Harper's Boating Book*, and over decades provided plans upon plans, directions upon directions, always focused on the boy with hammer, handsaw, and—most important—measuring tape. By 1952, however, *The Boy Mechanic* had begun to emphasize plywood boatbuilding, not so much because plywood made a simpler or even a more durable material but because the wide pine boards of the Depression could no longer be had cheaply, if at all.[13] The editors of *Popular Mechanics*, moreover, had noticed a marked change in their audience and no longer assumed that most young boatbuilders lived near the water.

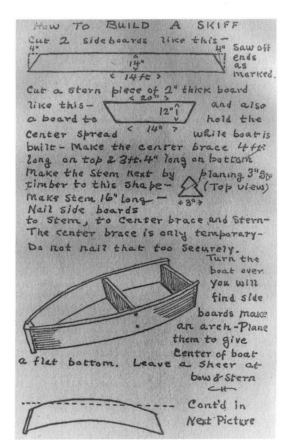

A newly built rowboat, even a sailboat, weighs remarkably little. Not until it has floated for some days will it stop leaking, as the wood swells with absorbed water. But by the end of summer, when its seams are long swollen tight, it is immensely heavy, often far too heavy for boys to trundle far. Until the perfection of the car-pulled boat trailer in the 1950s, therefore, young builders assumed that their boats would winter upside down in some boatyard, or behind some building adjacent to the waterfront. Before boat trailers, boats rarely moved far beyond high-tide mark, and then only on rollers, and the writers of so many how-to-do-it books knew it. Late spring found dozens of boys hard at work at the fringes of boatyards, each scraping and repainting a dried-out rowboat or sailboat built a year or so—or twenty or thirty years or so—before, and early summer found the boats afloat. Yet the boats furnished the land often nine months of the year, their hulls advertising the presence not so much of the shore but of amateur alongshore activity.

Everything about the how-to-do-it books smacks of amateurism, although not in any pejorative way. The vast literature aimed at boys includes few technical terms beyond the simplest and frequently eschews nautical vocabulary. The twisted-rope method of bringing together side boards makes perfect sense in book after book, for example, but never do authors employ the technical term, *Spanish windlass*. Moreover, directions for boys and grown men alike address not so much the novice builder as the one-time builder, the teenage boy building his first and only boat, the newlywed man determined to impress his bride, the father looking to please wife and four children—and a dog, too. And so almost none mention the tricks professional boatbuilders use to hasten their work, say their incredibly efficient technique of banging once on the point of each nail to flatten it, and so keep it from splitting wood when driven, or the building of jigs to speed repetitive assembly. Tricks of the trade were wonderfully useful to professionals, of course. At least one early twentieth-century five-man crew built a thirty-one-foot whaleboat in two days, and one late 1950s–early 1960s lone boatbuilder turned out a ten-foot plywood skiff a day, albeit unpainted. But decade by decade, boy-oriented how-to-do-it books assumed professional boatbuilders to be increasingly fewer and less helpful, and the assumptions of adult-oriented guidebooks lagged only slightly behind. Drawings became more numerous and were often simpler, and after 1910 photographs become important instructional accessories, perhaps substitutes for absent experts. As early as 1907, one writer extolling the many benefits that would accrue to adults who built their own boats concluded somberly that "the hardest thing about amateur boat-building is to find proper materials. The ordinary lumber yard seldom has the kind of stuff on hand needed for boat-building purposes; the ship yards seldom care to bother."[14] By the Depression, when many would-be small boaters saw do-it-yourself efforts as their only option, many savants dismissed boatbuilders as a vanished breed and urged straightforward experimentation. In "Yachts for All of Us," a *Leisure* article of 1934 on building sailboats, Edwin S. Parker explained that he had begun to build boats when he was fourteen, that they "have been plebian craft, dressed in sails I have sewed myself; sometimes I have even felt they verged on the proletarian," that building cardboard models makes plans more intelligible, that all of his boats leak.[15] Most guidebooks and magazine articles make clear in fact that amateur builders worked in gross ignorance of such professional skills as how to move completed boats to and from the water.

Few detailed, seemingly honest descriptions of turn-of-the-century adult boatbuilding survive in libraries, and perhaps none is as candid as that published in 1910 by W. R. Bradford in an issue of the *Rudder*. In the backyard of a Philadelphia rowhouse, Bradford built a twenty-five-foot cabin cruiser, in which he, his wife, and their young son could go gunkholing, "cruise in, live in, sleep in, cook in, and go where no hated railroad train could ever follow with its smoke and cinders." Full of hope, confident that "any amateur who was fairly handy with tools could build a good boat from these plans," the newspaperman quickly moved from his cellar to his backyard, constructing a rough-board platform and erecting a tent. "By this time the neighborhood was aroused, as only a Philadelphia neighborhood can be, at the sight of the unusual." The one professional boatbuilder who visited the scene was insulting, then grudgingly helpful in pointing out initial, massive errors caused by Bradford's disregard for measurements. Bradford broke twelve planks in the same place trying to bend them around misplaced frames, and soon the neighbors began to complain about his hammering and "pounding." Shortly after, the whole set of frames collapsed, and his wife became frantic. "How I kept out of the divorce courts I never knew," Bradford relates of the times when noise-conscious neighbors had begun to send anonymous letters of complaint and threaten arrest, and when nothing in the backyard effort, nothing, seemed to go right. Eventually, however, he did complete his vessel—and discovered he had no way to remove it from his yard.

"Here was a serious problem; hemmed in on all sides by brick houses, with only a narrow passageway between fences." His neighbors forbade his removing any of the fences, and Bradford quickly abandoned his scheme to build a railway over them. Then another friend "with the boat bug" suggested hiring riggers to lift the boat and its cradle over the house, into the street. The riggers erected two massive masts, a maze of stays and tackle, and actually raised, swayed, and lowered the boat, all before an immense crowd of disbelieving onlookers.

From beginning to end, W. R. Bradford confronted extraordinary difficulties in building **Dubbalong**. He had to build a tent to shield his project and himself from harsh weather, his neighbors complained about noise, and he faced almost insurmountable problems in getting **Dubbalong** from his backyard over his house and into the street. In the end, however, he had a fine craft that pleased his wife and son—and himself—no end. (Harvard College Library)

Six miles away, a derrick lowered the boat into the water. "Yesterday a fool! Today a hero!" [16] Of course, the engine would not start because of an unfinished gasoline line, but everything else seemed perfectly and amazingly right.

Bradford built in a time of building, an era of happy and sad mistakes, an era that lasted through World War II, an era that subtly transformed the seacoast environment. The boys who built rowboats and sailing sharpies might be hidden in barns and disused carriage sheds, but most men like Bradford built in the public view, even if in a Philadelphia backyard, simply because they had no indoor space. After 1900, however, more and more of them built along-shore, beside cottages rented for a summer or in some vacant lot hired expressly for the purpose. Although no one compiled statistics, editors of boating magazines understood that many small boaters had become small builders, partly to obtain a particular boat they were unable to afford to have built, partly to extend the pleasure of boating year-round, and partly to enjoy the pride of doing something many boaters considered extremely difficult. In the first years of the twentieth century, boating magazines featured how-to-do-it articles on small catboats, power launches, rowboats, even yacht-tenders, and these continued into the early 1960s. [17] Amateur building efforts often spanned years, and intermittent visitors to the coastal realm learned to look for the boat in the field, the boat in the yard, the boat that dwarfed the sheds around it, simply to see how much progress had been made by some amateur builder working part-time, in fits and starts, as money came to hand. "I myself have been guilty of building a small cutter under a shed, built up against the back of the house," wrote one *Rudder* author in 1906. "To get her to the water necessitated cutting a section out of the back fence, carrying her down an embankment, across a vacant lot, up another embankment to a wagon on the next street, waiting to carry her half a mile to the Bay."

The half-mile haul and others like it meant nothing as expensive or dramatic as Bradford's over-the-house lift, but they often proved tremendously difficult, even at the beginning if the amateur builder had neglected to build the boat high enough off the ground for beams and jacks to be easily inserted. Egyptian-like, often aided by dozens of friends and three or more teams of horses, amateur builders used rollers and pry bars to move their boats toward salt water, or else hired rigging companies, which used winches and other special equipment—even oxen. Boatbuilding articles make clear the extraordinary postcompletion efforts undertaken by men who built too far from shore, who bargained unsuccessfully with glum boiler and furniture movers, who endured the heartbreak of moving boats over too-soft ground that twisted hulls and split planking, who dismantled farm wagons to build giant hauling dollies. Many builders spent days heaving and shoving, learning that "yacht-building for pleasure is a rather strenuous mode of getting it, but nothing like so bad as football." Through the 1950s, extreme examples of boat-moving effort appear in newspaper stories like the Cambridge *Chronicle* article that detailed Frederick Butler's exertions. "A shiny new cabin cruiser floated handsomely in the Charles River off Cambridge last night," begins the story on June 23, 1952. "But to launch it, the builder had to chop down six trees, tear up a wire fence, saw three inches of moulding from two houses, and rip up some shrubbery."[18] Blisters, lame backs, and light wounds accrued to most amateur builders, but accrued fastest when finished boats began the long creeps shoreward.

Not surprisingly, therefore, amateur builders sought out sheds, barns, and vacant lots as near as possible to launching places. Weekend after weekend, evening after evening, they worked at their "second homes," struggling with one joinery problem after another, and becoming a picturesque element in harbor scenery. By the 1930s, they had long surpassed professional boatbuilders in number and visibility, and many landowners rented out waterfront sheds and vacant lots to entire streams of amateur builders too wise to build far inland. Harborside locations meant proximity to marine hardware stores and professional help, too, of course, but almost equal in importance to nearness to water, the locations offered nearness to other amateur builders, men, and sometimes women, who had already learned how to spile a garboard strake or were will.ng to lend a hand in lifting a heavy frame. While wartime slowed amateur boatbuilding dramatically, there was an amazing resurgence in the immediate postwar years, along with a burgeoning interest in repairing one-of-a-kind or mass-production boats.

By the 1950s, many boatyard owners had long abandoned any attempt to keep boat owners from working on stored boats, especially in spring, in "fitting out season." Instead, boatyards became the early spring resorts of dozens of owners eager to repair and repaint their wooden boats, or even to continue working on boats under construction. "The work of removing covers, scraping bright-work and deck, burning off old paint and preparing new, scraping and varnishing spars, and otherwise preparing the hull of the boat for the water, is a delightful diversion after a winter in an office," opined one writer of his favorite time of year. "The whole spring may be consumed before the boat is made ready for the water."[19] In the face of such dedicated do-it-yourself efforts, professional boatbuilding declined, rushed on not only

From the 1880s on, the first hint of spring brought pleasure-boat owners to boatyards to begin "fitting out," often by burning off old paint. (author's collection)

by the force of Lyman and other big-firm competition but by the success of amateurs who learned to use plywood, Weldwood glue, and other postwar products as unfamiliar to professionals as to amateurs but cogently described in books like S. S. Rabl's *Boatbuilding in Your Own Backyard*, written in 1947. Many boatbuilders turned to repairing hulls beyond the capacity of amateurs, to maintaining boats owned by folks who disliked handwork, or to part-time jobs from carpentry to installing overhead garage doors.

Into the era of fiberglass boats, small harbors stood surrounded by concentric rings of boats. Along the very beaches and high ground just inland lay the boats of boys, often nested one atop the next all winter, usually covered with tattered tarpaulins. Further inland, often just

Marine railways endure as one of the oddest applications of railroad technology to transportation. Here the carriage, complete with scaffold, awaits its next assignment. (JRS)

a few hundred yards in, rested a dense ring of boats more or less ready for launching, perhaps lacking only painting and varnishing. Under better cover all winter, that ring stood between the boys' ring and the builders' ring, the most inland ring of boats under construction, often immense ketches longer than fifty feet, or thirty-foot-long cabin cruisers made of plywood and powered with rebuilt Ford automobile engines. Crammed everywhere from late autumn to late spring, the boats gave a peculiar character to harbor neighborhoods, a character different from the late nineteenth-century one given by the boats of fishermen and lobstermen. As Spectre correctly remembers, by the 1960s, pleasure boats dominated many harbors even as Lymans dominated pleasure boats. Small boaters en masse had learned the folly of keeping boats far from hard-sand beaches, far from launching ways, far from the boaters' favorite launching device, the marine railway.

"Nothing looks more helpless and awkward than a boat out of the water being hauled along like a freight car," mused Gladys Taber in the 1950s, as coastal people accustomed themselves to boats moving on rubber-tired trailers. "Then when this same bulky craft is lowered into the water a miracle takes place."[20] But until the 1950s, between the rollers or dray or horse-drawn "truck" and the miracle of flotation stood the marine railway, the limicole element at one time so essential to boatbuilding, boat repairing, boat launching.

American histories of technology dismiss marine railways, often burying them in longer articles about docks, especially dry docks. Although these railways worked well as early as the 1840s, at least around New York Harbor, their origins are obscure. Only inquiry into British documents produces a clear-cut chronicle of origins, along with the implied charge that American shipbuilders stole the invention—rails, cables, and carriage—wholly and rudely violating patent rights. The marine railway proved too good an idea to pay for, and by the 1850s many observers filed it as one more example of Yankee ingenuity. In 1852, *Appleton's Dictionary of Machines, Mechanics, Engine-Work, and Engineering* coyly explained that "the American Marine Railway, as it is commonly called, or the Railway Dock, as it is styled in the patent, was brought forward about twenty years ago," and said nothing about its inventor.[21] Yet by 1880, when Englishman Ernest Spon argued in his massive *Dictionary of Engineering* that the device resembled sixteenth-century Venetian rails, few experts seemed certain of its modern inventors.[22] The exact origins of the oddest of railways lie hidden from retrospective scrutiny, just as one end of its track lies always underwater, beyond the view of anyone but the most curious of children—and the rare diver hired to unsnag the cable block.

Well into the 1960s, marine railways clattered and shrieked as they had a century before,

launching pleasure vessels every spring, retrieving them every autumn, and in the meantime "hauling out" boat after boat for inspection and repair. Even now, whenever one works it attracts attention. No mechanical device speaks more clearly of the whole notion of harbor, of safety from the sea, of the curious mix of water and shore, ships and buildings, but when not working it passes unnoticed by most casual visitors. Only when its carriage emerges from a boat shop and slowly, almost deliberately rolls down the suddenly noticed rails do tourists pay attention. And only when a sloop or lobster boat emerges from the water and rides high up the beach do they grab for cameras and stand amazed.

A marine railway—its name certainly hints at British origins, for no one calls it a marine railroad—is just that, a railway. Parallel steel rails often twelve feet apart lay spiked across great beams of wood, beams everyone calls sleepers—another British term—never ties. On the rails rides a carriage—not a car—with a cable attached to each end. Inland from the carriage, the cable runs between rails to a winding machine, often securely housed in a building and usually anchored to massive rocks or blocks of poured-in-place cement. Seaward runs another cable, snaking between the rails into and under the water to a block—what a landsman calls a pulley—anchored far out in the depths, usually to a massive stone but sometimes to a piling, and then around, back inshore, and up beneath the carriage, parallel to the landward cable, to the winding machine. Like an old style apartment-house clothesline on pulleys, therefore, the cable enables the winding machine to pull the carriage seaward and then under water, and then to pull it back again onto land. At first glance, the mechanism appears extremely simple, and undoubtedly mid-nineteenth-century boatbuilders and shipbuilders found it simpler than many other techniques.

Before the marine railway, mariners and shipyard workers rarely hauled out large vessels, especially oceangoing ships. Here and there naval dockyards maintained dry docks, great bathtublike containers into which a ship floated at high tide, then slowly lowered onto blocks as doors behind it closed and pumps removed the water. Extremely expensive to build and operate, dry docks rarely appeared away from great cities, and smaller harbors everywhere survived without them. In such places fishers and owners of schooners and other small trading vessels careened their craft on hard beaches, gently running them onto the shore at high tide, then working on their bottoms as the tide dropped. "When the night tide served, the young men turned to with a will, and warped the vessel across the harbor to Uncle Darius Kentle's tiny beach between the rocks, where she was soon grounded in the desired position," wrote George S. Wasson in 1903 of the time-honored small-harbor practice. "To insure her listing in as the tide ebbed, a couple of hogshead tubs were moved to the inshore side of the deck, and filled with salt water, while as a further precaution, both booms were swung broad off inshore also, and guyed in that position, so that no one felt the least doubt that the schooner would incline in as the tide left her, and be found accessible for work early the next morning."[23] Any vessel left to "dry out" on the hard sand lay in a dangerous position, liable to damage in sudden squalls, of course, but also likely to shift slightly and perhaps crack frames or ribs. Owners and shipwrights alike, moreover, knew well that only one side of the bottom lay open to repair and that the part next to the keel—and the keel and its adjacent garboard strakes generally needed repair most often, at least on old craft—stood open only to men willing to crawl belly-down in the cold sand and roll on their backs to reach the planks above. Marine railways, even the simplest, made maintenance and repair so vastly easier, quicker, and cheaper that everyone pronounced them wonderful.

Horses provided marine railway motive power well into the twentieth century, at least in the smallest harbors. Often only a single horse walked around and around a capstan, a massive vertical winch set securely in the ground. Jumping over the chain or cable once each revolution, the lone horse and massive reduction gear pulled carriage and schooner from the water and inched them up the slope.[24] In the 1900s, steam engines replaced horses, and after the 1920s, large, slow-turning gasoline engines coupled to winches replaced steam power. In the years after World War II, army-surplus heavy trucks often stood half-dismantled, their rear axles driving ancient windlasses and cable drums, their diesel engines roaring, their exhaust pipes blasting skyward funnels of blue smoke. Marine railways followed the horse-steam-diesel progress of inland railroads, but in little else did they resemble their counterparts.

Of primary and immediate visual importance, of course, the constant incline struck many neophyte observers as extraordinarily odd. Each marine railway consisted of a single track running up from the water and terminating at its winding equipment, the track never deviating from its angle. No hills, no valleys varied its short route, and its terminus, usually outdoors but sometimes within a giant ramshackle shed, stood alongside the angled track. Any vessel hauled out stopped at precisely the same angle it took when it left the water, and everyone from caulkers to painters worked on the vessel while it sat at this angle. Simple structural mechanics explain the everyday appearance of vessels pointing gently skyward. Although the railway carriage could easily negotiate changes in grade, its flexing would spring or even snap the keel of any wooden vessel. And so the carriage had to behave as substitute water, offering constant support to the very bottom of the uplifted craft.

Moreover, marine railways rarely curved and never diverged at switches. Straight and constantly sloped, they displayed the extreme limitations of cable-towed carriages hauling immense weights. A curved route overtaxed the towing cable, and although cables could be rigged to tow a carriage around a bend, the additional equipment struck few boatyard owners as worth the expense and trouble. Switches involved not only the diversions from straight routes implicit in curves but the expense of building moving rails and frogs, balky mechanisms that might easily fail. Marine railway operators could move more than one vessel by hauling up one carriage and vessel, and then slipping free the cable, snaking it beneath the carriage and vessel blocked in place on the inclined track, dragging it down to the water, swimming it down to the sunken carriage waiting at the underwater end-of-track, and attaching it, a job so awkward few attempted it. Only U.S. Navy and Coast Guard installations experimented frequently with railways involving switches and turntables, the Coast Guard struggling as always to insure the safe launching of motor lifeboats in storms that could damage railways and wharves.[25] In small civilian boatyards, no one bothered with curves or switches; instead they simply duplicated routes.

A prosperous small boatyard often had two or three marine railways running almost parallel or perhaps radiating slightly from a nearly common underwater point. With its own carriage, its own winding mechanism, and its own route, each marine railway stood nearly invisible except when its carriage passed, either loaded or unloaded. At other times it rested unobtrusively, its nearly sunken rails almost invisible, its rusty cable stretched from structure to someplace unseen. In autumn the carriages clattered back and forth all day, pulling up boats to be slid sideways from the carriages on massive wooden cradles, themselves often equipped

High tides often float cradles away from boatyards, and low tide strands them everywhere, making them perfect for postcard photographers seeking alongshore views. (author's collection)

with steel wheels to ride on the rails that ran at right angles to inclined tracks. A team of heavy horses strained to pull the cradled boats along the tracks that terminated at the edges of the yard, then returned for the next pulled-up vessel. In spring, after repair and painting, the team pulled each cradled vessel back to the inclined track and onto the carriage. Unlike dry-land railroads, on which crack passenger trains bypassed sidetracked slow freights, the marine railway worked strictly according to first-in, last-out rules. The first boat hauled out for the winter and dragged to the edge of the yard had to be the last boat hauled back to the railway and launched the following spring. Only vast effort could pull a boat out of line and enable other cradled boats to move around it, and such vast effort, as many boat owners learned to their amazement, cost vast sums.

Attempts to escape the inflexible schedule of boatyard-owned marine railways challenged pleasure boaters until well into the 1950s. Many, perhaps thousands, built private marine railways, diminutive and fragile but serviceable. After excavating shallow troughs through salt marsh—good summer exercise for boys—or scraping a constant grade across an inner harbor or estuary beach, they followed instructions like those provided in 1906 in a *Rudder* article, "How to Build a Small Marine Railway." First warning readers that only one in a hundred boat owners lives in a spot where a small marine railway might be built, Charles G. Davis describes how to build a small version of commercial railways, focusing on the underwater section. In emphasizing that a "high-tide" railway is best for the amateur, because it eliminates the need to sink rails and return blocks far out underwater, he makes clear that his design will work only at high tide. But his drawings and text demonstrate the utter practicality of the project, as well as its relative cheapness. A few scrap rails, several small flanged wheels, and a cable and hand-cranked winding mechanism are about its only metal parts, and any carpenter or determined amateur—perhaps one having built a boat already and eager for another saltwater project—can build it.[26] Davis largely ignores the actual operation of the home-built, home-property miniature marine railway, but other authors emphasize hauling-out techniques. In *The Ship's Husband*, written in 1937, Harold Augustin Calahan warns that "by railroad standards, the best yacht yard railway is a joke" and teaches his reader not only how to evaluate yacht-yard railways but how to insure that professionals operate them carefully.[27] His advice, gleaned from years of watching satisfactory haul-outs and disastrous ones, encapsulates traditional hauling-out lore, the sort of lore Davis simply assumes, the sort of lore thousands of families must have had, to judge by the derelict railways that linked summer homes with seawater.

Family railways worked seasonally until the perfecting of boat trailers in the 1950s and the simultaneous proliferation of war-surplus four-wheel-drive vehicles, especially Jeeps. Once boats could be trailered reliably across sand and cobble beaches, tiny marine railways fell into disrepair, and many families ripped out the underwater rails and sheave-boxes, having condemned them as obstructions to navigation. Trucks and trailers, moreover, all but elimi-

Boats remain babies, at least out of water. They sit on cradles that must be carefully blocked, and they, and their cradles, are moved to marine railways with utmost care. (JRS)

nated the inflexible boatyard schedules that so maddened boat owners, and soon boatyard operators began to abandon their marine railways.

Stark, impossible-to-miss vertical elements replaced many marine railways in the 1960s when boatyard owners began to buy used construction cranes. A heavy crane, usually diesel-powered and moving bulldozer-like on self-laying tracks, dramatically quickened boatyard launching and retrieval. Yards that owned more than one railway ripped up one and dug a wide, deep ditch—technically a dock—in its place, using the sand and mud to build a high, dry pad on which the crane could work. Hauling out a boat involved merely running the boat into the dock, slipping two slings under it, and lifting it from the water. The crane then pivoted and lowered the boat gently onto a waiting trailer, for transport to some inland backyard or for storage a few hundred yards away. Fast, cheap, and almost as reliable as the railway, crane-hauling blossomed in the 1960s and 1970s and in time produced specialized equipment, the Travelift. Essentially a moving gantry crane rolling on its rubber tires along two parallel finger piers, the Travelift works quickly and seemingly effortlessly to lift boats from the water before driving them to waiting trailers or cradles. Although it is far more expensive than a used crane and cannot remove and insert sailboat masts, the Travelift drives speedily around boatyards, eliminating the need for four-wheel-drive trucks to jockey trailers and cradles. No

Boatyard cranes come in many forms. Some almost a century old still work occasionally, providing onlookers with a glimpse of the nineteenth-century rigging work that lifted Bradford's Dubbalong over his house. (JRS)

Until the trailer revolutionized the building and storage of pleasure boats, every harbor stood surrounded by boats stored over the winter. (author's collection)

The Travelift Crane, its slings ready to retrieve another boat from the dock between its narrow piers, revolutionized boatyard storage and launching, too. (JRS)

longer are boatyards low-silhouette complexes of sheds, stored boats, and marine railways, serviced by rusted-out vehicles ranging from fenderless Dodge Power Wagons to farm tractors. Now they strike skyward, their one or two cranes thrusting booms sixty or more feet in the air, their Travelifts dwarfing waterfront sheds.

Marine railways linger in the glim of nineteenth-century technology, battered not only by the usual winter storms but by the insurance industry. Beginning in the 1980s, insurers frowned on marine railways. After all, they argued as they increased premiums, a fishing trawler or sailboat might topple from the carriage, crushing boatyard employees or spectators. Arguments raised against the scenario, arguments buttressed by harsher and harsher questions about when such catastrophe had happened, made little impact. Common sense made little impact either. After all, a boat seems less likely to fall from a carriage than from slings that suspend it twenty feet in the air, and the boom of a crane may always fall. Perhaps state licensing laws somehow affected the industry; crane operators must be licensed, but not so marine railway operators. Perhaps the uncertain age of so many railways disconcerted an industry increasingly at ease only with new equipment. Whatever the reasons, having decided that marine railways are dangerous, the insurance companies drive more and more boatyard owners to cranes of one sort or another, by raising premiums higher than cranes raise boats. But the railways endure.

Fishing and lobster boats so big that no crane, Travelift included, can lift them keep some railways in business, but a peculiar symbiosis between wooden boats and marine railways keeps many trundling between land and sea. Owners of wooden boats, especially owners of fine yachts, quickly learned that slings sometimes strain frames and planking beyond repair. Boatyard talk slips around and around the issue, for no one experiments, of course. But some hull shapes seem to be damaged more easily than others, and owners of wooden yachts now often demand railway hauling only. To be sure, the railway offers other advantages, particularly its wonderful ability to launch a boat ever so slowly into the water so that experts crouched in its bilges can identify the source of every leak—or the source of one stealthy, incredibly troublesome leak—and so save its owner large repair bills. Yet the continuing concern about slings damaging wooden boats keeps many railways in operation and causes many boatyard owners to regard the resurgence of interest in wooden boats with hope.

In the meantime, of course, fiberglass boats have transformed the edges of so many small harbors that the mid-twentieth-century alongshore landscape is all but gone.

So long as marine railways offered the only avenue for launching and retrieving boats, they lined the shores of every harbor and, with their respective yards and sheds, made a sort of collar snuggled against the water. Boatbuilding and boat-repairing efforts focused on the immediate edge of the water simply because marine railways could scarcely curve, let alone shift inclines. Even boat storage extended only slightly further inland. Right next to the water, therefore, stood a densely packed concatenation of spaces and structures and vessels, a concatenation enlivened from late spring through early autumn by furious activity, for even in midsummer wooden boats rumbled up marine railways for repair and repainting, especially for the bottom repainting that helped fast boats win races. The ring of land jammed with cradled boats all winter marked the visual high-tide zone, a zone in from the actual high-tide limit but in which vessels occupied space comfortably, almost "naturally." Crowded, active, and instantaneously announcing itself as the boating edge of the land, the harbor collar attracted everyone from boys building or repairing a rowboat to fishermen examining a hauled-out trawler to tourists savoring a salty scene.

And then, in the 1960s, the collar frayed. More and more families each year trailered fiberglass boats, and boatyard owners watched launching and storage income diminish almost as rapidly as repair income. Many met the challenge by building webs of floating piers and calling their yards by a new foreign name, *marina*.[28] In the transition years, between about 1965 and 1985 along this length of New England coastline, boatyard owners struggled to maintain inventories of wooden boat supplies as well as the resins and cloths needed to repair fiberglass boats, to provide electricity and fresh water to each of the "slips" in their webs of floating piers, to build cleaner restrooms and sewerage dumping facilities while repairing marine railways and musing whether to buy Travelifts. They confronted novel and difficult situations, perhaps chief among them the owners of a fiberglass boat furious that the boat stored next to theirs had not only a wood hull and superstructure but crazed owners who enjoyed working on it in early spring, strewing around not just pieces of wood but tools, extension cords, and sanding dust. Within ten years they confronted even more novel situations, say the owners of suddenly "classic" wooden boats complaining about the stink of sanding dust and wet resin that drifted from fiberglass boats under repair. Some boatyard owners banned wooden boats, ostensibly to reduce fire hazards, and so watched demand for their marine railways slacken even faster. Others insisted that all repair work done in the yard be done by their own employees, and so learned the high cost of retaining skilled employees for whom work did not always exist—and the fury of boat owners who learned that the laid-off employee who did know how to fix the diesel fuel filter or retrieve the hollow-mast lifting cable had long gone and that a "specialist" had to come from some other boatyard-marina some miles away. Still others confronted a growing skein of environmental regulations. Those that restricted dredging caused many yard owners to lose out on the sailboat-owning boom that followed the early 1970s energy crisis, simply because their wharves jutted out over water too shallow for deep-

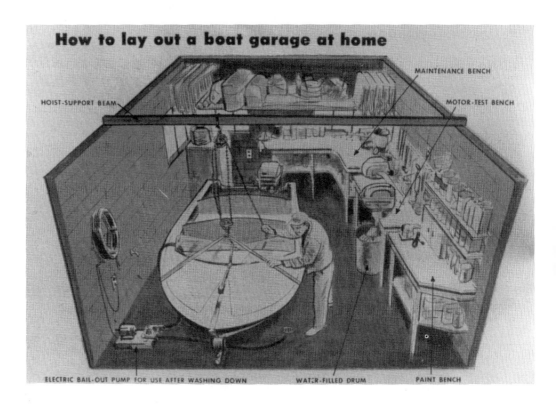

How to lay out a boat garage at home

HOIST-SUPPORT BEAM
MAINTENANCE BENCH
MOTOR-TEST BENCH
ELECTRIC BAIL-OUT PUMP FOR USE AFTER WASHING DOWN
WATER-FILLED DRUM
PAINT BENCH

keel sailboats. Regulations that restricted wharf building delayed some yard owners intent on installing finger piers to carry Travelifts so long that the new equipment went into service just as wooden boats became trendy, and well-to-do wooden-boat owners demanded marine railway service. Hard-pressed to choose and harder pressed financially, in desperation boatyard owners sold off their seemingly surplus real estate, especially the high land once used for winter storage.

And by the mid-1980s, magazines aimed at small boaters, and even some aimed at blue-water yachtsmen, began to publish warnings about harbors without marinas, without boatyards, without boatshops. In 1986, David J. Damkoehler attempted to answer one question: "Are real estate developers and government agencies killing Cape Cod boatyards?" Essentially, he argues that boatyards and marinas are worth more as vacant lots suitable for condominium development than as operating businesses. Moreover, although all marina operators favor environmental protection legislation as good for business—as Damkoehler points out about boatyards and marinas, "it is in the operators' commercial and personal interest to keep them clean"—the permitting process is so incredibly time-consuming and expensive, always taking at least six months and often more than a year and frequently requiring fees in excess of five thousand dollars and sometimes three times that figure, that changes to marina facilities, even ones made necessary by storm damage, are made hesitantly, if at all. Damkoehler asserts that most marina operators now admit that they cannot sell their businesses to successors who intend to operate them as businesses. A boatyard, much like a coastal-realm farm, is far too valuable to sell to anyone expecting to make mortgage payments from working it traditionally.[29] A boatyard is merely a condominium development yet to be.

Boat repair is something tourists love to see, even adjacent to their fancy inns. It is salty, peculiar to the coast, and intriguing to watch— unlike, say, auto-body work. (JRS)

The most casual survey of boatyard owners confirms Damkoehler's findings, but the most casual drive around almost any harbor confirms it faster. Everywhere stand the condominium buildings, often several stories tall, offering their occupants the most prized view of all, that of a harbor fronting the sea. Now and then the developments come complete with "dockominiums," slips and floating piers owned as real estate as part of the dry-land property. And as one boatyard blossoms into condominia after another, condominium owners and purchasers discover that the harbor activity, particularly the summertime small-boat activity, that should make the condominium units so attractive has somehow vanished. In many harbors the fringe structures no longer include a single boatyard, not even one sporting a Travelift.

Instead boatyards have moved inland, far from the sea but near highways, often in low-rent outer-suburban districts. "An inland boatyard may sound odd, but owner Tom Coughlan says there's a good reason for locating his boat-engine repair shop eight miles from the nearest open water," begins one business-section story in a coastal newspaper. "This is actually a very good area because I get the Rhode Islanders coming up Interstate 95 as well as the Boston crowd."[30] Inland, away from the sea, away from harbors and marine railways and wharves, away from skyrocketing land values and lengthy environmental-permitting processes, glisten the new shops. And the small boater, fiberglass boat comfortably perched atop a high-speed trailer, begins his or her first voyage by driving from backyard to boatshop, then to some seaside launching ramp. Around every harbor the new condominia glisten, too, as the final fruits of the fiberglass boat, high-speed boat-trailer inventions, and the small boater with a leak or a balky engine learns how forbidding so many harbors are, the harbors lacking any boat-repair businesses at all.

Every harbor now is merely a node of converging courses plotted from a thousand houses, each the retreat of a fiberglass boat on a high-speed trailer. No longer are harbors filled with Lymans or other "popular" models, let alone with local boats built in adjacent shops. Instead harbors exist merely as parking lots in which boats temporarily dock, midway in odd passages from garages and inland boat dealers to garages again. And tourists stand on wharves, charmed by the neatness, by the quiet, oblivious to the economic and social change so manifest in the fiberglass boats and the condominia.

Contemporary maintenance of fiberglass boats is far removed from the traditional care of wooden boats. Here several teenagers use a boat-launching ramp to jump-start their boat engine. (JRS)

7 WHARVES

*As this Rudder photograph of November 1902
indicates, even flimsy wharves attracted tourists
eager to get nearer the sea than the sand.
(Harvard College Library)*

No one, not even a summer person, stands on docks. A dock exists only as a space, an open area adjacent to a wharf, a pier, a landing stage, even a bit of marsh. One can swim in a dock, unless the dock dries out at low tide. Then one can stand *in* the dock, walk about on pebbles or hard sand, or slog and slip through the muck. No one can be on a dock, only in it. And in the dock, especially at low water, one can scrutinize the landing adjacent to it, see deep into the shadowy spiles that support the tourist- or fishermen-peopled deck above, see the cut-granite stones slick with rockweed, see the underside of things. Trolls lurk beneath bridges, but no supernatural creature, not even a Nereid, dwells among the mollusk-encrusted spiles.

Out in the sunlight lies merely a patch of water, or patch of ebb-uncovered ground, the parking space of skiffs and speedboats, trawlers and dude-carrying schooners. Between two wharves thrusting seaward—or merely into some small harbor or estuary—a dock is most clearly defined. A rectangle wide or narrow, closed at one end by the shore—or some structure—and open at the other, open on adventure, on freedom. Every dock is a jail of sorts, a cell in which vessels lose their beauty, their independence. To be in a dock is almost like being aground.

Experienced seagoers call it "taking the ground," at least sometimes. Yachtsmen tend to say "running aground," perhaps because they so often sail or motor in haste, and almost no one along this stretch of coast—or elsewhere in the Republic—uses the Anglicism "stranding." When any vessel bigger than a yacht-tender or other small craft intended to run ashore routinely on beaches or clam flats "grounds out," all the peculiarities of alongshore language parade themselves immediately.

Seen from the shore,
from the end of a path
that wanders downhill
through the pines, an old
wharf is the jumping-off
place for swimming or
modest exploration.
(JRS)

Viewed from the estuary at low tide, the same wharf is a nearly unreachable goal, a ruin, something detached from its long-lost landing float. (JRS)

" 'Taking the ground' is the professional expression for a ship that is stranded in gentle circumstances," wrote Joseph Conrad in his memoir, *The Mirror of the Sea*. "But the feeling is more as if the ground had taken hold of her." In ruminating about the professional expression, the sailing-ship-master-turned-novelist gnawed at the implications in the turn of the phrase. "Stranding is, indeed, the reverse of sinking. The sea does not close upon the water-logged hull with a sunny ripple, or maybe with the angry crash of a curling wave, erasing her name from the roll of living ships." Instead the ship sits in plain view, usually bows-on to the ground. Nine in every ten mariners wish themselves dead at that moment, in Conrad's experience, for "after all, the only mission of a seaman's calling is to keep ships' keels off the ground. Thus the moment of her stranding takes away from him every excuse for his continued existence." Given the staggering importance of such collision, Conrad defined his terms very precisely, concluding that "stranded" in his book "stands for a more or less excusable mistake," then remarking on two additional phrases: "A ship may be 'driven ashore' by stress of weather. It is a catastrophe, a defeat. To be 'run ashore' has the littleness, poignancy, and bitterness of human error." As an experienced master in the British mercantile marine, and as a devoted writer in his adopted, beloved English language, Conrad insisted on accurate terminology. "To take a liberty with technical language is a crime against the clearness, precision, and beauty of perfected speech."[1]

Tyehee no more "took the ground," then, than did Thacher's pinnace. Both vessels wrecked, and although Conrad might argue that *Tyehee* was run ashore and the pinnace driven ashore, no one fluent in alongshore English would say now that either went aground. "Grounding" in all its contemporary variations, including Conrad's "taking the ground," connotes quick reversal, either instantaneous, say with the reversing of engines, or soon, say with the next rising tide. Whatever the wreckage of its master's pride, a grounded vessel usually suffers no greater damage than scraped paint, and it moves off, away from the ground.

A dock, therefore, is something intriguing indeed, something so odd that vast numbers of otherwise plain-spoken Americans mistake it for the wharves adjacent. A dock permits the loading and unloading of a vessel without the vessel taking the ground, although the vessel touches land. Or does it? Is a wharf land? Or is it something else, something marginal, something in-between?

Dictionaries offer no immediate help. Attorneys and other jurists quickly learn that the terminology of real estate law frays badly at the edge of the sea, and worse just beyond the edge. In United States jurisdictions, a wharf and a pier are essentially the same thing, built forms erected for the efficient unloading of ships stopped in docks adjacent. Architects dismiss wharves as too impermanent for notice, and consider piers the property of engineers concerned merely with building, not with beauty.[2] But casual conversation reveals nuances beyond the needs of law and architecture, nuances that easily serve the needs of locals all alongshore. *Wharf* typically designates a structure set on wooden spiles and carrying a wooden deck, whereas *pier* designates a stone, earth-filled, or steel-and-concrete construction carrying a deck made of anything but wood. Local language eschews the word *quay*—along this stretch of coast it vaguely defines landing places in French Canada, and few users know that the term applies only to landing places constructed parallel with the shore, not jutting into it—and relegates *jetty* and *mole* to a jumble of nouns useful in describing breakwater arrangements.[3] And since most towns nowadays boast only a single landing place built of anything but wood, usually the main place built at public expense sometime between the heyday of the China Trade and the destruction of the latest northeast storm or hurricane, *wharf* remains in general use except when someone refers to the *town pier*. No one uses *pier* to identify a recreational structure, say with a casino or restaurant or amusement park, for almost no one knows —or cares—of their existence, either in Great Britain or in Atlantic City.[4] Along this stretch of coast piers are for business, and whatever amusement they provide is strictly accidental.

Wharves stand everywhere away from beaches exposed to ocean surf, thrusting into every navigable creek, every harbor. Flimsy, silvered by sunlight and salt spray, barnacle-encrusted and often lopsided or sagging, wharves attract landsmen. Fishers perch on their

PORT JEFFERSON.

By the end of the century, tourists had learned that stage or staging meant flimsy wharves used mostly for boarding small boats. (author's collection)

Stage, Bridge, and Harbor, CHATHAM, Mass

Only prosperous small towns boasted stone piers, but by the early twentieth century, such piers stood as relics of long-vanished overseas trade. (author's collection)

Cohasset Harbor, Cohasset, Mass.

The flimsiest of salt-marsh wharves often took the name "staging," in part because everyone, including the builders, knew them to be temporary structures. (Society for the Preservation of New England Antiquities)

edges, loners shading their eyes stand at their outermost reaches, tourists throng them, staring at fish forked up, at lobsters heaped in baskets, at immense, utterly private yachts undergoing emergency repairs a foot or so away. Wharves let poor landsmen venture to sea, to get out over the water, if not exactly on it, to shed the land, or nearly shed it.

Wharves lose themselves in mystery, especially in wet weather or at night. "The wharves stretched out towards the centre of the harbor, and, in this inclement weather, were deserted by the ordinary throng of merchants, laborers, and sea-faring men," writes Nathaniel Hawthorne in *The House of the Seven Gables*, "each wharf a solitude, with the vessels moored stem and stern along its misty length." What a place for suicide, Hawthorne continues. All one need do is "bend, one moment, over the deep, black tide," take one step, or slightly overbalance.[5] Hawthorne presumes readers familiar with something that contemporary tourists forget, forget until confronted directly, immediately. Wharves, unlike bridges, have no rails,

nothing but a six-inch-high timber—if that—to mark their edges, nothing to confront the would-be suicide—or giggling toddler—with an obstacle to jumping or falling. Tourists often walk jauntily to the edge of wharves, then literally rear backward, test their equilibrium, and mince forward to gaze down over the unfenced edge. Parents suddenly conscious of the sheer drop sprint forward, arms outstretched toward children running toward the precipice exposed by low tide. Somehow every wharf becomes what no sign ever announces. A dead end.

And what of the people of the wharves, the figures half-hidden by twilight, by fog? Are they honest landsfolk or seagoers, or some marginal, criminal half-breed marooned from an old

Shipowners sometimes repaired their vessels in docks, careening them over so that their tophamper rested on an adjacent wharf. (New Bedford Whaling Museum)

Bogart or Brando film set on disused steamship wharves, on some crime-ridden urban waterfront? A wharf traps the unwary, the innocent. Might not the criminal—the wharf rat—follow a victim onto the wharf, and so trap her or him? Might not criminals lie in wait at the outermost edge of the wharf and pop up suddenly, surprising their prey? Who walks the wharves in the wee hours of the morning, when unsuspecting tourists lie asleep in their fragile craft, moored stem to stern? Maybe it is better to anchor, to anchor in the outer harbor where open water surrounds the little boat and makes pilferage and murder less likely. Wharves, after all, may be humansized weirs.

Here Melville makes creepy reading. His "Loomings" chapter introduces *Moby-Dick* in ways modern readers miss, miss beyond any inkling of looming, for modern readers see nothing evil, nothing uncanny in Melville's remarks about Sunday visitors thronging quiet New York wharves. The people "leaning against the spiles," the people all week pent up in lath and plaster are, after all, gazing outward at ocean adventure, having walked as far on land, and as far over water, as they can. On any "dreamy Sabbath afternoon" the thousands of wharf visitors look outward into light, toward looming. Innocent, virginal, almost simple, their heart-rending scrutiny of openness opens Melville's romance, directing attention to the nautical meaning of "looming." But Melville presents something else, too, something ghastly, something nowadays forgotten.

When Melville wrote *Moby-Dick*, the wharves of New York swarmed with criminals. Even on Sundays, while thousands of law-abiding citizens stood in quiet reverie atop wharves, thousands of waterfront criminals slept under or "worked" the wharves. Beneath the law-abiding citizens, literally under their feet, pulsed another way of life, a way as dark, as evil, as pitiable as any imaginable. In telling his readers to walk "from Corlears Hook to Coenties Slip, and from there, Whitehall, northward," Melville provided an itinerary above, if not through,

Even at crude cribwork wharves, unloading oysters, lobsters, or scallops enlivened the day, flavoring the harbor with a trace of big-port excitement. (author's collection)

some of the worst poverty and criminal activity known in the Republic.[6] His introduction, then, looks beyond loomings, for in so deftly implying the evil beneath the feet of the casual visitors it introduces the evil that ballasts the whaling voyage.

Thirty years after he published *Moby-Dick*, when Melville worked as a customs officer along the New York waterfront, the under-wharf situation had grown only worse, as Charles H. Farnham discovered firsthand, pulling beneath the wharves in a small boat rowed by a member of the "steamboat squad," the waterfront police. Farnham explored a criminal realm of "thieves, wharf-rats, smugglers, and their kind" within inches of bustling legal activity. From beneath the wharves, thieves routinely stole goods piled atop wharves, often by such simple tricks as cutting a hole in the decking and looting piled crates. Sometimes they bored holes through the deck into hogsheads of whiskey and drained the liquor into barrels sitting in their rowboats. Others cut holes beneath heaps of fruit and collapsed the heaps so quickly that no one noticed. Foreign seamen frequently contrived to smuggle goods to the thieves, who then stored them in caches all along the waterfront. "The tramping of the passengers overhead drowns any noise these men make in operating down here, and they can live in safety here till discovered by some boating expedition. The passengers overhead don't imagine that they walk over a robber's cave, but some strange things are done under ground as well as over it," the policeman explained while pulling beneath a busy ferryboat wharf. "We lately found $3,700 worth of velvets and silks on such a ferry-float in Jersey City." Stolen goods went to

Mass-circulation periodicals routinely depict boys swimming among wharf pilings, but almost never do they show girls. Even in the late 1920s, docks remained part of the male realm. (Harvard College Library)

licensed pawnbrokers and junk dealers so anxious for items that they and their crews rowed about beneath the wharves, protected from arrest by their affirmations that they had paid for everything in their boats.

Farnham gradually figured out what every small boy growing up alongshore learns so quickly. Wharves are hiding places made to order, a sort of constructed forest whose crown lies just under the oh-so-sunny deck. Almost every large wharf consists of rows of spiles marching waterward from the shore, each row capped with a massive timber, lying parallel to the shore, too. Joists bridge the timbers, running at right angles to the timbers and the shore. The deck planks run at right angles to the joists, parallel to the rows of spiles, the heavy timbers, and the shore. Although it is wholly commonsensical, wholly ordinary, wholly in the American common tradition of building in wood, wharf construction typically produces an extraordinary nesting area atop the spiles. The space between the top of the spiles and the bottom of the deck planking, often two or three feet high, provides not only a grand place to climb but a perfect place to hide—or to hide things—and as Farnham learned, even a place to live.

"This underside of the city is a shadowy world even at high noon, and its structure, as well as its seclusion, makes it as good as a forest for hiding." Almost desperate to impress his *Scribner's Monthly* readers with the scale of the places, Farnham argued that "the under side of a pier can hold a good sloopload of packages, and a box on a stringer is invisible to any one passing under the pier, unless he passes very close to it." Near the criminal-infested region called Hell's Kitchen, the policeman pulled Farnham into "a long, narrow passage between the back of the new sea-wall on the right and rows of spiles on the left. The faint light shone through the forest of spiles and lighted a dismal, slimy cavern, bounded by the green ledges of the wall, the rocky shore full of dens and holes, and the heavy timbers close overhead." Staring at rats staring at him and at sewers "vomiting filth," breathing air "unendurably loathesome,"

Many adventures, including fictional escapades, begin on wharves. (Harvard College Library)

Farnham suddenly realized that people lived here, perched atop the spiles. "A few boards lay across three stringers, and a bed on these was made of a heap of rags and papers from the gutter. A few vials of medicine completed the picture of utter misery with suggestions of sickness and death." The so-called nook-and-corner men found perfect nooks beneath such wharves, making "a vast region of secrecy right under the busiest part of New York." Only the police, probing and probing, disturbed the nook-and-corner men—and their women and children.

In spite of police assurances that the organized gangs had been broken up, that the steamboat squad followed thieves into every hole, Farnham boggled at the vast numbers of under-wharf criminals, and at their achievements. Because so many wharves stood on such densely set spiles that "only the smallest skiff can pass through the narrow, tortuous openings," the police could rarely probe in forces large enough to make mass arrests. And the police let slip that until recently, the under-wharf criminals had used a channel they cut through the spiles "from one end of the city to the other, by which they could travel nearly the whole distance without showing themselves." For all his fascination with the on-wharf activity, the frantic loading and unloading of ships and lighters and boats, the jams of drays and small wagons, the frenzied work in shops that made everything from figureheads to small boats, the under-wharf world fascinated more and more deeply. "So here is a vast region of secrecy right under the busiest part of New York," he concluded, knowing something about which most casual visitors to the waterfront, most law-abiding, preoccupied people running for a ferry or picking up some item or cargo guessed nothing.[7]

Underfoot but hidden, secrets lurk still. Low tide makes some under-wharf forests accessible to sloggers who are willing to brave the stink. Unlike healthy salt marshes and flats

Charles H. Farnham found all sorts of offal around the New York City docks. The garbage scow carried only one sort. (Harvard College Library)

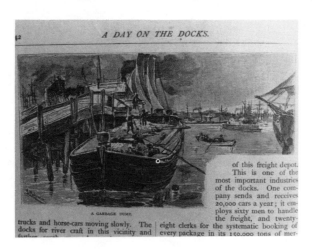

A DAY ON THE DOCKS.

A GARBAGE DUMP.

of this freight depot. This is one of the most important industries of the docks. One company sends and receives 20,000 cars a year; it employs sixty men to handle the freight, and twenty-trucks and horse-cars moving slowly. The | eight clerks for the systematic booking of docks for river craft in this vicinity and | every package in its 150,000 tons of mer-

of sand or mud, the ever-shaded under-deck zone exudes odors that call miasma to mind, odors belched from every new-made footprint gaping in muck. No sunlight purifies the air, let alone the mud, but enough light penetrates the cracks between deck planks to illuminate the voyeur's way. High above echo the calls of tourists, the scuffing of feet, now and then the rumble of a fish truck, if the wharf is busy. Vibration dislodges grit and splinters of long-dried, creosoted wood. Litter falls through the cracks, dropped or shoved by daylight visitors unconcerned with the dank below.

Bushels of broken glass lie beneath every old wharf. Permanent, never decaying, the shards and bottles offer mute evidence of long-ago picnics (ketchup bottles), hard drinking and soft (whiskey nip bottles and bottles once filled with ginger ale, Coca-Cola, and Simpson Spring), and seasickness (medicine bottles, often blue). The shards remain sharp, as sharp as any barnacle, any broken shell. A vast slicing field, the last place of barefoot wandering, lies beneath the wharf-forest crown.

And in the slicing field lies treasure, albeit minor. Thousands of fishing lures adorn some places, or heaps of lures and hooks and sinkers from upset tackle boxes, especially if the wharf is old and the spaces between splintered deck planks are large. Dozens of hand tools, usually screwdrivers but sometimes pliers, remind the slogger that much marine repair work occurs not aboard the fishing boat or sloop or fiberglass motorboat, but just alongside, up on the wharf deck, where engines and other recalcitrant mechanisms can be upended in sunlight, stability, and proximity to auto-mechanic toolboxes. Steel cable, sometimes stainless but usually long rusted, and bits and pieces of chain twist through the mud, often half-attached to some waterlogged timber that once meant a secure mooring, a safe grip on a safe wharf. Everywhere monofilament fishing line glistens, and everywhere the detritus speaks of tourism, in layers extending upward from ancient bottles packed solid with sand and long-dead mollusks to tonic-can "flip tops" to slime-covered metal and plastic 35-millimeter film canisters to ice-cream-bar wrappers and other scraps of paper. Magnets may prove useful to adolescent treasure seekers but usually yank up only rusted steel cans, fishing hooks, and hundreds of bent nails—almost never the hoped-for Swiss-made pocket knives. And criminals throw revolvers from bridges, not wharves.

With the incoming tide all vanishes into deeper mystery, the shadowed water masking the larger items, say the crumpled supermarket shopping cart, the beer keg long discarded as a mooring buoy. The tide warns away the rare slogging explorer, preferably before the slogger turns swimmer caught among dozens of mussel-covered, razor-sharp trunks, and now and then it welcomes a novice boater, someone lured into the wilderness of spiles locals now call pilings or piles.

Only rarely does a local steer a boat from dock or channel into the maze beneath the wharf deck. Most small boats are simply too wide to fit easily among barnacle-encrusted spiles, where oars become as useless as outboard engines. Sometimes a boat can be poled along, but most often it is moved by "handing," by pulling it from one spile to another. Even youngsters hear of the dangers. The boat can become jammed between two off-vertical spiles, moving into the apex of the angle as the tide rises or falls, or it can become twisted among four or more spiles, unable to budge. Even worse, the boat can be trapped by the incoming tide, held against the beams that top the spiles and carry the deck joists. Stories move through junior high schools of adventurers crushed or drowned against the bottoms of old wharves.

Of course, adventurers do navigate the wharf forests, especially now that sea-kayaks need so little width to move. Adult kayakers discover what small boys in old rowboats discover, not only the mysteries of the under-wharf zones but the voyeuristic pleasures of floating just under the feet of unsuspecting wharf-deck visitors, especially tourists half-consciously thrilled to be standing over the water anyway and ripe for startling. A length of straw or reed, perhaps a

As this illustration from the June 16, 1980, New Yorker demonstrates, artists still savor the forestlike darkness beneath New York City wharves. (Harvard College Library)

stiff piece of rockweed pushed through the crack and rubbed against some bare ankle elicits a very satisfactory shriek, followed by an even more satisfactory scolding from a relative convinced that everything is imagined, that nothing can be down there, reaching up. Unlike a bridge, a wharf ought not have anything under it but its spiles, no trolls, no nook-and-corner men, no spies, playful or otherwise. Wharves have no channel ways, for no navigation is expected, welcomed, permitted. Indeed many wharfingers, public and private, fear under-wharf exploration, not for reasons of safety but for reasons of discovery.

Nook-and-corner or nook-and-cranny men worked small-harbor wharves along this coast, too, though few remember them. Long after the East India trade died, long after even the West Indies trade collapsed, a few scavengers waited for fresh fish and other modest treasures to fall between schooner and wharf. Called "gaffin' fish," the night-time, low-tide occupation sometimes irritated, sometimes enraged fishermen and wharfingers, who found it difficult to detect, let alone catch the men and boys who poled briskly in narrow boats called punts, boats perfect for scooting into the dark safety of wharves.[8] In the smallest harbors no nook-and-cranny men worked for long, even among the fishing boats routinely docked without watchmen. After only one theft, the close-knit communities watched too closely. But always the under-wharf shadows seemed a likely lair for anyone up to no good, from the tramp sleeping rough to the local stealing lobsters from a floating pound to the armed robber seeking a getaway boat, and wharfmasters concerned about their reputations and clients warned away everyone suspect.

Yet wharfingers warn away explorers for another reason, too. They know what explorers may find beneath the smooth deck of the wharf.

Under the old small-harbor wharves of coastal New England the spiles stand in near-chaotic proximity, in every stage of decrepitude. Above them, the deck is often well maintained, with a handful of new planks laid down each spring, in advance of tourist season. But under the new planks so prim in their pressure-treated, rot-resistant greenness, the spiles almost always need work.

Repairing piles means summoning a pile driver, a weighty machine that stirs up secrets and memories even if no longer powered by steam. In *Wharf and Fleet: Ballads of the Fishermen of Gloucester*, Clarence Manning Falt caught the turn-of-the-century magic of the "spider and the fly," the massive iron weight dropped again and again on the fly, the wooden pile. "W'en the iron rings its sounder / Birthin' piers," even grown men stopped work, and children gathered, at a distance, hoping to see the rare sight of the immense weight jerk free of its cable and crash into the sea.[9] Little has changed today. Still expensive, still heavy, still

Derelict wharves, even the small wharves abandoned on silted-in salt creeks, are dangerous. They collapse, sometimes with no more warning than a groan. (JRS)

pounding away with a throbbing that pleases many listeners and gives others headaches, the pile driver is terribly expensive to rent, and arrives only for major repair work, not to replace a rotted spile or two. Its immense weight means that every spile between shore and worksite must be sound, for a wharf weak enough to scare truck drivers and fishers can scarcely carry the pile-driving rig. So the expensive, welcomed, detested machine begins to pound in new spiles nearest shore, its rhythmic ramming carrying far across open water and marshes, its vibrations stirring up mud for yards. So impressive is the driving that few observers catch the half-done job. Only rarely does the repair crew try to extract rotted spiles. Instead, new

Winter ice and spring gales batter even well-kept, pleasure-use-only wharves, splintering diagonal bracing and sometimes knocking over entire rows of pilings. (JRS)

piles are driven alongside the old, making over the years a haphazard, twisted arrangement that challenges carpenters who have to place cross timbers. And sometimes, usually for lack of funds, the wharfinger replaces only the inshore half of the wharf, closing the outer end to trucks, then a year or so later to cars, then to tourists, then finally to local fishers hitherto trusted enough to avoid the most rotten spots.

Rot alone fells few wharves, however. Attack by marine animals is the greater menace. So bizarre is the assault, so seemingly unfair, so successful, that many wharfingers give up in despair. And nowadays, concerns about chemical poisons cause many wharfingers to give up even faster.

Two types of animal attack wooden pilings. Neither is lovely enough to attract amateur marine biologists, but because some grow almost four feet long, the lack of amateur interest will surprise anyone who tramps around harbors at low tide, pokes about in a rowboat, and—especially—hacks open an ancient spile. Of the class Crustacea there are three genera of importance, but *Limnoria lignorum*, nicknamed "the gribble," is the most destructive for all that its quarter-inch-long body looks like a woodlouse. Of the mollusks *Teredo*, *Bankia*, and *Martesia*—all three genera are distantly related to the clam—*Teredo* is perhaps the most destructive. Known as shipworm, the many species first came to prominence when they attacked the ships of Europeans exploring the Caribbean, but they appear to have worldwide distribution. Some perhaps moved north in the wood of European ships, in time infesting the warmer harbors of Europe and both coasts of the United States. Although the paths of migration cannot be retraced and indigenous species cannot always be differentiated from new arrivals, the ubiquity of the animals astounds any inquirer. *Bankia* range from Kodiak in Alaska to Gulfport, Mississippi, for example, and early in the twentieth century, wharf-deterioration experts began to learn that marine zoologists were not familiar with the species of *Bankia* discovered in infested pilings.[10] Nearly a century later, even sophisticated littoral field guides are of no use to the dedicated amateur determined to explore the marine ecology of a somnolent Massachusetts Bay or Cape Cod harbor.[11] Animals of salt marshes and rocky shores, estuaries and sandy beaches receive much attention, but the little nasties that eat away the spiles on which the naturalist stands get none.

Anyone who pokes about in a small boat nowadays discovers a low-tide world of abandoned wharves and pilings. (JRS)

In the late nineteen-teens, however, the animals precipitated an extraordinary crisis, one nearly forgotten now even by marine ecologists, one that must intrigue anyone propped in the sun atop a wooden wharf. Beginning in 1914—and masked from widespread public attention by the start of World War I—the wharves of San Francisco Bay began to collapse. Shipworm, probably *Teredo* species, had been noticed in the bay since 1849, perhaps brought north by Gold Rush vessels following the Cape Horn route that took them through the infested Caribbean and then the warm waters west of the Isthmus of Darien. But in 1914 the U.S. Navy noted massive, seemingly overnight damage to the wharves at its base on Mare Island, and by 1917 wooden structures began to fail further up the bay. By 1921 almost every wooden structure touching the waters of San Francisco Bay had failed, and replacement pilings were failing rapidly, sometimes within a year of being pounded in. Railroad cars simply plunged through weakened wharves, and on one breezy night, the sailing vessel *Cyrus Wakefield*, properly moored to a wharf, went adrift, pulling the wharf deck along with it.[12] New species, especially of *Teredo*, may have arrived, or perhaps indigenous species somehow "bloomed." Whatever the cause, the destruction nearly panicked the Navy, steamship firms, railroad companies, and other owners of wooden wharves, and the local committee of engineers, biologists, and chemists quickly asked for national help.[13] The National Research Council, a nonprofit research group funded by member firms, established a Committee on Marine Piling Investigations, which learned after the most cursory of introductory inquiries that similar destruction was occurring everywhere along U.S. coasts. Within months, railroad companies from the Bangor & Aroostook of Maine to the San Antonio and Aransas Pass of Texas, wharf owners from the American Sugar Refining Company to the Port of New Orleans, and steamship firms from the Cunard Line to the Atlantic Steamship Lines had contributed financing for a large-scale, exceptionally sophisticated research-and-development effort.[14] Not until 1924, however, did much make sense to the researchers, and even now the tomes reflect an appalling uncertainty about the end of an era the researchers wholly missed.

Worldwide economic recession at the close of World War I ended the momentary last hurrah of wood-hulled sailing ships. So desperate were shippers during the war that shipyards constructed many wooden freight vessels, some of them sailing vessels so large that they carried steam engines to power sail-raising hoists or auxiliary steam engines to power propellers.[15] But steel-hulled sailing vessels and steamships had replaced most wooden vessels before the war, as early as the 1890s on some routes. Coastwise cargoes that moved in wood-hulled schooners and other small vessels until just before the war had begun to move in steel barges towed by tugboats, and although the wartime boom brought many coasters out of retirement—and caused Maine builders to build many new small ones—"the tows" quickly grabbed away trade after 1918. And in countless harbors, especially in New England, small sailing vessels, especially fishing schooners, were moored for good, or else run aground and left to decay.[16] The fate of these wood-hulled vessels turns out to be critical in any study of wharf deterioration, for the long historical backward glance reveals something the National Research Council investigators were too close to notice. By 1914, practically every harbor in the Republic had at least one, and sometimes many, wood-hulled vessels rotting in place.

And in the hulls lived the Crustacea and Mollusca picked up in far-off places and brought to a final place to mature, breed, and hatch. And once hatched, the young attacked the nearest wood, the wharf pilings.

In a timber-rich country few wharfingers had ever taken pains to preserve spiles, and then usually only from long-term rot. Creosote, the tried-and-true preservative railroads applied to cross-ties, slowed rot in saltwater-immersed spiles, but its effectiveness against boring animals depended in part on its concentration in the wood and on the location of the spiles. In warm-water ports like Mobile, a creosoted spile might last only sixty days before borers honeycombed it to the point of collapse, and even in cold-water harbors, swift currents seemed to leach away the creosote quicker than still water dissolved it. While the U.S. Forest Service remained convinced that high-quality creosote could forestall damage by borers, by 1920 experts knew that any damage to a spile, especially before insertion into the ground, might provide an opening for borers. Suggestions for protecting spiles before and after creosoting became almost comical; researchers urged that no spile ever be lifted by chains, ever shoved about with a sharp pike pole—even that no men wear cleated shoes when walking over heaps of uninstalled spiles or when climbing set ones. Experiments rapidly confirmed researchers' worst fears: "Shipworms and *Limnoria lignorum* will seek out and attack through the smallest defects, and after the shipworms have become established and their development has proceeded beyond the larval stage, they seem to work only little less freely in creosoted than in uncreosoted timber."[17] But wharfingers accustomed to cheap spiles, and to replacing them every twenty-five to fifty years, rebelled when researchers offered creosoting as the best solution. Every nick, every deep splinter in a driven spile essentially negated the preserving power of creosote, and as any wharfinger—or observant child—knew, pile-driving damages every spile, even if slightly.

Desperate, nearly broke in a difficult recession, and yet oddly willing to try anything new, anything "scientific," wharfingers and researchers begged the U.S. Army for help. And so at Edgewood Arsenal in January 1923 the Chemical Warfare Service began to learn how to poison the enemies of wharves.

Within six months, the service began field tests at Beaufort, North Carolina, experimenting with Bureau of Fisheries biologists and naval experts on anti-fouling paints intended to keep American warship hulls free of the plant and animal growth that so slowed vessels only a few months in service. Army experts proceeded on the basis of battlefield expertise, quickly concluding that "several chemical warfare compounds" were immediately effective and that "the best all around specific toxic found was chlorvinyl arsenious oxide, a modification of the well-known war gas 'Lewisite.'" But although impregnating sections of new railroad ties—hastily and enthusiastically supplied by the Pennsylvania Railroad Company, itself a wharf owner—certainly proved that Army poisons killed any creature dangerous to spiles, even the Army admitted that the chemicals soon leached from the test pieces. Creosoting seemed to bind the poisons into the wood, and the Army concluded that "there is no doubt that the addition of a specific toxic to creosote will give protection for a longer period than creosote alone." Otherwise, every spile under every wharf would need a cloth poison-dispenser along its entire submerged length; fresh poison could be inserted every two weeks or so. The poisons were cheap—the Army estimated a cost of two dollars per pile for installation and a one-year supply of "the cheaper toxics." Another method was also proposed, but shooting copper plugs into existing spiles in the hope that salt water would erode them and by electrolysis make chlorine gas did not seem particularly worthwhile, and certainly was expensive to install, requiring the services of a diver for existing wharves. A permanent, or at least long-term, solution seemed technically feasible but politically impractical, though the Army itched to try it. Poisoning the water around existing infested wharves should work, but "so little is known of the dispersion of small quantities of difficultly soluble materials in extremely large bodies of water that it did not seem feasible to try any large-scale poisoning of the water in Beaufort harbor on account of the fishing industry there, and at this writing no place seems available for a test of this kind." Yet the Army, knowing that its poisons would kill all forms of marine life even "in an additional region, depending on the local harbor conditions, such as tide, current, wind, etc.," had merely begun to follow lines of argument already worked out by federal Department of Agriculture scientists, who were beginning to argue for the elimination of all insects from a farm field infested with any noxious ones.[18] In a busy port dependent on its wharves, clams and flounder might have to die in the all-out war on *Teredo*.

How far the Army's argument went no one knows or cares to remember. The military well understood the secondary costs of poisoning only part of a harbor, and even the Navy never tested the idea at its large bases.[19] But private industry began to produce all manner of poison paints and other spile coverings, and certainly experimented with impregnating creosote with additional poisons, too. National Research Council investigators found none of the patent preservatives especially useful but remained convinced that some firm might stumble onto something cheap, durable, and immensely effective.[20] In the meantime, since some mollusks attacked even cement piers, the council recommended creosote, admitted temporary failure, and looked to experimental techniques like armoring spiles with everything from copper sheathing to floating barnacle scrapers.[21]

Yet the investigative reports reveal extraordinary oddities. First and foremost, the overnight effort to catalogue spile-destroying creatures now suggests that the wharf destruction of the 1920s occurred not only because infested wooden vessels were laid up but because canals were opened between infested and clean places. The Cape Cod Canal, opened in 1914, linked Cape Cod Bay with the waters of Buzzards Bay, which the National Research Council discovered in 1923 was infested with shipworm and *Limnoria*. The problems of infestation in Buzzards Bay had troubled the lighthouse service—creosoted, white-oak spiles were "completely destroyed in two years"—but Cape Cod Bay was relatively untroubled until about 1920, when the slightly warmer waters of Buzzards Bay had begun to have an effect on marine pests.[22] Second, as more municipalities and agricultural irrigation districts drew off fresh water before it flowed into the sea, the rise in salinity allowed borers to move further into harbors and estuaries, exactly as happened in San Francisco Bay. In a fierce article in the *Pacific Marine Review*—excerpted in 1920 in an issue of the *Literary Digest*, for by 1920 the West Coast *Teredo* problem had attracted national public attention—one writer blamed the entire infestation on the increased harbor salinity that resulted from diverting fresh water to irrigate rice farms and concluded that "the present condition should show irrigation engineers the advisability of research work to determine just what effect their reclamation programs will have in upsetting the nice balance of the ordinary processes of nature."[23] Third, as cities built sewage-treatment plants and so purified the effluent discharged into harbors, boring animals multiplied in the cleaner waters and became more destructive. But issues of water flow and water quality simply escaped scrutiny, as did the issue of mass importation of destructive marine animals.

Larger steel-hulled vessels collected extraordinary amounts of hull-attached organisms, which were routinely dumped near whatever shipyard scraped them. In August 1909, the

Navy scraped six hundred tons of barnacles and other animals from the hull of the cruiser *South Dakota*, which was dry docked at Hunter's Point in San Francisco Harbor. Although the load of barnacles attracted some attention—*Harper's Weekly* ran an illustrated story pointing out that barnacle-covered warships cruising in tropical waters experienced increased fuel consumption and reduced speed—no one glimpsed the potential ecological effects of scraping.[24] Five years later, when naval officers first reported the San Francisco Bay infestation, no one remembered the *South Dakota* barnacles or the animals scraped from other warships, especially those returned from serving in the Asiatic Squadron. Indeed, researchers on the international movement of agricultural infestations were concerned mostly with ship cargoes and decks, not hulls, and rarely considered the long-term effects either of beached hulks in which species could complete life cycles or of dumped hull scrapings. Since some species of *Teredo* lay 100 million eggs per animal, infested hulks and mounds of scrapings ought to have drawn some attention.[25]

Although council researchers ignored some major issues, often for reasons that are now inexplicable, they did try to interpret the influence of harbor pollution. Sewerage alone did not discourage borers. Indeed, spiles that protected the outfall of some large urban sewers deteriorated rapidly, and the four-year destruction of wooden accessories to the New Jersey Passaic Valley Sewer outfall so worried researchers that they attached an emergency appendix to their *Report* of 1924. But where sewerage did not immediately flow away with the ebb and where chemical pollution existed in high concentrations, wharfingers noticed long-lasting spiles.[26] In upper New York Harbor, anywhere tidal currents impounded sewerage and chemical pollution for days at a time, many spiles remained in exceptionally sound condition, and wharfingers hoped they would remain so. Indeed, the pollution Farnham witnessed pouring from sewers beneath wharves actually lessened the ravages of borers. After 1870, the depredations of *Teredo navalis*, thought to have arrived in the harbor at least as early as the sinking of a British frigate destroyed in the Revolutionary War, began to decrease. Within five years, wharfingers remarked on the lessened activity, and while a wharf at Greenport failed in 1898 due to teredine borers, much of the harbor had become almost borer-free.[27] By 1925, researchers and wharfingers knew that eliminating harbor pollution would trigger infestations of borers, and advocates of clean harbors encountered subtle opposition.

In the 1920s, however, wharfingers had more than borers to worry about. As the postwar shipping slump decreased profits, wharf owners made fewer and fewer repairs, until many wharves became so splintery that their heavily creosoted deck timbers turned into kindling ready to blaze at the slightest spark. Once the National Fire Protection Association adopted its "Regulations for the Construction and Safeguarding of Piers and Wharves" in 1924, city governments and insurance companies began to frown even more deeply at wooden piers saturated in highly flammable preservatives.[28] In urban ports, building codes began to specify steel-and-concrete piers, and only in small fishing harbors did the spile-supported wood-deck wharves endure. After all, everyone familiar with them understood that wharves exist as essentially temporary, marginal structures, always within grasp of the sea.

Then, too, large-port city governments finally understood something Farnham had noted half a century earlier. "The view of the piers from the water is singularly out of keeping with the grandeur of the city," he determined from his rowboat, deciding that only the "dignity" of the great harbor and the vast collection of fine ships keep the city front from being contemptible. "The docks are a tattered, dirty fringe to the city," he suddenly realized, the fringe that welcomed every transatlantic steamship passenger. In 1879 no one cared much about the waterfront fringe, for every other urban terminal stood equally shabby, but by 1920, when most cities boasted fine, ultramodern railroad terminals, the wharves looked awfully seedy by comparison.[29] The shabbiness of wharves everywhere, even in the smallest ports, quickly became a minor theme in period writing. Stately sailing ships and oceangoing steamers attracted sustained attention, even in docked masses, but almost no one wrote about wharves without remarking their shabbiness. Quite literally, the eye of arriving passengers and crewmembers jumped from the docked ships to the tall buildings in from the waterfront, skipping wharves as uninteresting if not worse. One Maine schoonerman, recalling his first arrival in New York Harbor in 1882, remembered the "mighty panorama," the "spectacle" of the harbor that burst before him as his schooner swung to port into the East River: "It was not so much New York City. There was no skyscraper skyline in those days as the highest buildings were some ten stories, which is just a portico now. But the river! Docked on both sides, the lofty spars and yards of ships laying hull to hull, stretched away for miles until they were lost in the distance." Albert Averill enumerated the types of vessels, including ferryboats and "splendid great harbor tugs," even scows, but the wharves, even the piers by then covered with cobblestones, rate hardly an oblique mention.[30] And nowadays, in so many small harbors, they rate less.

Wharves mean only vantage points to most summer visitors, even to many year-round residents drawn on Sunday afternoons to the harbor to look out at the breakwater, to look down on the fishing trawlers, to look at the colorful fiberglass sailboats moored everywhere except in the channel just beyond the wharves. From the small boat rowed just alongside, however,

Some small turn-of-the-century wharves boasted a crispness that reflected
prosperous local industries, especially fishing. (Society for the Preservation
of New England Antiquities)

small-harbor and salt-marsh-creek wharves remain what Yankee localism still calls them—landing stages. The term applies now to small wharves that totter atop wooden spiles still routinely replaced in continuing warfare against borers, but it connotes much more than newcomers imagine. Essentially a corruption of "staging," the term implies that locals view small wharves as kin to the carpenter's staging, something temporary to provide access, something that lets a vessel off-load without taking the ground. But the smallest of wharves still harbor something darkly mysterious, something that suggests forest gloom, something that reminds any viewer of nook-and-cranny men or worse. One of the last extant forms of traditional or common construction, these wharves, with their beam-atop-spile, joist-atop-beam, deck-atop-joist construction dating from the eighteenth century, still speak to anyone who is willing to pull alongside at low tide, to splash about in the mud.

In 1885, Sarah Orne Jewett did just that, finding in the deck-topped spiles of a salt-marsh estuary village something picturesque, quaint, yet vaguely sinister, as impenetrable as salt marshes. Ebb tide revealed the spiles, the darkness beneath sunlight. "The tide was going out; the foundation of the village seemed to be insecure piles and slender sea-bitten timbers, between which one could look, as if they were great cages for long-since-escaped marine monsters."[31] Decked in olive and brown seaweeds, standing out over mud and shallow water, the flimsy village wharves might well resemble cages, cages big enough for sea monsters, small enough for people. But nowadays, long after the demise of coastal steamship companies, only pleasure boaters glimpse the darkness, the twisted spiles, the wreckage below deck.

They glimpse them only momentarily, as they step from a fiberglass speedboat onto the floating pier they call a float, not a landing stage. Ahead of them, angling down from wharf deck to float, the steep-pitched, wheeled ramp shifts slightly with the chop, or rests rock-still. Almost always the adults hurry up the ramp, clutching neon nylon duffel bags, coolers, fishing tackle. Only children pause to stare into the forest of barnacle-encrusted pilings, to look deeply into the shadows, to point out vague shapes in the dark, to glimpse the lair of inshore sea monsters, the lurking place of pirates. And no one, not adults, not children, thinks of wharves as the entryway of invading armies, armies too clever by far to land on open beaches.

8 SMUDGE

Cubelike, gritty gray against the winter sky and gritty gray against the summer sky, too, the visored monoliths march along almost the whole United States Atlantic coast, south from northern Maine to Cape Hatteras in North Carolina. Indestructible, almost invariably prominent from any landward vantage point, the towers stand exactly placed, each barely in sight of two others, one at either side. On hilltops squat, sometimes just twenty feet high, the towers change shape in low-lying places, reaching five stories into the air. By land, close-up scrutiny of any or all reveals little, no more than three nearly uninterrupted concrete walls facing inland and alongshore, the inland one broken only by a massive steel door, often welded shut. Only the view from the sea offers any clue, unless the wanderer of deserted beaches finds one of the immense steel doors vandalized, wrenched open.

From the sea, the towers are distinct, marked against sky and inland scenery as clearly as on charts that locate them as seamarks useful for coastwise navigation. Always the mariner notices the horizontal slits near the top of the seaward-facing facade, slits usually shaded by visorlike slabs of concrete. The towers appear to squint into the rising sun, to watch through narrowed eyes, to serve some sinister purpose unlike that of lighthouses, buoys, and other brightly painted navigational aids.

Trespassing is wonderfully educational, if sometimes dangerous. Unlike lighthouses, long unstaffed but precisely automated, functioning, and so protected by warning signs, immaculate white and red paint, and unannounced visits by the Coast Guard, the towers stand so evidently abandoned that only their grotesque ugliness deflects hikers and picnickers. Now and then, however, the trespasser finds one with its steel door pried open, the spot welds shattered by some adventurer's sledgehammer. More rarely the hiker along the winter beach

So ugly, so charged with unsettling associations are the World War II observation towers that many tourists and many scenery photographers ignore them, even when the towers stand adjacent to accepted salty structures like lighthouses. (JRS)

hears the freed door booming in the gale, slamming shut then crashing open, its deep bang echoing from the top of the tower. Curiosity prompts a change of direction through the dunes, then a long stare at the gloomy tower wreathed in blowing sand. Prudence dictates wedging open the door, perhaps with a piece of driftwood, and then the most careful of explorations into the boxlike ground-floor chamber. A match, better a candle, better yet a flashlight or kerosene-fueled boat lantern becomes necessary on all but the brightest summer days, unless the steel door can be propped wide open. In one corner a concrete, steel-railed stair leads up, into a second room unlit by any window but offering the trespasser another stair. In the gloom the trespasser climbs the concrete-and-steel stairs, candle guttering in the drafts, and emerges into the topmost chamber, into light, into awareness. Into weapon.

Certainly the tower stands at attention, a timeless military construct. In World War II it served two purposes. Always it sheltered coast watchers, soldiers or sailors charged with scanning the inshore waters, watching day and night through binoculars, watching for U-boat attack, spies and saboteurs, invasion. And in the event of attack, its men acted as spotters, recording the fall of shells fired by the coast artillery, the immense guns that protected most major and second-ranked ports. So ruggedly built that it could withstand near misses from naval guns, the tower was almost impossible to demolish at war's end, let alone to reuse when purchased as surplus real estate by summer-cottage seekers.[1] Sentinel-like, it stands at the beginning of the twenty-first century as a mocking reminder of incidents conveniently, cleanly forgotten except alongshore.

German U-boats shelled Cape Cod beaches. Admittedly, the German seamen had targeted merchant vessels inshore from their submarines, and the shells that struck the beaches struck by accident. But all alongshore, old-timers recall the closeness of naval warfare in 1942 and 1943, the ships burning near the horizon, the lifeboats creeping painfully toward land, the periscopes glimpsed by fishermen. Naval historians detail the exploits of convoys, the stunning maneuvers in the Coral Sea and elsewhere, but only rarely do they remark the intimacy of U-boat assault. Tanker after tanker burned off of Cape Hatteras simply because U-boat commanders positioned their submarines so that the tankers would pass between periscopes and navigation buoys left illuminated well into the war. Silhouetted against the white light, the tankers made perfect, utterly perfect targets. Not for months did anyone think to extinguish the buoy lights.[2] Only the alongshore people saw the night-time fires, watched the oil slicks come ashore, gathered the charred bodies from the surf. From Cape Hatteras north to Canadian waters, old-timers muse still about the days of darkness, the days of watching

Far beyond a barrier beach exposed at low tide stands a submarine watchtower, a solitary sentry scarcely noticed by tourists. But even many locals know nothing of the hill adjacent to it, a wholly artificial, wholly hollow hill that once shielded a great cannon from enemy attack. (JRS)

warships steam closer and closer until they could make out the American flag or the white-and-blue Cross of St. Andrew that identified Royal Navy vessels, the days of wondering if the Allied warship steaming at top speed meant to attack some U-boat between it and land or if it steamed to escape some U-boat, some pocket battleship, some Q-ship camouflaged as a rusty tramp steamer.

On winter nights around the boatshop stove, at the Masonic Lodge, in some lobsterman's kitchen the light-hearted rumrunning stories turn somber. Old men talk slowly of what they saw, the periscope stationary just off the harbor, the bomber dropping depth charges half a mile ahead of the unsuspecting fishing trawler, the throbbing roar of the Navy blimp diving in the night, the rattle of machine-gun fire in the fog. Younger men wonder, shake their heads, catch each other's eyes, try to sift truth from exaggeration. But they do not smile. The old-timers know too deeply, know something beyond the high school history books.

And on pleasant summer days the wartime coldness intrudes, too, sharply, like a knife. As his yacht-tender lies aground on the sandbank, the historian scans the busyness of a July afternoon behind the barrier beach, in the shallow bay. Children splash each other, adults fly kites, a hundred motorboats and sailboats and kayaks course about, vacationing Brazilians learn to windsurf, families fish from the wooden bridge atop brand-new pilings. He talks casually with the old-timer bent over in the mud, raking clams. The old man chuckles, pleased with his pickings, and says the clams are making a comeback. Comeback? Intrigued, the historian inquires, and the old-timer straightens. In the war, torpedoed oil tankers sought refuge in the little shallow bay, steamed in until they took the ground, their ruptured tanks spewing oil, the oil that killed the clams. Why did the ships come here? the historian asks, wondering how an oceangoing tanker could enter such a narrow inlet, even at high tide. Why not Boston, thirty miles north, the port of dry docks, shipyards, tugboats? The old timer grunts and bends again to his clam rake. "It was come here or die, young fella. The U-boats was between here and Boston, the U-boats was everywhere, for a while."

The towers stand as monuments to something everyone alongshore tries very hard to forget but simply cannot. They are monuments to what so few young people now know. In the Battle of the North Atlantic, the front line lay all alongshore, and the enemy cruised just beyond the outermost hazards, inshore from the lobsterboats, the fishing trawlers. Revisionist history began in earnest during the early Cold War, it seems, when the military advised first lines of defense far away from the continental United States. The forces of the North Atlantic Treaty Alliance, the nuclear-powered submarine fleet, the Dew Line in Canada, and Camp Thule in Greenland protected Americans. Never would the Commies get close. Perhaps the Cuban Missile Crisis struck so deeply at Americans not because the missiles were poised only ninety miles offshore but because adults and children had begun to believe the military-spawned fiction. Somehow in Cuba the Soviets had come close, almost as close as the U-boats. But then the Soviets blinked and shipped home their missiles. Everyone relaxed, and children learned anew the schoolbook lesson.[3] The U.S. Navy and Air Force will keep any enemy at a far distance.

Nowadays casual visitors to the barrier beach only laugh at the historian when he asks his question. Even his students, fortified by chowder and ready to walk the winter beach, look askance as they trudge up the dunes toward the sea. What a ridiculous question. Suppose as we crest the rise we see a thousand men landing in inflatable boats, grim-faced men in battle gear wading ashore, deploying into positions near the dunes, waving ashore amphibious tanks from the landing craft beyond the surf. And further out, warships. And suppose the flags are yellow and green, or red and black?

How do you report a war?

In his **Picturesque Views of the Southern Coast of England,** *J. M. Turner routinely included the towers erected against potential French invasion. (Houghton Library, Harvard University)*

And everyone laughs or grins, even the graduate students assuming a sort of joke. War no longer starts like that, does it professor? War is a big bang in the night, a flash of incineration. War is not a shelling by battleships, not a surprise invasion. War is not a submarine poking its way inshore among the hazards, laying mines, its torpedo tubes loaded and waiting, its hatches opening as it surfaces to disembark boatloads of terrorists.

War is something far off and usually out of mind. War is something no American encounters on home ground. War is something filed away, something that has nothing to do with a summer day at the beach, nor even a winter day when the squat tower looms gray against a gray sky.[4]

Contemporary warfare is something so awesome, so horrible that it makes the visored monolith totally invisible, or—if someone does notice it—anything but quaint, nothing like a lighthouse or weathered fisherman's shack or tottering wharf. War does not belong on the beach, does not belong on vacation, does not belong on the postcard. So no one sees the tower. No one *realizes* it. However precisely modernist it is in outline, no devotee of international-style architecture admires its purity of form. However bold against the dawn sky, no painter, no photographer records it.

But once upon a time war did mean alongshore attack, and the diligent observer of seacoast turns up evidence everywhere. Often no more than a vaguely rectangular hollow, something like a giant cellar hole grown over with dune grass and bayberry, marks the colonial artillery emplacement, usually on some headland adjacent to a harbor entrance. Near larger harbors, the evidence is often stone, sometimes merely the sand-drifted foundation on which a more or less permanent fortification stood, perhaps in the era of the War of 1812. But often the stone-work is half-buried in concrete, suggesting a rapid, haphazard modernization of obsolescent emplacements. And now the concrete, half-masked by sumac, bayberry, scrubby pine trees clinging to the high ground, erodes in the frosts and sand-filled gales, a fitting emblem of the interest in coastal fortification that comes and goes, almost tidelike. When far-off enemies threaten, coastal residents and shipowners demand protection, but in times of quiet peace, a frugal federal government abandons the fortifications to weather and sheep.

For almost a century after Independence, in fact, officialdom decreed that proper coastal fortifications provided surer—and certainly cheaper—protection than a strong navy. However isolationist in theory, and however frequently lambasted by strong-navy advocates, the concept of coastal defense endured in American political and military thinking. To be sure, the higgledy-piggledy fighting in the War of Independence biased civilian thinking for decades, as people remembered the brief hit-and-run attacks waged by blockading British forces, attacks that often had no purpose greater than to procure a supply of firewood for gal-ley stoves, some farm animals for fresh meat, even some ripe vegetables. Such minor attacks on shallow harbors, even against farms behind barrier beaches, frequently became Ameri-can victories, as determined farmers gathered behind rocks and trees to shoot at the Royal Marines who protected scavenging seamen. Each longboat or jolly boat that returned to some sloop or frigate just offshore became a cause for celebration, even if everyone knew that the raiding party would have returned to sea immediately anyway, fired on or not. Although the British intermittently occupied some large ports and burned shipping in smaller ones, far too much late-eighteenth-century colonial property—shipping included—lay protected be-hind coastal hazards and narrow inlets Royal Navy officers found too dangerous to navigate in deep-draft warships or at night in ship's boats.[5] Throughout the long War of Independence, the stark dangers of gunkholing contributed mightily to the peace and quiet enjoyed by so many coastal people.

Even into the nineteenth century, coastal warfare, especially hit-and-run attack, remained remarkably eighteenth-century in method, as one locally known incident makes clear. Along this stretch of coast, schoolchildren are indoctrinated—and tourists entertained—with the War of 1812 heroism of Rebecca and Abigail Bates, the lighthouse keeper's daughters. On grade-school bulletin boards and harborside restaurant placemats, local teachers and vict-uallers reinforce the lessons of their childhoods, the lessons of quick-wittedness, bravery, sneakiness, and a harbor of ships saved by two adolescent girls.

The attack began with the utmost professionalism. At dawn the local farmers and craftsmen discovered a British frigate sailing inshore, out of the sun, angling northeast, toward the mid-point of the great barrier beach that separates Scituate Harbor from Cohassett Harbor. The militia of Scituate assembled immediately and marched north, faster and faster, determined to assist the citizens of Cohassett. But as the exhausted militia reached Cohassett Harbor, the frigate captain put about and sailed rapidly downwind, far more swiftly than the men could hope to march. Off Scituate Harbor—he clearly worried about moving his deep-draft war-ship near the narrow channel—the captain furled sail and embarked marines and seamen in his ship's boats. Everyone who remained in Scituate understood the intent. The British had tricked the militia and would burn not only the shipping but perhaps the boatyards and other buildings, too—and destroy the lighthouse erected in 1810 at the harbor entrance, the very symbol of the proud new federal government. But Rebecca and Abigail Bates, their father gone with the other men, grabbed an extra fife and an extra drum, ran madly along the inland side of the dunes, and then marched back, playing the fife and drum exactly as the militia played. The landing party hesitated as its officers listened to the music, then put about and pulled back to the warship. Perhaps fearing the arrival of a horse-drawn cannon, perhaps for reasons still unknown, the British captain made sail, leaving Scituate unscathed and the Bates girls perpetual heroines to generation after generation of children and adults.[6] After all, they not only outwitted the Royal Navy, but they played loudly enough to carry their tune above the rote of the sea, no small feat in itself.

The local victory cheered the locals, but far-off defeats signaled a changing art of warfare. The British put ashore nearly eight thousand troops at Washington and burned the new capital buildings. Indeed, Royal Marines landed almost at will anywhere. And in 1814 they accomplished something almost beyond belief, a grand descent at New Orleans.

By "grand descent," military experts of the time understood something vastly greater a raiding party, even one as large as sacked Washington. "Grand descent" designated a major landing party intended to remain more or less permanently on a site, to invade a region and hold it, to secure not simply a landing spot on a beach while more troops arrived but a deepwater harbor, perhaps a city. Although American irregulars defeated the ten thousand troops landed at New Orleans in a battle fought before news of peace arrived, the decades-long celebration of that victory masked the demonstrated inability of the Republic to protect its coast against invasion. To be sure, American irregulars—and especially Kentuckians with their long rifles—under the leadership of Andrew Jackson had defeated a superior British force.[7] But the victory that launched Jackson toward the White House became a national legend, a vastly expanded version of what happened at Scituate. Superior bravery, quick-wittedness, and the right equipment carried the day for the Americans. The connection between the derring-do of the Bates girls and the wily bravery of Jackson and his fellow Southerners may not be mere coincidence, for American theater audiences soon delighted in such songs as "The Hunters of Kentucky" and plays like *The Champions of Freedom*, the work of Scituate native Samuel Woodworth. In New York City, Woodworth succeeded wildly, not only in writing nation-sweeping songs like "The Old Oaken Bucket," about his Scituate boyhood home near the salt marshes, and in establishing the New York *Mirror*, but in keeping alive the popular belief that American virtue and bravery had defeated the British at New Orleans. His "The Hunters of Kentucky" became perhaps the most popular American song of the decade, drowning out the few voices that recognized the New Orleans triumph not only as luck but as victory snatched from the first grand descent on American soil.[8]

Not until the 1840s did educated Americans begin to understand what some American military officers had long known. Just as the war with the Barbary pirates made Timothy Dwight and his contemporaries more aware of seacoast geomorphology, so coastal attacks by European imperialist powers made educated Americans suddenly uneasy about coastal defense. More important, by the 1840s historians had made clear that the ten thousand men landed at New Orleans were part of a combined operation involving the landing of ten thousand more in Canada. Retrospect made crystal clear that the British had been thinking in terms of a continental-scale pincer movement, something as stunning as the nearly one hundred thousand troops they had landed in 1809 in the European Antwerp expedition.[9] By 1856, however, when an anonymous writer, probably a military officer, published the path-breaking article "Our Sea-Coast Defense and Fortification System" in *Putnam's Monthly*, the impact of the Crimean War had begun to unnerve thoughtful Americans, in particular those who lived along the coast. Britain and France had transported by sea an army of two hundred thousand troops, had made a successful grand descent in the Russian Crimea, and had supported their troops month after month.[10] Warfare had changed.

Thoreau glimpsed the change, mentioning it in *Cape Cod* in an aside almost unintelligible now but every bit as clear to his contemporaries as were Melville's snippets about wharves. In quoting George Anson, a British major-general, on "the comparative advantages of bituminous and anthracite coal for war-steamers," Thoreau makes plain his understanding of the military significance of Anson's assertion that " 'in America the steamers burning the fat bituminous coal can be tracked at sea at least seventy miles before the hulls become visible, by the dense columns of black smoke pouring out of their chimneys, and trailing along the horizon.' "[11] In what now appears to be a passage about the glim, about staring into vastness, about the ever-shifting horizon, about looming, Thoreau demonstrates his understanding of state-of-the-art military technique. Vastly more floats in the *Cape Cod* paragraph than his understanding of ocean-watching as experiment in physics and aesthetics. Thoreau here speaks to his contemporaries, telling them that he, too, knows the significance of the ocean steamship in the reinvented art of war, in the new scale of grand descents, in the survival of the Republic. He understands the connection between the steamship and the grand descent, the fact that no massive plume of smoke advertises the advent of British warships and merchant steamers. Americans will have no warning. None at all.

Thoreau merely nodded toward a subject that perplexed many well-read Americans, especially those who feared not merely a civil war but the potential for foreign intervention during or after such a war. "The great advances in ship-building—the rapid expansion of commercial transportation, and especially the general introduction of steam-power, both for military and commercial operations, have removed much of the difficulty incident to sending expeditions across the Atlantic," the nameless expert warned *Putnam's Monthly* readers. Disunion might well lead to foreign invasion, a grand descent by French and British troops transported largely in British steamships, a descent that would focus on a major seaport, a descent that the tiny U.S. Navy—its ships betrayed from afar by miles-high plumes of black smoke—could do little or nothing to stop, even if not engaged against rebels at home. No American could hope for some New Orleans–like victory. "Imagine a hostile fleet coming up the Narrows, and no forts, no batteries to obstruct its progress; all New York, taking to muskets and field pieces, sparrow shot and shillalahs, pitchforks, and tape scissors," he sneers. "Verily, of little use are the million 'strong arms and stout hearts' in such a case." The warships will stand off and bombard the city, setting it afire and driving off its scantily armed defenders; then the troops will pour ashore, onto beaches if necessary, most probably onto wharves.

The anonymous authority wrote in a context hard to imagine now but utterly straightforward in his era. "The defenses of Quebec are still progressing." [12] In one phrase, he acknowledges what his contemporaries knew, what even Thoreau in *A Yankee in Canada* states. [13] North of the American border, the British military presence grew stronger year after year, and although American plainsmen cast covetous glances at western British North America, while they began mumbling, shaping the words that became "54–40 or fight," the *Putnam's Monthly* writer understood that the United States would lose *any* war it fought with Britain, and that Britain and France might well decide to combine and divide up the Republic, securing not only the extant financial investments of their citizens but the holdings of gold and silver rumored to exist somewhere west of the High Plains. And the United States would be unable to strike either at British and French home ports or at their colonial bases, for both Britain and France, partly in mutual fear, partly in fear of Germany, not only had built steam-driven warships far, far superior in number and power to anything that flew the American flag but had fortified their home coasts and the coasts adjacent to every important colonial harbor, including Quebec. No longer did Daniel Webster's assertion during the War of 1812 that he "was for doing something more with our navy than to keep it on our shores for the protection of our coasts and harbors" make any sense. [14] The United States Navy would either suffer prolonged blockade and destruction in its home harbors or, in the unlikely event that its little ships sailed or steamed unscathed to Europe, would encounter defeat by superior ships or

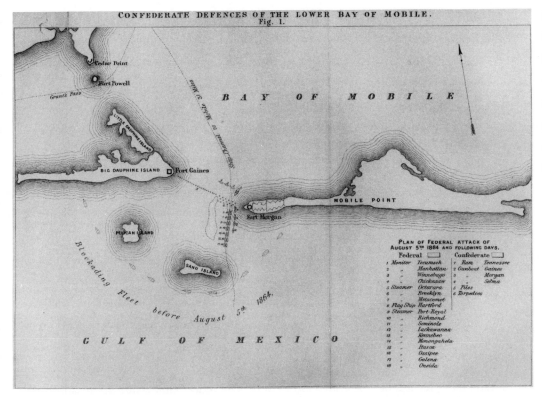

CONFEDERATE DEFENCES OF THE LOWER BAY OF MOBILE.
Fig. 1.

In his Treatise on Coast Defence, *Viktor Von Scheliha devoted much attention to the Confederate fortifications that protected Mobile, Alabama. (Harvard College Library)*

state-of-the-art coastal fortifications. However gallant its efforts, the Navy would fail, and its conquerors would descend as they had on Sebastopol.

Only coastal fortifications offered a sound guarantee against such a likely, indeed probable scenario, so the *Putnam's Monthly* essayist explained how such fortifications could be built at low cost.

"A long array of instances might be cited, in which forts have beaten off fleets; and, in many of these, the disproportion of strength amounts to the grotesque," he asserted before calling to mind local examples remembered locally just as Scituate remembered its patriotic, quick-thinking, strong-limbed girls. "Fort Moultrie, Fort McHenry, Mobile Point, and Stonington Point, are good illustrations in our own history, though in each of these cases the works were small and of weak profile." Simple earthwork forts, properly outfitted with high-powered, long-range cannon, would not only prevent an enemy fleet from entering a harbor but could actually prevent a fleet from steaming near enough to launch an effective, accurate shelling. Of course, in 1856 such forts needed more than thrown-up sand ramparts. Most needed some sort of barracks and some sort of magazine, both protected from newfangled mortar shells that fell almost straight down—when they didn't burst in the air, inspiring national anthems. Moreover, the forts needed reliable land access so that reinforcements could rush to them, indeed so that troops might arrive to man them if a frugal federal government left them empty most of the time. In the time-honored tradition of American military thinking, the essayist assumed that carefully trained local militia would serve the massive guns. "The manual of heavy guns can be quickly learned by intelligent men, who, under cover of walls and parapets, can be relied on to serve them well in action." As in the War of Independence, citizen-soldiers would protect the coast, even from the most modern, most professionally manned enemy battleship. "A well-armed fort, served by the spirited and quick-witted population of one of our New England towns, would give such formidable battle as no fleet could long withstand. A nucleus garrison, thoroughly trained in defensive service, would give a right direction to the entire local force." Steady cannon served by steady men would sink any enemy fleet tossed by the ocean swell and firing into massive ramparts of shock-absorbing sand.

Forts on particularly exposed headlands, and on islands, needed additional protection, not so much from shelling as from night attack by landing parties, however, and such protection cost money. Sheer stone walls high enough to stymie troops equipped with ladders and grapnels, sally ports for infantry to outflank attackers, and other battlements would add to the original cost yet would still be cheaper than building a massive navy. But no forts, at least in peacetime, struck Congress as cheaper still, as the essayist lamented before he terrified his readers.

Readers of *Putnam's Monthly* found more than the short stories of Herman Melville and other writers. They encountered stunning predictions of wholesale slaughter, massive property destruction, and certain military defeat. "New Bedford, the third port, in amount of enrolled shipping, in the United States," the anonymous expert argued in withering prose, "the very port whence war would send forth the most efficient privateers, has no protection worth mentioning." Indeed almost the entire Atlantic coast stood unprotected in 1856, nearly naked of forts. "A single armed vessel could make a clean sweep of most of these ports," he concluded, after listing harbors as important as Salem, New Haven, Georgetown, and Galveston.[15] Destruction lay just beyond the horizon, under almost invisible smoke, waiting for the right moment, for the upstart Republic to act too self-righteously, too independently.

Civil War distracted many educated Americans from the threat of foreign invasion, but as the naval successes of the Confederacy mounted almost beyond Union belief, and Brit-

ain and France grew remarkably hospitable to Confederate overtures, coastal Unionists began to wonder about the potential for allied attack, for Britain and France to combine with the Confederate states and defeat what remained of the Republic. To be sure, the czar sent two Russian fleets, one to visit New York, one to visit San Francisco, but few Northerners considered these fleets more than goodwill gestures, certainly no match for combined British-French squadrons. Prolonged, agonizing attempts by the Union navy to enter fortified southern ports merely demonstrated the correctness of the *Putnam's Monthly* line of argument, and left many war-weary Americans carefully eyeing the post-1865 possibility of a British-French grand descent.[16]

Throughout the last decades of the nineteenth century, Americans worried increasingly about the real possibility—indeed growing probability—of British attack. The year 1882 confirmed the worst fears of civilians and military officers alike. After the British fleet stood off Alexandria and reduced obsolescent Egyptian coastal defenses to rubble, then landed a massive military force to protect British investment, everyone understood that modern battleships firing modern high-explosive shells could level a city unless the city surrendered at once.

In 1885, United States Corps of Engineers officer Eugene Griffin published his up-to-date analysis of the Alexandria demonstration. "A hostile fleet lying in the upper bay of New York would have within reach of its guns about two billion dollars' worth of destructible property in New York City alone, and including Brooklyn and Jersey City, over two and a half billions," he argued in *Our Sea-Coast Defences*. British battleships fired shells filled with seventy-five pounds of gelatine explosive, vastly more powerful than gunpowder. "The result would be terrible beyond description." No building in New York could withstand a hit by such a shell, for each hit would demolish any structure, no matter how big. Fire would break out immediately and sweep the entire city, blazing instantly beyond the capacity of any fire department. *"New York would be doomed."* [17]

No hysteria produced his italics. A year earlier, another United States Army officer, Edward Maguire, had vividly explained how shellfire would ignite conflagrations worse than those started by accident in Boston and, more recently, in Chicago. Maguire's *Attack and Defence of Coast-Fortifications* raised the image of people fleeing fires that started in many places, fleeing on foot because all public transportation routes, especially railroads, had been severed. Manhattan, Maguire reminded his readers, is an island, an island trap. [18]

High-speed battleships mounting long-range cannon destroyed all pre–Civil War notions of citizen soldiers aiming coast artillery. The battleships would steam at very high speeds— Griffin explained matter-of-factly that a battlefleet could arrive off Boston only thirty-six hours after leaving Halifax—and maneuver extremely rapidly. So suddenly would they appear, and so massive would be the destruction on land immediately afterward, that volunteers would have no time to man forts, even if Congress had appropriated money to build them and fit them with up-to-date heavy guns. In 1885, when a congressional committee heard testimony about the impact of ultramodern battleships, the "dreadnoughts" that enforced British imperial rule everywhere, it and the public learned that almost every port on all three coasts could be shelled from points three to five miles offshore. [19] In fact, only the port of Mobile, well inland along a difficult channel, seemed safe from British naval bombardment. "There is not a harbor on our coast that cannot be captured with comparative ease by an iron-clad fleet properly armed and equipped," Griffin concluded. "There is not a single important power in the world which does not possess such a fleet." Britain, France, Germany, Italy, and even China had battleships with ten-mile-range heavy guns, and although Italy and China, and perhaps Germany, could only hope to destroy Boston, New York, and Washington in the same week, Britain and France could capture and hold such port cities. [20]

Fear and frugality might have warred on into the twentieth century had the Spanish-American War not intervened. By merest chance, the U.S. Navy encountered the Spanish Atlantic fleet at sea. Schoolbook histories ignore the proto-panic that swept Boston and New York at the outbreak of war, the half-mad questioning of insurance companies and the furious telegrams sent to congressmen. Even now, serious histories downplay the terrible chance—and the stunning eastern-states bias—implicit in the frantic dispatch of the USS *Oregon* from California around Cape Horn to the Atlantic coast even as a second Spanish fleet, based in the Philippines, threatened cities on the west coast. Only military historians remark how everyone, particularly everyone in the form of daily newspaper editors, in New Orleans suddenly remembered Griffin's calculation, that a Spanish fleet could steam from Havana to New Orleans in a mere forty-five hours.[21] One of the forgotten wars of the Republic, a sort of odd naval prelude to the Great War sixteen years later, the Spanish-American War still whispers in its very title that Spain had superior force, perhaps even right on its side.

War with Spain confirmed four decades of warning, prompted the upgrading of existing fortifications and the building of new ones, and vindicated military theorists who had argued that not every harbor, not every beach should be fortified. Nowadays the astute walker of beaches, the educated puller of rowboats notices the something that is not there in so many coastal places. Almost nowhere, except near the mouths of important harbors, does the beachcomber find traces of fortifications built between 1865 and 1940. Almost everywhere along the American coasts, the smaller harbors, even harbors like New Bedford, stand bereft of late nineteenth-century fortifications. The beachwalker who strives to understand the absence of fortifications needs more than a cursory knowledge of congressional frugality and the diminishing political clout of towns that ringed shallow-water harbors cluttered in 1900 with rotting wooden ships. He or she needs a copy of Artillery Corps major John P. Wisser's *Tactics of Coast Defense*, a treatise of 1902 best read with a telescope in one hand and a pocket calculator in the other. Whereas the anonymous writer in *Putnam's Monthly,* Maguire, Griffin, and other military and civilian experts warned of the consequences of not building coastal forts, of not installing heavy guns, Wisser quite simply explained how to site the forts and aim the guns. For Wisser, emplacing coastal artillery meant using science, the science that would rout even the biggest dreadnought, the science already employed "in our own waters" by the British, who had fortified Halifax, Bermuda, and Victoria.

Wisser reasoned bluntly that dreadnoughts would not attack fishing villages, even in passing, but would instead bombard nine important targets distributed along the three coasts of

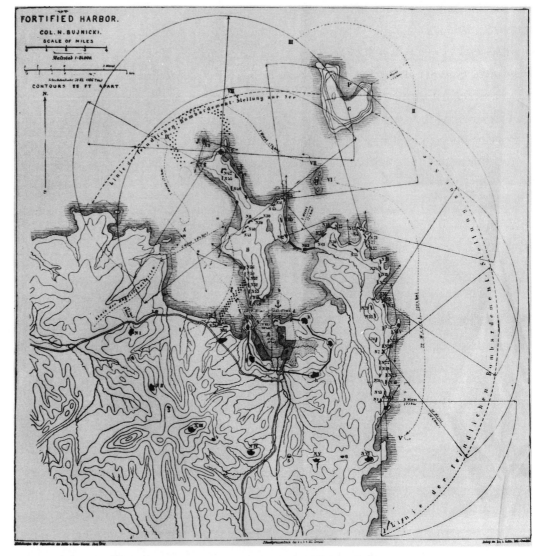

the nation. From the northeast, where Portland formed the terminus of the Grand Trunk Rail-road, "the natural outlet of the Dominion of Canada and the key to the wedge of Maine pro-jecting northward into British territory to within a few miles of the Inter-Colonial Railway" to Key West, "controlling the narrow channel out of the Gulf of Mexico into the Atlantic," across the Gulf Coast dominated by New Orleans, and from San Francisco to Puget Sound, "a base for operations in the Pacific Ocean, covering the northern Pacific Coast," only nine ports needed massive fortifications and the emplacement of great guns. But the same ports must be defended by mine fields that would force attacking fleets to move inshore along narrow mine-free channels raked by heavy and light guns. The deadly accurate fire from the heavy land-based guns would keep the battleships at bay while equally accurate fire from shorter-range artillery would destroy smaller vessels trying to deactivate mine fields. Bombardment and invasion both would become impossible at the nine chief targets along the U.S. coast, the wealthy target cities that boasted commanding military locations, great riches, and massive wharves perfect for disembarking tens of thousands of troops.

Everywhere else, Wisser reasons, would be subject to raids, but not to grand descents, simply because dreadnoughts and other steel-hulled warships could not approach beaches and shallow harbors. "A few men of the Coast Guard Corps" would be able to patrol vast stretches of coast, linked to local Army commands by telephone, and able to summon troops along railroad lines. Should an enemy fleet attempt to raid or temporarily invade and hold a shallow-water harbor, the defending troops, with the assistance of local fishermen, must activate the already placed mines, and so delay the landing until regular troops arrive, preferably with light artillery. But Wisser does not explain what would happen then, when infantry confronted the gigantic shells fired from enemy warships just offshore.[22] However strongly worded it is concerning the defense of New York, Boston, Hampton Roads, and San Francisco, *Tactics of Coast Defense* shrivels as it confronts the near impossibility of defending much of the United States coast from surprise raids by powerful dreadnoughts that could send men ashore in fast, motor-driven boats bristling with machine guns. Its mathematics and geometry demonstrate the utter practicality of defending major ports and the hopeless expense of fortifying barrier beaches and small harbors.

Nowadays both impassioned pleas for coast defense and impeccably structured guidelines for emplacing coastal artillery seem quaint, almost laughable. The walker along some headland often misses entirely the earthwork ruin from the War of 1812 or misjudges the purpose of the early twentieth-century emplacement. Attempting to understand the triangulation of shellfire, the position of never-placed mine fields, the old form-column-and-attack method of closing the coast requires not only a solid understanding of trigonometry but a willingness to imagine an era before airplane reconnaissance, before airplane bombing. But just as the nameless expert fitted his article so perfectly into the pages of *Putnam's Monthly*, so Maguire, Wisser, and other turn-of-the-century authors fitted their books into the literary context of educated readers who learned something, however vague, of international politics and foreign military prowess simply by reading European travel writing.

Readers of Rudyard Kipling's *American Notes* in 1899 grasped the military implication at once, for Kipling rubbed it in their faces. "San Francisco is a mad city—inhabited for the most part by perfectly insane people, whose women are of a remarkable beauty," Kipling asserts at the beginning of his first chapter. What evidence of insanity does he have? "When the *City of Pekin* steamed through the Golden Gate, I saw with great joy that the block-house which guarded the mouth of the 'finest harbor in the world, sir,' could be silenced by two gunboats from Hong Kong with safety, comfort and despatch. Also, there was not a single

American vessel of war in the harbor." Of course, Kipling bore a grudge born of copyright infringement, and the grudge prompted him to point out how easily an angered Britain might take the port. "This may sound bloodthirsty; but remember, I had come with a grievance upon me—the grievance of the pirated English books."[23] Kipling's comments are perhaps the rudest, the boldest of any made by turn-of-the-century Britons arriving in the Republic, but they mirror what so many Britons noticed from the decks of ocean liners. The great cities of the United States lay ripe for the taking, open to bombardment, to a grand descent, to tribute.

What warning would Americans have? Perhaps only a smudge on the horizon, the smudge that horizon-watchers knew marked the passage of an Atlantic steamer.[24] And beneath the smudge, what? The Royal Navy convoying French transports? Perhaps the imperial German fleet?

Germany offered no real threat, Wisser assures his readers, for it lacked the navy to deliver its crack army to American shores. Britain and France would eventually cause trouble, Britain almost certainly massing ships and troops at Halifax, perhaps at Bermuda, certainly somewhere along the coast of British Columbia, then descending against Portland, Boston, Hampton Roads, and San Francisco. Though he mentions briefly that "submarine boats" could be useful in harbor defense, Wisser clearly saw coast defense as a refined version of mid-nineteenth-century tactics and strategy, likely to change only as heavy artillery exceeded a twelve-mile range.[25] Within two decades, however, coastal New Englanders understood Germany as the threat, as a very different sort of threat indeed.

World War I submarines demonstrated an amazing ability to close American waters and do battle with vessels typically ignored by battleships. German submarines surfaced in New England waters and shelled American fishing schooners, usually after allowing crewmembers to leave in dories, and they crept inshore at night to shell tugboats and barges moving in very shallow water. Submarines threatened shallow-water shipping, small harbor towns, and even barrier-beach property owners not so much because they moved about underwater and fired torpedoes but because they made nonsense of established coast-defense thinking. Every German submarine operating off the coast of New England announced that a long-distance, powerful warship could also be a shallow-draft warship, one eminently capable of threading its way into shoal-water harbors and wreaking destruction from below or on the sea. Its powerful deck cannon threw large shells at such great range that any which missed tugboats and other small vessels might easily land onshore, and should the submarine commander so choose, the cannon could shell any of a thousand shallow-water ports undefended by coast ar-

War with Spain prompted intense interest in coast artillery, and particularly in rapid-firing long-range guns like this one depicted in Henry Houghton Beck's Cuba's Fight for Freedom. (author's collection)

Early twentieth-century ocean-liner posters frequently emphasized the massiveness of the steamships by juxtaposing the vessels against sailing ships or harbor steamers. (author's collection)

HAMBURG-AMERICAN LINE
CRUISES

In Service next May—S.S. Imperator
World's Largest Ship

WEST INDIES and PANAMA CANAL. 8 cruises during January, February, March and April by S.S. Moltke and S.S. Victoria Luise. 2 cruises from New Orleans, Jan. 23, Feb. 10, by S.S. Kronprinzessin Cecilie.

AROUND THE WORLD
Sixth Cruise sailing from San Francisco Feb. 6, 1913, by S.S. Cleveland (17,000 tons), duration 110 days, $650 up, including all necessary expenses aboard and ashore.

ORIENT From New York, January 28, 1913, by S.S. Cincinnati (17,000 tons), 80 days, $325 and up.

Hamburg-American Line
41-45 Broadway   New York City

Boston   Philadelphia   Pittsburgh   Chicago
St. Louis   San Francisco

tillery.[26] Cape Codders immediately understood the significance of freighters burning just off their coast, of burning vessels run aground on sandbars, of stray shells hitting the beaches.

After the war, however, coast defense became a very muddled issue in the popular imagination. Britain and France seemed permanent allies, Germany permanently defeated. Moreover, airplanes, dirigibles, and blimps appeared likely to defend coasts against submarine and surface warships, and coastal artillery suddenly looked hopelessly inflexible and need-lessly expensive. Now and then some worried military officer published scenarios of enemy warships firing poison-gas shells into New York City, but during the 1920s and 1930s, inter-est shifted curiously away from coast defense and toward a more powerful naval offense.[27] Even Hollywood dismissed the coast artillery as antiquated, as stupid as the "horse marines." One mid-1930s comedy film, a Three Stooges short subject, depicted the Stooges firing an old coast-artillery cannon first into an admiral's battleship, then into land targets.[28] Coast artillery had become obsolete, ridiculous, and clearly incapable of handling new threats to coastal peace, particularly those posed by massive battleships mounting guns that could lob shells twenty-four miles.

As war with the Axis powers became likely, American military experts embarked on their program of coast defense, the militarization of the Coast Guard, the emplacement of har-bor booms, and especially the building of watchtowers. No one expected the towers to be of much use against German battleships like *Bismarck*, which could destroy cities and harbor shipping from afar, although near major harbors the towers might shelter coast-artillery fire-control spotters. But the towers did offer some hope of scanning inshore waters. In a time before radar, telescopes and binoculars provided the best bet of enhancing the time-honored utility of the keen-eyed lookout. From the towers the watchers might spot submarines, might spot clandestine landing operations, might spot an air attack.

What 1940-era airplane could attack the Atlantic American coast? After all, transatlantic flying was still an infant undertaking, and even the new Pan-American transatlantic service depended on flying boats that made Irish and Canadian refueling landfalls. In 1940, no one in authority said much, but clearly the military understood two possibilities more likely than transoceanic bombing runs. First, some battleship or cruiser might launch catapult aircraft that would bomb harbor installations, or second—and albeit it less likely, even though the U.S. Navy had owned such aircraft since 1931—a dirigible might fly across the Atlantic and when near the coast launch the fighter planes it carried onboard, retrieve and refuel them, and relaunch them. About the third possibility, the grotesque idea that the Germans might own submarines like the massive French *Surcouf*, submarines so gigantic that they carried catapult-launched, derrick-retrieved aircraft, no one said much at all, at least not in public.

Airplane-carrying submarines became a nightmare to American military officers. As the war in Europe worsened, they remembered half-baked World War I efforts to launch tiny air-craft from surfaced submarines, then the postwar German experiments at building aircraft so tiny that they could be disassembled and stored in watertight cylinders attached to submarine decks, then assembled and launched after the submarine surfaced. The French experiments, which began with a successful launch in 1935, showed how easily large submarines could launch and retrieve aircraft, and although the Royal Navy had lost one aircraft-carrying submarine in an accident, it had continued to experiment. And every espionage agent knew that the German navy had already built very large, long-distance U-boats. But after the attack at Pearl Harbor, the nightmare became real, though never real enough.

In September 1942, the Japanese submarine I-25 surfaced a few miles off Cape Blanco, Oregon; minutes later, it catapulted a seaplane into the air. The seaplane dropped incendiary bombs on the thick forests of northwestern Oregon, bombs which the seaplane crew noticed ignited forest fires at once. After observing the damage, the crew returned to its rendezvous, landed on the sea, and taxied to the I-25, which retrieved the seaplane and submerged. The following night, the same crew flew the same seaplane on an equally successful mission over Oregon. While the U.S. Navy slept as soundly as it had when a Japanese submarine-launched reconnaissance plane flew a spy flight across the still-burning ships at Pearl Harbor, enemy aircraft bombed the U.S. mainland.

The Oregon air raids achieved their immediate mission, but of course they existed in a larger Imperial Navy effort, the full impacts of which became apparent only after Japan's surrender. Japanese submarine-launched aircraft flew over Sydney and Melbourne in Australia, and even over Wellington and Auckland in New Zealand, identifying potential targets and scouting shipping anchored in harbors. As reconnaissance aircraft, the submarine-launched planes performed splendidly, apparently far more splendidly than the some 125 helicopter-type aircraft that accompanied the long-distance German submarines as they cruised from Germany to Japan to fight alongside their allies. Moreover, the eleven Japanese submarines that carried aircraft in 1941 proved the viability of a daring idea.

By 1945, the Imperial Navy had twenty-seven airplane-carrying submarines, including the gigantic I-400 class boats, which operated as submarine aircraft carriers. The I-400 submarines existed to mount massive air raids on the American mainland in retaliation for the one-time Jimmy Doolittle B-25 raid on Japan the United States had launched in April 1942, the same month as the attacks on Oregon. Near war's end, the great submarine ships had become operational and moved in a battle group toward the most tender point in the Allied war machine. But the war ended before they could arrive where they could launch a force of some ten aircraft against a critical target.

A few days before the United States dropped two atomic bombs on Japan, the Imperial Navy was fully intending to blow the Gatun Locks to kingdom come, stopper the Panama Canal totally and perhaps permanently, and drastically retard the movement of men and matériel from the European theater of operations to the Pacific. The Imperial Navy had the intention— and the capability—exactly as U.S. military officers had feared in the early 1940s.[29]

By early 1942, blackouts and air-raid drills confirmed the possibilities hinted at by the towers built a year or two earlier, particularly the chance of air attack from the sea. All along-shore, coastal people took the possibility of air attack seriously, without knowing exactly why, but suspecting that any enemy whose submarines operated with impunity just offshore might have worse weapons at the ready.

And by 1955, certainly by 1965, the old threats seemed old indeed, as old as War of 1812 stories of young girls playing fife and drum. Hollywood emphasized the importance of air power, of sea power, and in time laughed loudly at coastal defense.

Alongshore, especially in small harbor towns, concern mixed with laughter. In one old the-ater whose front-row seats now and then stood awash when northeast gales forced the tide higher than usual, "The Russians Are Coming, The Russians Are Coming," played to massive mid-1960s applause and choking laughter, but not for reasons anticipated by Hollywood.[30] Everyone in the audience knew instinctively that the Navy would come too late, that a sub-marine could work inshore, that the locals, descendants of the Bates girls, would know what to do, and would do it. Children, even teenagers, who asked in subsequent days about the possibility of Soviet invasion, even near-accidental shore raids, heard the reassuring mes-sage told so oddly, as if half-believed. "It can't happen here. Not nowadays." No matter that the towers loomed empty and derelict. Airplanes, satellites, and electronic warning devices made such invasion impossible. Federal authority said so. Teacher said so.

In the high school the teacher turned back to the blackboard, having told the students the official federal-government explanation. The students, intimately familiar with stories of giant Soviet fishing trawlers and even more massive fish-processing ships, heard the explanation, and wondered. Maybe teacher was right.

But teacher was wrong.

9 TREASURE

T wenty years ago teenagers dated as their parents had dated. They drove down to the beach, to the marshes, then snuggled in the front seats of Fords and DeSotos, listened to the radio, and watched the submarine races, comfortably warm on a winter night in the lee of the dunes, the lighthouse, the coast-artillery watchtower. Not much happened during night-time proto-courtship. Every local girl knew the line tossed by every local boy, that the Minot's Ledge lighthouse flashing 1-4-3 signaled "I love you." Perhaps the last pre-beer, pre-drugs, pre-sex moment, the era of submarine-watching dating involved no one in much danger. Boys recalled the little advice given in grin, advice not about the girls but about the importance of stopping the car in gear, of securely latching the emergency brake lever. No need to roll into the creek, off the cliff. Girls recalled advice, too, usually about making sure the boy securely applied the emergency brake. After all, what else could go wrong?

No Soviets lurked about the inshore hazards, no Soviets crossed the surf in inflatable boats, no Soviets crawled through the bayberry. The roads ending in dunes, ending in salt marsh, ending at town landings unnavigable at low tide ended in quietude, in safety, in boredom. And besides, the local peace officer cruised the parking spots, too, flashing his spotlight, rotating the parkers, keeping casual watch over the harmless activity so vital to the town's happiness.

Today the image strikes teenagers as a fairy tale, some fiction of immaturity and innocence, some tale of another time, perhaps another galaxy. Teenagers know a different social life, of course, one fraught with perils unimagined in the middle sixties. But they know, too, that government, that teacher taught wrong. Under the great arc of satellites and spy planes, through the web of nuclear submarines and guided-missile cruisers crept and probed the old coastal scourge, penetrating the tiniest harbors, motoring inland along the estuaries, navi-

Hollywood movies reinforced the traditional alongshore understanding of pirates landing and burying treasure. In this image from The Sea Hawk, *the pirates pull ashore with their about-to-be-buried booty. (Museum of Modern Art, Film Stills Archive)*

gating the narrowest of gutters. Soviet spy ships and trawlers postured, then vanished, but pirates came again, seeking not tribute but wealth beyond the dreams of avarice.

Piracy and tribute lie deep in the American maritime psyche and in the national mythology. When the Barbary pirates demanded tribute from the new American nation, so legend goes, the new American nation sent the *President*, the *Constitution*, and other frigates to destroy their little xebecs and their little fortresses. "Millions for defense, not one cent for tribute." Somehow the cry endures, a faint echo in peacetime, a thunderous roar in wartime, the home-front roar of World War II war-bond campaigns. No one much cares that Congress tried first to pay tribute to every two-bit corsair chief, and only later ventured along Jefferson's cheaper-in-the-long-run but far more violent course.[1] But almost everywhere in the national memory, piracy and tribute lurk far away, somewhere off along the Marine Hymn "shores of Tripoli," on the Spanish Main, in the far South Seas, but never close to home. Only along the coast do piracy and tribute linger as present possibilities, present realities. Only alongshore do people recall that "the law" for most Americans ends at salt water. Only along the beach do readers of Thoreau catch the double meaning of his remark that even the most populous city cannot scare a shark far from its wharves. Not every shark is elasmobranch. As expressions like "card shark" and "land shark" suggest, *shark* is sometimes another word for pirate.

On the national scale, nowadays piracy often means something for children, perhaps a skull-and-crossbones flag flying jauntily from the stern of a skiff, a Halloween costume made from a bandana, a lath sword, and an eye patch. In the 1990s, pirates have become as cuddly as dinosaurs, and well-intentioned parents purchase Lego pirate sets with pleasure and confidence. Boys and girls assemble the countless pieces, creating forts and sailing ships, rowboats and palm-treed islands, enjoying above all the equipment peculiar to pirates: shoulder parrots, cutlasses and muskets, and—especially—chests filled with gold. So far no one attaches much importance to the peg-legged Lego pirates or the one-eyed ones, or those with tiny plastic hooks instead of left hands.[2] Pirates now are good-natured, bold and healthy, cute—and rich.

Until well into the eighteenth century, however, pirates enjoyed a wholly different reputation, one that lingered late into the nineteenth century. The appearance of Robert Louis Stevenson's *Treasure Island* in 1884 marks a sort of turning point toward the *Peter Pan* and Lego image, for although Stevenson set his novel in the eighteenth century, his pirates, especially Long John Silver and Ben Gunn, are scarcely wholly evil, wholly rapacious.[3] *Treasure Island* passed through dozens of subsequent editions, becoming a children's classic, enduring even now as one of the books most adults have read, or know through film versions. But few adults remark the pattern implicit in its illustrations, the pictures so important to young readers but insignificant to adults who purchase the book for their children. *Treasure Island* exemplifies the subtle significance of book illustration as a barometer of national thinking. The illustrations that changed from edition to edition reveal a definite reworking of piracy in both British and American popular culture, a reworking of troubling and timely import now, when the Lego-pirate image rules everywhere but alongshore.

Roughly nine in ten pirates from the mid-sixteenth century onward came from Britain and its colonies. France and the Netherlands produced some pirates, and black slaves sometimes turned pirate, but the vast majority of pirates spoke English as their native language, and—more important—looked English. Few Spaniards fell into piracy. Not only did Spain's imperial system offer a wide range of legal opportunities for amassing great wealth, but Spanish law punished piracy with death and rarely permitted privateering. Although many British pirates intermittently sailed as wartime privateers under letters of marque, almost never did the Spanish crown countenance piratical activity, or even letter-of-marque efforts against Britain.[4] Throughout the Caribbean and along the Atlantic coast north of Key West, *pirate* long connoted someone of English background who preyed on Spanish subjects, a usage that eighteenth-century Britons took for granted but that Victorian authors, writing half a century after the last spurts of Atlantic piracy, subtly masked. As late nineteenth-century British

Lego toys emphasize a dark-visaged pirate armed with massive gold coins, pistol, and shovel. (author's collection)

naval forces battled Malay, Dyak, and other pirates who tormented the fringes of imperial trade routes, authors like Stevenson began to rework the history of British piracy.

Certainly Stevenson wrote in a genre that includes Rider Haggard's *King Solomon's Mines* of 1885, Rudyard Kipling's "Man Who Would Be King," and, especially, John Masefield's *Jim Davis*, a genre that broadened and deepened the Victorian notion of the British hero while reworking British history. Just as Masefield redefined the eighteenth-century English smuggler, presenting him in *Jim Davis* as a nonviolent, gentlemanly, misunderstood patriot forced into smuggling by unjust laws, so Stevenson and subsequent writers emphasized the heroic deeds of law-abiding Britons who fought sometimes wicked, sometimes noble British pirates.[5] As late as the 1910s, pirates figured in *Treasure Island* illustrations as blond or red-haired and otherwise as distinctly English-looking as their English victims and adversaries. In Victorian hands, especially hands writing for young boys, smuggling and piracy became distinctly intramural squabbles, Anglo squabbles.

Sometime in the late 1920s, however, American illustrators of *Treasure Island* and other Victorian adventures began to transform the pirate, making him increasingly black-bearded, increasingly Latin in complexion.

Although the shift may have been speeded by the appearance in 1922 of *Captain Blood*, a best-seller about a gentleman pirate by the Italian-born English novelist Rafael Sabatini, the defining undercurrent was the warped enthnocentrism most clearly manifested in 1920 by an American historian, Lothrop Stoddard, in *The Rising Tide of Color against White World Supremacy*. Nowadays dismissed as mere racism when remembered at all, Stoddard's book clarifies the context in which appeared pirate novels like Jeffery Farnol's *Black Bartlemy's Treasure* of 1920 and adventure tales like Sax Rohmer's *Yellow Claw* and *The Golden Scorpion*.[6] The dark-visaged pirate-villain appealed mightily to adult and juvenile readers already jittery about the fragility of white supremacy, and soon he appealed to cinema-goers, too. *Captain Blood* was a spectacular success as a Hollywood film, an even greater box-office smash than another Sabatini novel, *The Sea Hawk*, produced as a film in 1924. By 1940, when

By the 1920s, inexpensive editions of Treasure Island *boasted full-color renderings of dark-visaged pirates. (author's collection)*

Warner Brothers produced a remake of *The Sea Hawk,* albeit it one quite different from its predecessor and from the original novel in which an Englishman joins the Barbary Corsairs, the dark-visaged but still sometimes noble pirate had become a stock figure. In the remake, Errol Flynn plays an Englishman who fights wicked Spaniards but is enamored of a beautiful Spanish woman, a Spanish governor's niece whom he finds aboard a treasure-laden galleon. First capturing a Spanish galleon, being in turn captured by the Spanish and condemned to galley-rowing, and then escaping and being rewarded by Queen Elizabeth for discovering Spanish battle plans, the pirate Errol Flynn is simultaneously wicked and noble, free and oppressed, disgraceful and honorable. At the same time, he is simultaneously enlivened by Spanish energy and emotion yet able to overcome Spanish sloth and coldness.[7] Whatever its oddnesses, *The Sea Hawk* and its successor, *The Spanish Main,* brilliantly represent Hollywood's confusion over the true "complexion" of the pirate, confusion that is implicit in Stevenson's original *Treasure Island,* then explicit in dozens of illustrations in post-1910 *Treasure Island* editions and in novels like *Black Bartlemy's Treasure,* then blatant in films like one starring Douglas Fairbanks, a movie called *The Black Pirate.*

For Stevenson and Masefield, pirates and smugglers represented that part of the noble English character which chafed nobly under Victorian restriction. Never revolutionaries, fictional pirates and smugglers nonetheless forced wide the narrowness of everyday life and so spoke boylike to many readers—and later to many movie-goers, too. Outside the law but, unlike the Wild West badman, able to reenter society in wartime under a letter of marque, the pirate inhabited an essentially lawless realm. Moreover, though he was often of a low social class, the pirate might temporarily—or finally—demonstrate the grace and courage of a gentleman, often by rising to command a ship, sometimes by finding himself a true patriot in wartime.[8] In the end, however, Victorian morality made difficult any accurate fictional depiction of piracy.

So long as Britons equated Spain—and Catholicism—with wickedness, British piracy could seem noble, or at least capable of being sandpapered and whitewashed into Anglo-

In 1935 Errol Flynn starred in a remake of **Captain Blood,** *surrounded by monkeylike pirates. (Museum of Modern Art, Film Stills Archive)*

Protestant nobility. A letter of marque and a knighthood make posterity remember Sir Francis Drake as a privateer, not a pirate, for example, and his defeat of the Armada and subsequent Caribbean depredations equally advanced British interests even if few historians said so plainly.[9] But no amount of Victorian sandpapering and whitewashing wholly masked the ugliness of British piracy post-1588, and gradually even diehard Anglo-centrists accepted the wrongs done to Spanish merchants and colonists by rapacious Britons, especially since late nineteenth-century historians had begun to detail those depredations. *Treasure Island*, at heart a tale of honest British adventurers fighting ordinarily dishonest British pirates, cleverly avoids the issue of Spain but in time fell prey to the new notions of piracy implicit in its illustrations.

By the early 1920s illustrators were often depicting seventeenth- and eighteenth-century pirates as Latin, sometimes preying on British or American vessels, sometimes attacking Spanish ships, but always having the black hair and beard of a Spaniard—and often the lavish clothing, too. Novelists and then Hollywood filmmakers rewrote not only history but Victorian fiction, fastening on piracy as essentially Latino, the product of hot-blooded Catholics afloat in tropic seas, something to be curtailed by Anglo, Protestant heroes.[10] Part of the change appears to have originated in strong beliefs that hot climates produce hot-tempered, violent, sexy men and women, people predestined by nature for such wildnesses as piracy—

beliefs that had just begun to prompt white Americans to get tanned at the beach, to capture some trace of tropical solar energy.[11] But whereas illustrators, novelists, and filmmakers increasingly depicted the pirate as a man of mingled British and Spanish ancestry and inclination, especially the inclination to carry off blond British heroines, historians understood the fallacy—indeed the foolishness—of such views.

Scholarly interest in piracy exploded in the 1920s, partly as an outgrowth of the burgeoning interest in the history of exploration, partly as an outgrowth of something more immediately newsworthy. As early as 1923, when George Francis Dow and John Henry Edmonds published *The Pirates of the New England Coast, 1630–1730*, scholars had made connections between piracy and smuggling, and between piracy and slaving.[12] From the 1920s on, therefore, piracy existed in the scholarly world as a particularly complex, particularly nasty activity that involved a range of wider issues, even as it existed in the popular imagination as something increasingly simplistic, increasingly exciting and noble, something filmmakers could use to advance their championing of the Latin lover, south-of-the-border sexual vitality, and high-seas freedom. Only along the coast did people still think of it as scholars thought of it, as something violent, dirty.

Treasure, of course, helped keep alive the old understanding of piracy, treasure buried in oaken chests, treasure brought north centuries before. But in the 1920s treasure meant

In 1926, just four years after the appearance of Lothrop Stoddard's **Rising Tide of Color,** *Douglas Fairbanks starred in* **The Black Pirate** *as a dark-visaged pirate battling grobian pirates depicted as African or Spanish. (Museum of Modern Art, Film Stills Archive)*

P.159

something else, too, something scholars had begun to notice. Treasure meant not only doubloons but smuggled liquor, the rum run in under the nose of the Coast Guard, the liquor that made fishermen rumrunners, that made them rich. It meant anything smuggled: jewels, drugs, white women, would-be Chinese immigrants. It meant trouble. And it meant wealth.

Coastal Americans still search for treasure, and sometimes find it. Even the barefoot historian walks particularly carefully along one beach after every autumn gale, knowing that every few years some beachcomber finds a piece of Spanish gold tossed up in the sand. For decades, so long as federal law forbade the private ownership of gold coins, the lucky finders kept silent and visited the great sandbar more and more often, and even nowadays the finds are announced only quietly, and then only among locals who already know that some hoard is sunk just offshore among the sandbars or else lies strewn deep under the sand. No one wants tourists to find the hoard or some urban millionaire to arrive with electronic equipment. And then again, everyone knows of the new laws that prohibit the looting of shipwrecks. Do the new laws obviate the old law of treasure trove?

Treasure trove, part of the British and New England common law, part of the law that children somehow learn, part of the law that echoes the children's "finders keepers, losers weepers," annoys attorneys. Yes, indeed, the backhoe operator gets to keep the chestful of gold dug up while installing a septic system. The owner of the land owns not the treasure. But how often do such cases arise? Here, rarely. Thirty years ago a mason reparing a colonial-era chimney foundation chopped out several frost-split bricks and precipitated a shower of gold coins. He split the trove with the woman who had hired him, and both kept remarkably silent, their find passing into the local lore of a small town in which similar stories circulate among carpenters and roofers, among plumbers and farmers. Attorneys mutter among themselves about the law of treasure trove, and mutter worst when the split is made immediately, without recrimination or dispute, without recourse to law.

This is the great coastal hobby, the little-remarked looking out for treasure that keeps coastal property owners home from work when backhoes dig up gardens, boatyards, and barnyards, that gives hunters pause when they discover bowl-shaped depressions in the woods next to the estuary, that prompts scuba divers to poke about in waters renowned for their unimposing natural scenery and marked scarcity of lobster. Pirates, every local knows, buried their treasure, often splitting their hoard and burying parts in several places, and sometimes they wrecked small boats while getting the chests to shore, and so dumped their gold on inaccessible bottoms. Liable to capture, trial, and imprisonment, liable to shipwreck, liable to boatwreck in the last few hundred yards of their passage from the Spanish Main to New England, pirates secured their old age by salting away at least some of their gold after each trip. And sometimes, perhaps often, they died off a-piratin', and their gold lies buried still.

Tourists dismiss the tales. How often did pirates move this far north? Everyone knows that pirates drank up their booty in Caribbean taverns rather than save it to secure their old age. Tourists scoff, as locals hope they scoff, as locals intend they scoff.

Locals know that long ago, well into the nineteenth century, some vessels putting out from the little harbors went out in the "irregular trade." And when they returned, if not from out-and-out piracy then from slave-running or smuggling or gun-running, their local crews buried their ill-gotten riches before running in and asking the customs officers aboard.[13] And sometimes, only sometimes, for all sorts of reasons they did not retrieve their treasure.

Early in the nineteenth century, American periodicals began to publish accounts of treasure-seeking, or "money-digging" in the slang of the time. One of the best of this forgotten genre is "The Antiquarians," a *Knickerbocker* story of 1842 set on "a lonely, unfrequented part of the coast" of Louisiana. From its beginning in the memories of an old-timer, the adventure quickly moves to a specific locale. "The situation was several miles up a bay, and near the bottom of a cove of that bay, with nothing but shallow water for half a league around; consequently no vessel of any size could come within miles of it. There was a narrow, crooked, uncertain channel running through the bay and cove, by which boats drawing two or three feet might approach; but the swamp, from the shore to the magazine, might be called impassable." The magazine is a solitary building, defended by "a barrier or out-work" consisting of a six-foot-diameter log jammed across the channel, then a ditch or moat about it, and having neither door nor window. Its masonry construction, of "immense strength" and "bearing an appearance of great antiquity," is half-masked by piles of driftwood and tall reeds. About the structure rumor abounds. Some argue that it is left from Spanish or French colonial days, a mere if massive powder magazine, but others, noting that it has "every appearance of having been located with a view to convenience of access from the ocean," determine that it is a pirate storehouse, or perhaps a vault of treasure erected by slaveowners fleeing the Santo Domingo revolt. In the end, the two well-educated men who are determined to enter the keep go adven-

turing not simply hoping for "bags of gold and silver" but out of some antiquarian impulse, an impulse similar to that which drove British explorers along the marge of the Syrtis. "Our equipage consisted of a clumsy skiff, none the better for its age and hard knocks; two men, an axe, a crow-bar, tinder-box, lantern, and boat-compass," relates the narrator, slighting the presence of two hired seamen but emphasizing the expeditionary nature of two experiences, one the gunkholing effort of finding the keep, the other the gunkholing-like penetrating of it.[14]

Much of "The Antiquarians" derives from the early nineteenth-century American literary effort to find on United States ground objects old enough and with enough "historic associations" to serve in stories that imitated the romances of English literature.[15] Its narrator frequently mentions the "deep-red, time-worn walls" and its "appearance of hoary age," reveling in the keep as ruin, something "so long buried in mystery and gloom, that old stories of magic and enchantment came to mind." Yet against the terminology of European romanticism pulses a wholly up-to-date Yankee pragmatism. Suppose they do find a "unmanageable quantity of gold" left "by the old Spanish government," one asks the other. The physician replies, winking both eyes and making a significant motion of the head, with "altogether the air of a man whose plans are cut and dried." The physician knows what to do, how to ignore hoary age. "'*Wouldn't* I smuggle it on board the fastest tow-boat, and then make her walk up the Mississippi and never stop this side of Louisville?'" Both men understand the antiquarian impulse, the literal digging up of alongshore history, and they feel the draw of romantic associations implicit in the hoary vault. But about treasure, especially treasure perhaps still legally the property of heirs or far-off government, they think differently indeed. Such treasure would surely attract official interest, perhaps confiscation. Such treasure will make them smugglers.

After much work, the seamen break through the masonry, although the doctor and the narrator strike the final blows from some "instinctive perception of the great principle of right and title, so venerable for its antiquity, having been admitted to settle beyond all cavil the right to greater possessions, such as crowns and kingdoms, from time immemorial."[16] Treasure trove drives them on, into the structure, into emptiness.

They find only a magazine, a massive military structure so heavily built that over the centuries it has sunk some ten feet into the muck below its foundation, its builders having disregarded the advice of Ive and other sixteenth-century quicksand experts. Clearly the estuary has changed course, or perhaps the marsh has risen after great storms thrust sand ashore. Nothing of value rewards their efforts, and they set off for home, becoming enmeshed for hours in the twisting creeks and sandbanks of the great salt marshes.

The tale is ultimately a let-down, a too-long description of gunkholing gone wrong. Extraordinary anticipation cracks as the masonry cracks, giving way to emptiness and fatigue. Yet "The Antiquarians" clearly defines not simply the treasure-hunting craze that so puzzled British visitors but the loneliness of so much of the American coast that inspired all sorts of tales.[17] Certainly slaveowners or pirates might have chosen such a spot, so inaccessible from the land and so difficult of access from the sea, to hide their wealth, although—as one character warns in the tale—they most likely would have buried it, not hidden it within a tower, masonry or not. The shallow inlet is the key topographical feature, for it offers entry to a ship's boat while the larger vessel stands inshore, but in deep water. *Any* shallow inlet might serve anyone at sea who wished to hide something onshore.

Just such an inlet figures largely in Edgar Allan Poe's "Gold Bug," a short story that appeared a year after "The Antiquarians." "A scarcely perceptible creek, oozing its way through a wilderness of reeds and slime," makes Sullivan's Island a true island, a perfect setting for

Seashore vacations traditionally turn many men into wholly selfish, wholly irresponsible, piratelike individuals. (Harvard College Library)

finding a mysterious memorandum, deciphering it, and using it to search for buried pirate gold. The map leads its finder, the hermit-philosopher William Legrand, along with his slave Jupiter and the narrator, into a skiff, in which they "cross the creek at the head of the island" to a very desolate part of the coast, "where no trace of a human footstep was to be seen." But more than romanticism drives Poe's story. Poe understands the mysteries of gunkholing, of chart-making, the mysteries that caused Stevenson, after drawing an imaginary treasure map, to so fall in love with his creation that he wrote *Treasure Island* about it.

Still a favorite among junior-high-school students and American literature anthologizers, "The Gold Bug" works on many levels, one being the mathematical-linguistic level out of which comes the verbal mapping, beginning "a good glass in the bishop's hostel in the devil's seat—forty-one degrees and thirteen minutes—northeast and by north," another being the orienting of the resulting map to coastal topography changed over time, and yet another the level of suspense and excitement sharpened by repeated navigational mistakes: "The error, however trivial in the beginning, increased as we proceeded with the line, and by the time we had gone fifty feet threw us quite off the scent." Poe succeeds mightily because his effort lies largely in the locating of the treasure spot, not in the post-finding gunkholing home that scars "The Antiquarians." Poe traces the act of forming a chart, of solving a problem by making a picture, and his brilliant story succeeds with adolescent readers perhaps because the readers have achieved their first real familiarity with the mathematical and verbal languages his adventurers use to find the right spot. It succeeds with adults perhaps because adult readers not only retain the childhood dream of finding treasure but because they perceive the complexity of the adventurers' final act, that of linking chart to landscape.

The adventurers navigate to the right spot, and after much digging they find the treasure in a great chest, "a treasure of incalculable value." A mass of gold and jewels—"there was not a particle of silver"—lies heaped in a great bulk. The coinage, all "of antique date and of great variety—French, Spanish, and German money, with a few English guineas, and some counters, of which we had never seen specimens before," adds up to "rather more than four hundred and fifty thousand dollars." Consulting currency-exchange tables, the adventurers learn that they are rich, not exactly beyond measure, but certainly rich in money alone, and definitely rich in jewels and sculpture, including "eighty-three very large and heavy crucifixes." The catalogue of treasure demonstrates Poe's mastery of imagination and his understanding of what might be found, for it is indeed a catalogue of Spanish wealth.

More than treasure trove guarantees ownership in "The Gold Bug." The adventurers have earned the treasure, not only by first honoring the possibility that the United States might have a deep-rooted romantic history but by deciphering the verbal message, converting it to a chart, and aligning the chart with nearby topography, and by glimpsing the importance of the original parchment scrap being written with heat-sensitive ink. Chemistry enables one adventurer to reveal not the drawing of a skull but the drawing of a goat's head or, more properly, the head of a kid, and antiquarianism enables him to understand the connection. The vellum belonged to Captain Kidd, and only he knew its secret, for as the adventurers later discover in their digging, Kidd killed the seamen who lowered the chest into the hole. Skeletons covered the chest, just as mystery obscured the scrap of parchment.

"It was lying half buried in the sand, a corner sticking up," recalls Legrand of finding the vellum while searching with Jupiter for nondescript insects, a perfectly correct, alongshore natural history activity. "Near the spot where we found it, I observed the remnants of the hull of what appeared to have been a ship's long-boat. The wreck seemed to have been there for a very great while; for the resemblance to boat timbers could scarcely be traced." At first Legrand thinks nothing of the scrap, then he casually establishes "a kind of connection." As he explains at the close of his tale, "I had put together two links of a great chain. There was a boat lying upon a seacoast, and not far from the boat was a parchment—*not a paper*— with a skull depicted upon it." Scrutiny and serendipity combine to convince Legrand that the cipher might lead to immense wealth, that the depositers of the treasure never left the beach. "'You will observe that the stories told are all about money-seekers, not about money-finders,'" Legrand concludes, remarking the coastal fascination of his era, what the narrator at first dismisses as "the innumerable southern superstitions about money buried."[18] Legrand argues logically toward fantastical ends.

All along this stretch of coast, every local twelve-year-old knows that Kidd's treasure lies buried somewhere, unless its finders secretly spirited it out of the country or otherwise concealed it. And everyone knows, too, that the death's head, the skull and crossbones, advertises the brutality of piracy and the likelihood that Poe imagined right, that atop every unearthed pirate chest lies a mass of pirate bones. Dead men tell no tales.

In the national mind, however, piracy became ever less important, especially when juxtaposed against the great exploration and settlement of the West. After the Gold Rush, unearthing gold and silver meant prospecting and mining in California, in Nevada, and finally in the Klondike, not deciphering scraps of vellum in some alongshore wilderness. Not until the twentieth century did Americans begin to notice piracy again.

Piracy reemerged in the 1920s with the resurgence of smuggling, of rumrunning, of the dawning public awareness that the U.S. Navy and Coast Guard could not prevent liquor smuggling and so might be equally unable to protect the United States from military attack as from the rumrunners who threaded some vast salt marsh or the nook-and-cranny men who worked beneath an urban wharf. Moreover, rumrunning twisted American notions of legality and even sovereignty. Nowadays almost any attorney, even one devoted to sailing or fishing, will fall silent when questioned about maritime law, about the old British law of the admiralty. Just offshore in the glim, out in the offing where miles are nautical, not statute, the law of the land wastes away like sand, and maritime law takes hold. Law schools teach little of it, and only specialist attorneys know much of it, let alone practice it. Piracy, mutiny, crimes of property committed by stateless persons—such are the stuff of specialists now, the stuff that Prohibition made so necessary to understand, the stuff that in 1929 produced books like William E. Masterson's *Jurisdiction in Marginal Seas with Special Reference to Smuggling*. In the 1920s, educated Americans confronted legal mysteries like the so-called hovering laws, which applied to foreign-flag steamships that sat motionless just beyond the three-mile legal limit, waiting for nightfall, for fog, waiting to off-load liquor into high-speed motorboats or fishing boats.

As Masterson and others patiently explained, the three-mile limit originated in the longest possible range of eighteenth-century cannon, and whereas artillery had improved exactly as Maguire and Wisser knew, the three-mile range had become codified, static, a limit for fishing regulations and for customs actions, the limit of territorial waters. But as Masterson argued, Prohibition forced the "littoral state" to reencounter age-old problems, to understand the limits of a state's authority and range of action. "It has not as high a concern with a collision between foreign vessels 2.9 miles out as it has with an armed and powerfully manned foreign smuggling vessel hovering 12 miles out, seeking an opportunity clandestinely to run ashore or to unlade merchandise whose importation is forbidden."[19] Who became pirates and how? When did smuggling occur? Who owned contraband? What cause did the state have for stopping, let alone searching small craft? How could the state defend its coasts?

Nineteenth- and even eighteenth-century rulings quickly enmeshed law-enforcement authorities, attorneys, and diplomats; antiquarian interest in piracy and smuggling soon became full-tilt research into legal precedents; and young readers quickly fathomed the changed connotations of coastal life. In *The House on the Cliff*, the second Hardy Boys detective novel, Edward Stratemeyer, author of *Rover Boys on the Ocean* and writing under yet another pseudonym, Franklin W. Dixon, set his 1927 adventure in the thick of alongshore legal confusion.

This 1950s dust-jacket illustration (left) shows the Hardy Boys finding their way to an old smugglers' hideout converted to a modern smuggling operation. A cover illustration from the 1980s (right) still emphasizes the 1930s-era inboard motorboat but perches the house atop a sea-facing cliff, not one facing a hidden harbor. (author's collection)

No legal treatise or contemporary nonfiction account better delineates the twentieth-century rediscovery of piracy and smuggling.

At the end of the novel, Frank and Joe Hardy and several high-school-age friends rescue their detective father and—with the help of the Coast Guard and state police—round up a gang of drug smugglers. The smugglers, bringing in drugs and other illegal items from ocean-going vessels that move inshore toward a deep-water harbor, use not only an ancient clifftop house as an observation point but a web of tunnels dug by pirates and smugglers centuries before and hitherto forgotten. Most important, the smugglers use motorboats to bring the drugs almost up to the cliff, right among rocky hazards, before transferring them into rowboats they pull silently through the shallows and then into a narrow opening in the cliff. Although much of the story focuses on the kind of adventurous motorboating decried in the *Rudder* and other period magazines, much also involves the Hardy Boys wading and swimming the knifelike passage into the secret harbor, then figuring out how pirates and smugglers long ago used the tiny wharf that juts from the tunnel mouth. Yet the key to the significance of *The House on the Cliff* lies elsewhere.

At the beginning of the story, the boys set up a powerful telescope atop the cliff, using it to watch for smugglers. "'On a clear day you can make out human figures at distances of twenty-four miles,'" Frank Hardy tells his friends. But his brother amends the statement. Even with such a fine telescope, one can identify a person only "'about two and a half miles'" away. Instead of coast artillery, the Hardy Boys—like the rumrunning-enmeshed federal government —use a telescope to sweep the sea. And like penny-pinched federal authorities, they never employ an airplane. Much of the plot depends on scrutinizing, of gazing into the vastness for one motorboat after another, and of peering from a motorboat at one piece of cliff after another, trying to see how a rowboat might disappear. *The House on the Cliff* is about surveillance, about *watching* the coast, looking outward from the land and looking landward from the sea.

And the Hardy Boys descry dark-visaged men. While *The House on the Cliff* makes clear the wholehearted Americanness of boys like Tony Prito, who roars about in his motorboat, the *Napoli*, and Phil Cohen, who scouts the coast road on his motorcycle, the boy detectives and their friends discover the wickedness of Ali Singh, a crewman aboard the *Marco Polo*, a smuggler in league with an American mastermind.[20] Somehow, in a vague, mixed-up way, *The House on the Cliff* implies that America is at war with dark-visaged smugglers based perhaps in Asia or the Caribbean, some place away from the "areas of white settlement" Lothrop Stoddard mapped so precisely.

Skin color has a twofold meaning in many ocean-focused 1920s and 1930s juvenile and adult novels, even in *Saturday Evening Post* stories like the Guy Gilpatric tales about the adventures of the tramp steamer *Inchcliffe Castle*, or the John Marquand stories that feature the impassive Japanese secret agent Mr. Moto. The Caucasian American heroes, even the Hardy Boys, are tanned from outdoor activity, flushed from continuous exertion, and are "as lithe as Indians" when swimming, climbing, or wading.[21] Their skin advertises their vitality, their virility, their manliness, their easy familiarity with the outdoors, especially the seacoast. In contrast, the smugglers and other modern-day pirates hail not from Europe, not even from southern Europe, but from warm, stinking ports filled with dark-skinned men married to vice. Shanghai and other cities along the China coast, Havana and other Caribbean ports, Pembeni and any other place east of Suez produce the Ali Singhs and others who think nothing of killing Americans.[22]

Nationalism, not racism, not even straightforward ethnocentricism explains the motifs in the simplest juvenile tales like *The House on the Cliff*. Neither dark-complexioned Americans nor law-abiding foreigners offer the slightest danger. But criminal foreigners, almost invariably dark-skinned in post-1920 alongshore-focused fiction, not only intermittently invade U.S. territory but entrap Americans in their activities. Ordinarily law-abiding Americans, especially fishermen and other men whose alongshore or inshore occupations provide intimate local knowledge of shallow inlets and creek-crossed salt marshes, succumb to the temptation of rum, jewels, illegal immigrants, even drugs.

Late twentieth-century juvenile fiction often depicts the pirate as vaguely Latino, and certainly dark-haired. (author's collection)

Local complicity in international smuggling clarifies the intensity with which Masterson argues in *Jurisdiction in Marginal Seas*. Law enforcement authorities rarely encountered deepwater vessels close inshore, directly landing contraband. Deep-draft vessels simply could not work among the sandbars and other hazards, especially at night, in fog. But they could and did hover, waiting for locals in high-powered boats to reach them offshore, take on cargo, then speed toward any number of shallow inlets.[23] Locals had no need to gunkhole, to watch for buoys and other navigational aids, even to wait for good weather. Laden with whiskey, gin, or vodka, they ran inshore, then into creeks barely navigable at flood tide. At some long-disused landing fit only for gundalows they off-loaded their cargo into trucks driven by big-city gangsters, took payment in cash, and motored off to some local mooring, secure not only in their understanding of shallow-water navigation but in the growing dislike of the Volstead Act.[24] Although it was not an honest way to earn a living, to many people in the Depression, rumrunning seemed necessary, seemed almost a lark, seemed a way to get some cash quickly.

All along this stretch of coastline echo the rumrunning tales, although more softly each year, for each year fewer and fewer participants remain to retell them. But locals, especially locals who ran no rum, know the result of rumrunning. Mortgages paid off, children sent to college, boatyards or fifty-acre woodlots purchased, rental properties acquired—such were the fruits of smuggling payoffs. Treasure stands all around, in plain sight, still paying dividends, as when the woodlot becomes a condominium complex and the grandchildren of rumrunners become millionaires. And treasure lies sunk, too.

Here and there, at the lowest of low tides, the fisher or swimmer spies the heap of broken bottles and—if the fisher or swimmer is a local—knows that here a rumrunner dumped his cargo when the Coast Guard chase grew too hot. Sometimes, rumor insists, some bottles lie still sealed. So rare high-schoolers go a-foraging in salt creeks and under abandoned wharves, and find only shards and whole bottles with long-rotted-out corks. But now, seventy years after Repeal, a word still pronounced with a capital R in many coastal towns, few high-schoolers care to forage for old bottles, even if they know of rumrunning, of smuggling.

And smuggling is different now. Never does it smack of a lark, and always it smells filthy. Running in drugs, running in illegal immigrants, running out stolen automatic weapons destined for foreign terrorists turns out to be a remarkably chancy activity, something fraught with violence. And everyone knows that the smugglers include locals, locals often in over their heads, locals as dangerous as any pirate landing gold. As John Casey points out in his superb novel *Spartina,* a drug smuggler can be a fisherman making a one-time effort to get out of debt, to bring ashore a shipment of heroin beneath his iced fish, to drop the shipment into some small speedboat that waits in the mist.[25] Nervous, almost crazed with fear, the one-time smuggler or five-time smuggler now orders all sorts of thrillers.

Just as in the 1930s Phoebe Atwood Taylor set her Asey Mayo mysteries in a Cape Cod tainted and thrilled with rumrunning, so contemporary drug smuggling has spawned an extraordinary number of mysteries, thrillers, gothic romances, sea stories, and local-color tales focused on "the drug trade."[26] Wharfside bookstores carry dozens of titles in dozens of copies, perfect for summer people to read while lazing on the inn porch, while sprawled beneath beach umbrellas. How tourists can be pleased to learn that among the locals lurk drug smugglers, thieves, and murderers remains a deep unreached by dipsey lead, something the barefoot historian ponders frequently between Memorial Day and Labor Day. Do the well-read tourists look askance at every load of fish swung up from hold to wharf? Do they stare at lobsterboats, wondering about the ones that glisten in new paint? Do they expect bodies to wash up from under wharves? Do they wonder about the sailboat making land at four in the morning, the one with no navigation lights?

Whole series of books avidly read by men and by women—and by many teenagers of both sexes—revolve around the landing of drugs. In the late 1970s such books centered on the Florida coast and nearby archipelagos, especially the Bahamas, but by the late 1980s authors determined that New England inlets sheltered dozens of loathesome characters, vividly described in books with titles ranging from *Dark Nantucket Moon* and *Billingsgate Shoal* to *Death below Deck* and *The Body in the Kelp*.[27] They may be varied, and sometimes ridiculous, in their details of gunkholing, offshore boating, and rendezvous techniques, but the thrillers correctly conclude that few Latin Americans actually land drugs. Instead they move them north in international waters and hire locals to bring them in the last few miles. Every powerboat, every large sailboat can be a drug-runner, and every inlet, especially every inlet filled with sandbars and rocks, can be a suitable landing place. And most important, any smuggler can become a murderer, especially when money and a reputation is at stake.

Although they are more sophisticated and violent than *The House on the Cliff*, contemporary thrillers emphasize the 1920s notion of the dark-visaged man as evil. Almost always the long-distance drug-runner is Latino, and if not, then he is an Anglo well-tanned from the Caribbean sun. Moreover, the long-distance drug-runner is not only remarkably fit, perhaps from sailing his schooner or yawl so far, but remarkably willful, able to convince his shallow-water, part-time accomplices to do his bidding, to lure them in deeper and deeper. From the

Dawn finds a large sailing yacht anchored in the cove. Are the occupants honest seafarers or smugglers at the end of their run? (JRS)

old Spanish Main comes deep-sea trouble that spills itself inshore, corrupting locals already corrupt, corrupting even honest men convinced that one trip will end their financial woes, perhaps corrupting even high school students into making a run or two in the fiberglass speed-boat. The drug-trade alters the balance of things—and alters the view.

Now and then the authors of thrillers connect almost perfectly with the quiet, half-stated local knowledge, and in repeating that knowledge give their books an eerie semblance of reality. " 'Something phony stuck out all over. Those people made sailing mistakes,' " remarks one character in J. S. Borthwick's *Bodies of Water*, a New England coast–based thriller. The man explains how inexperienced long-distance drug-runners make all sorts of errors as they close the coast. " 'Two boats, one fetched up on a ledge, the other grounded out in shoal waters,' " he remarks at the beginning of a list of things done that arouse suspicion. " 'A crew member spoke Spanish in public and that suggested southern connections, and then there was the cash,' " the flashing of which is the " 'one sure way to make a New Englander's hair stand on end.' " But odd behavior gives away even the careful smugglers. " 'They tried to avoid contact with the harbor master. Always anchored at the outskirts of the harbor, and any fool knows that being tied up to a fixed mooring is the way to rest in peace.' "[28] In the first pages of her thriller, Borthwick encapsulates what all alongshore locals—except those deliberately unwilling to hear or repeat the truth—accept without question. Everywhere, especially in summertime when thousands of powerboats cruise everywhere, pirates move among honest men and women.

Part-time pirates, full-time pirates, pirates from the northwest coast of South America move stealthily or boldly, changing the local view of the seacoast, its vessels, its people, even its scenery. Does the salt-marsh gundalow-only landing used by rumrunners still serve the grandsons of rumrunners, and who if not the grandsons of rumrunners would know the tricks, say of using a boat small and light enough to drag over a barrier beach and into the bay just inland, and so eluding the drug enforcement agents? What does one make of the immaculate wooden sailboat nosing in at dusk, a wooden boat showing nothing electronic, nothing metallic, nothing plastic? Does one think of it differently when one realizes that a wooden sailboat, especially one without an ugly radar reflector, is a sailboat the Coast Guard cannot track with radar, a sailboat owned either by a traditional small boater or a very competent drug-runner? Is the out-of-town, voluptuous, bronze-skinned woman asking directions in the harbormaster shack simply a woman utterly at ease in her sunglasses and half-askew bikini or a drug-runner who knows that almost every man will remember her tanned, taut body but not her face? Is

Before entering the tower and keeper's house, tourists read the sign extolling the courage and quick-thinking of two along-shore girls, whom local schoolteachers still encourage local boys—and girls—to emulate. (JRS)

1636   1976

OLD SCITUATE LIGHTHOUSE

DURING THE YEAR 1810 THE U.S.
CONGRESS VOTED 4000 TO BUILD A
LIGHTHOUSE AT SCITUATE HARBOR.
DURING THE WAR OF 1812 ABIGAIL AND
REBECCA BATES YOUNG DAUGHTERS OF
THE LIGHTHOUSE KEEPER PREVENTED
A BRITISH NAVAL FORCE FROM SACKING
THE TOWN BY PLAYING A FIFE AND
BEATING A DRUM. THEY HAVE GONE
DOWN IN HISTORY AS "THE ARMY OF
TWO" AND THEIR COURAGEOUS ACT
HAS BEEN RECORDED IN MANY
TEXTBOOKS AND STORY BOOKS.

SCITUATE HISTORICAL SOCIETY

the sixteen-foot fiberglass speedboat putting out for offshore fishing or for some other activity that requires eight extra gasoline tanks and no beer? While the tourists read their thrillers, the locals look around.

Of course, the locals take action, official and unofficial, remembering the federal failure to protect the harbors against U-boats and other obvious enemies. In March, long before the onset of the summer boating season, they gather in town meetings to debate arming the harbormaster, to appropriate money for a high-powered rifle or an infrared-sensitive night-use telescope or a new radio to connect police officers with the Coast Guard. In summer, newspapers report drug busts involving Florida schooners and Massachusetts speedboats, bales of marijuana and bundles of cocaine, pistols and Soviet-made submachine guns. And the locals notice the gray sedans parked before the harbormaster shack, and the summer-suited, short-haired, binocular-holding young men who stare at the mass of boats in the harbor or at the immense wedge of boats that heads harborward at sunset.[29] Now and then on a summer morning one local nudges another and gestures at the Coast Guard wharf across the channel, at the low gray speedboat with "Coast Guard" barely discernible on its hull, a gray speedboat utterly unlike the white-and-orange lifeboat made fast alongside it, a speedboat that comes and goes with the darkness. And the barefoot historian, telescope at his eye, stands on the dune watching the Coast Guard roar off toward the horizon and no longer thinks automatically of rescue efforts but instead ponders the odd word *interdiction.*

And the smallest boaters of all see things in the shallows. In daytime little arouses their suspicion, usually a deep-water yacht working far up a bay, nudging into the creeks that run from estuary into salt marsh, its crew suddenly silent but ostensibly friendly toward the rowboat that materializes from some grass-roofed gutter. But at night, the kayakers, the canoeists, the pullers of skiffs all navigate the shallows in engineless silence, in centuries-old silence, and in centuries-old darkness, too. No lights announce their coming, for the law still exempts muscle-powered craft from carrying anything more than a light that must be shown to prevent collision, not shown continuously. The romantic midnight row, the before-dawn paddle to a favorite fishing spot opens on all sorts of things mysterious. In the misty, moonless silence the rower feels a presence and discerns the gray Coast Guard boat, lights off, anchored in the shallows, its crew listening, its radar unable to detect the wood skiff, the wooden yacht-tender. Or the rower hears voices whispering in the salt-marsh grass and figures—for he or she does not change course to look closely—that an engine boat lies nestled at the mouth of some gutter. Or on the coast road, the narrow sand-drifted road that might have given Franklin W. Dixon the idea for the first Hardy Boys novel, *The Shore Road*, the predawn sportsman after striped bass discerns a pickup accelerating from the old lane that ends at the gundalow landing, a pickup accelerating without headlights.

Teenagers are careful now where they park, where they neck. After all, after the necking, when the boy flicks on the headlights, who knows what might be revealed, what might start inland, toward the car, knife or silenced pistol in hand? In the dark everywhere alongshore lurks the old pirate evil come again, come through the most sophisticated of military defenses, come to stalk the quaint villages that so please the thriller-reading tourists.

Aerial adventure begins just offshore in Virginia of the Air Lanes and quickly moves to the beach as Virginia Suarez crashes her flying machine into the dunes. (author's collection)

Along the Gulf Coast, from Perdido Bay to Mobile, Herbert Quick focused the action of his futurist novel *Virginia of the Air Lanes* (1909). Set in the middle of the twentieth century and organized around twin plots, a love triangle and a fierce competition between venture capitalists and inventors, the novel thrusts deeply into the cultural implications of human flight. Over the barrier beaches and bays glide the immense "aerostats," the dirigibles doomed to extinction by helicopters, "aeronefs," and other creatures of a disturbingly complex technology. And intertwined in the technical advance is the new woman, the woman as unsettling as the fluttering aircraft that taunts the zeppelin owners.

Virginia Suarez is no typical turn-of-the-century heroine. Her half-playful, half-desperate escape from a noxious admirer aboard her uncle's aerostat slams into danger. Her experimental helicopter falls from the larger craft, then zooms upward, seaward. "She studied the machinery, trying to apply her picked-up knowledge of engines." Growing calmer, she begins to master the "demon machine," and eventually jumps from it as it skims over the isolated barrier beach on which the hero has his secret workshop and flying machine. Rescued by him and his aged, sea-captain friend, Virginia gulps down some whiskey, becomes first amorous then drowsy, and then determines to return home. Her love for the young engineer builds throughout the novel, and finally she saves his life. When his aircraft, which she has learned to fly as well as he, becomes tethered to a small submarine piloted by ruthless competitors, the inventor hands her his revolver and climbs outward to cut the cable. "She took the pistol with the air of one who knows how to use it, and nodded her head," writes Quick. "She braced herself against the rail, aimed conscientiously at the middle of the mark presented by the villain below, and fired, fired with the curious certitude the marksman feels when he is making a good

By 1913, illustrators had begun to project a new sort of ocean liner, one destined for greatness. (author's collection)

Lighter-than-air flight entranced early twentieth-century observers. In this image, made on June 6, 1908, on Long Island, an intrepid flyer experiments above a crowd. (Queens Borough Public Library)

shot." In the "clean, unsullied place in which to meet the end," Virginia masters technology and men. Beyond the reach of custom and restriction, beyond the summer cottages along the bay, Virginia plunges into a new world, savoring excitements unknown to inland women.

Virginia of the Air Lanes presents a remarkable woman in a remarkable undertaking, and for all its futurist, science-fiction orientation, it emphasizes the freedoms of women at the seacoast. Of course, Virginia is somewhat special, as her last name, Suarez, indicates. Filled with tropical vitality and energy, and flirtatious by turn-of-the-century standards, she is a perfect match for the dark-visaged pirates she encounters and from whose clutches she rescues her engineer-lover.[1] At the beach—or just above it—anything can happen.

Quick's novel is surely not fine literature, nothing comparable to Henry James's inquiries into women on the loose in Europe, but it is more than B-level or pulp fiction. It deals with three intertwined issues, and does so surprisingly well. It addresses the social implications of edge-of-the-land, cutting-edge technology, implications that range from the effects of mechanical power on women to the relation of inventors to venture capitalists and to monopolists threatened by technical innovation. It scrutinizes the role of women in a future society in which suffrage and other turn-of-the-century demands have been met but in which men have become even more rapacious, almost piratical, and women have become quick-thinking and physically strong. And it constantly addresses the seacoast as the place of technical and social experimentation, the place where land, sea, and air mix best.

At the back of Quick's novel stands the achievement of Wilbur and Orville Wright at Kitty Hawk, North Carolina. By 1909 everyone knew that these Ohio brothers had flown over and over again from their dune-top headquarters, that they had progressed rapidly from glider flights to powered flights and, from 1903 onward, to more and longer flights powered by heavier engines mounted in more complex airplanes. At Kitty Hawk the Wright brothers demonstrated not only the success of their original idea but the pace of technical innovation that entranced and worried both expert and casual observers. The Wrights flew over the ocean, almost to make flying seem even more marvelous than anyone had dreamed, for the ocean was vast and dangerous, topped by waves over which the aircraft purred so reliably that at first only naval officers fretted about the future of dreadnaughts and only urban cranks wondered about aerial bombing. At Kitty Hawk and other coastal places, by about 1909, Americans who already feared British naval attack and French grand descents learned that someday, perhaps quite soon, given the pace of the Wright assault on previously insurmountable barriers, foreign powers would fly across the oceans and lay waste American cities.[2]

So Quick wrote in a rather odd, brief era, one chronicled in *Scientific American* and in *Youth's Companion*, in half a dozen other periodicals committed to keeping adults and children absolutely up-to-date about coast artillery and flying machines. His jump from technical experimentation to social and financial envisioning, however dramatic in retrospect, may have meant less in the amazing years of seashore flying. On the other hand, given the simultaneous outpouring of seashore writing, B-level and otherwise, *Virginia of the Air Lanes* may have struck some readers as stunningly different from traditional writing about seashore space.

Henry James's nonfiction effort of 1904, *The American Scene*, offers a brittle counterpoint to Quick's adventure story. After a long residence in Europe, James returned to marvel at a changed United States, a modernized, mechanized, urbanized, speeded-up United States that figures badly in his book. But as he remarks early in *The American Scene*, he encounters "a supreme queerness on Cape Cod," a vague and disquieting sense that time had stopped all along the shallow-water New England seacoast, that the life of each coast town "was prac-

tically locked up as tight as if had all been a question of painted Japanese silks." At first on a local train that stops at harbor after harbor, then in a buggy driven by a local, remarkably taciturn youth, then finally on "an earnest stroll, undertaken for a view of waterside life," James struggles to understand the timelessness of places characterized by "crooked inlets of mild sea" and "low extensions of woody, piney, pondy landscape, veined with blue inlets and trimmed, on opportunity, with blond beaches." He knows the painterly terminology, knows enough to use phrases like "a mere brave wash of cobalt" to describe the color of the sea, but he admits finally that he "pursued in vain the shy spectre of a revelation."[3] Along the shore of Cape Cod, James finds nothing of the novel geology that entranced Dwight, nothing of the activity that entranced Thoreau, nothing of the salt-marsh farming that fascinated Jewett. What he finds is a thin, ribbonlike region somehow bypassed by progress, a place marked by poverty as well as poverty grass, a place poor in history and scenery both.

James walked in search of something old, something that could boast "historical associations" in the way the British conceived such associations, in the way that authors like that of "The Antiquarians" sought something hoary in the young Republic. For at least a century before James walked the Cape Cod coastline, British readers had been instructed in ways of looking at things littoral. Indeed, a handful of writers and painters had created a sort of pedestrian coastal aesthetic, something resembling a pair of spectacles to be worn when landscape merged into the coastal realm.

In 1775, for example, William Gilpin published his *Observations on the Coasts of Hampshire, Sussex, and Kent, Relative Chiefly to Picturesque Beauty*. Filled with snippets of historical information blended into topographical description, *Observations* is an essentially didactic work. "An object of this kind is by no means picturesque," Gilpin explains of a three-mile-long moor abutting the beach, "but it is *grand* from its *uniformity*, and *striking* from its *novelty*." The diligent reader learned how to look at the coastal realm in a precise, sophisticated way, to speak of different views accurately, in specific language, and, of course, to understand better paintings and drawings of coastal-realm subjects. But just as Gilpin ventured to Solway Moss to examine quicksand simply because quicksand belongs in any encyclopedic vocabulary of landscape observation, so he paid close attention to modern elements in the coastal realm.

"Everywhere as we approached Portsmouth, we saw quantities of timber lying near the road, ready to be conveyed to the King's magazines," he commented of Royal Navy matériel. "This is both a *picturesque* and a *proper* decoration of the avenues to a dockyard." In his final analysis, new things, even freshly hewn timber, might be picturesque because they are proper, correctly peculiar to the coastal setting. Rowing about the inlets and harbors of the coastal realm, he noticed the ornate carvings that decorated so many British warships, and condemned the carvings as stupid: "The *impropriety* and *deformity* of these ornaments, I think, are great. The *impropriety* of them consists in *decorating* a machine with carved work, which is professedly intended to be battered with cannon."[4] What is best fitted to the coastal realm, therefore, is that which the coastal realm *requires*.

The depiction of the proper constituents of coastal-realm scenery long entranced British artists, who saw in their depicting something utterly patriotic in an island nation. Sometimes they sought the antique, but equally often they scrutinized the new, following Gilpin's precisely outlined aesthetic. In *Picturesque Views of the Southern Coast of England*, a profusely illustrated work of 1826, J. M. Turner emphasizes elements as modern as ropewalks and shipyards. But especially Turner emphasizes up-to-date coast defenses, particularly the martello tower, "one of those defensive erections established to guard against invasion from the opposite coast of France." He described how the towers marched along the coast, each one-half to three-and-a-quarter miles distant from the next, and explained that each mounted two or more small cannon behind thick walls—and under bombproof roofs.[5] Martello towers, essential to coastal safety, struck Turner as utterly picturesque, and certainly proper, for all their vertical newness.

James walked the sandy beaches of Cape Cod in a wholly different aesthetic tradition, of course. Although the American tradition derived from Dwight, Thoreau, and other educated Americans of the early nineteenth century who visited the coastal realm not for its historical associations or its up-to-date seacoast defenses but to examine its natural curiosities and beauties, by the first years of the twentieth century it conflicted with the newer tradition of finding "local color." In out-of-the-way areas "unsullied" by such modern contrivances as railroads, telegraphs, and urban dailies, educated Americans hoped to find traces of an earlier, less troubling time. In the Tennessee mountains, in the upper peninsula of Michigan, in northern New England, they traveled alert to "simpler" ways and definite dialects, and alert to pick up the earliest of the old items they had learned to call "antique."[6] And almost by accident, they discovered the seacoast away from great port cities.

Steamships completed the impoverishing of many small harbor villages begun by the Embargo Acts and subsequent War of 1812. Harbors that had dispatched sailing vessels to Canton, Rio, and other far-off places no longer boasted even one full-rigged ship, not even a brig.[7] While Thoreau noted the turning inward of so many Cape Cod towns, only late nineteenth-century authors analyzed the quiet. In *Cap'n Simeon's Store* and other novels, George S. Wasson captured the dialect of long-beached deep-sea captains confined to Maine harbor towns, and in *The Country of the Pointed Firs*, *A Marsh Island*, and other books, Sarah Orne Jewett studied the same men and the women who lived wraithlike, almost haunting the houses filled with Oriental treasures and other proofs of long-lost world trade, and the tumbledown shacks filled with children and want. Into such villages came the first of the summer people, the warm-weather-only residents James understood as the sometime inhabitants of frame hotels and "alignments of white cottages."[8] But no Gilpinsque aesthetic theory, no Turnersque delineation guided them to and around the ordinary coastal realm.

The first summer people arrived essentially because the late nineteenth-century seashore offered "rustication" as cheaply as any other rural place discovered by middle-class families eager to vacation away from cities. Instead of vacationing on a farm, or at some mountain-valley resort or at some northern lakefront hotel, some families simply chose the seashore. As early as the 1830s, gazetteer-authors noted the rising industry of "boarding" summer visitors, and seashore families—even ship captains' and merchants' descendants living in large houses—began to accept paying "guests."[9] Between the 1830s and the turn of the century, however, such guests apparently valued highest the sea breezes that cooled their sleeping rooms and, most of the time, cooled verandas and quiet roads. Away from sunbaked manure, disease, open-window racket, and other summertime urban ills, the seacoast summer visitor enjoyed a week or month or season with scant benefit of aesthetic theory or historical association—and surely no instruction in marveling at nonexistent up-to-date coastal fortifications.

After the Civil War, however, the seacoast visitor visited according to—or in spite of—an ever more clearly defined web of instructions, though never as explicit as Gilpin's aesthetic. Newspapers, magazines, and books enjoined every visitor to embrace the coastal realm on several levels, all at once. The visitor ought to understand the wilderness beyond the porch, just off the pier, counseled many writers eager to advance zoological and botanical understanding of the ocean. The visitor ought to know something of the history of even the most ordinary harbor town, say something of the War of 1812, or the long-lost West Indies or China Trade. And the visitor should know something of the contemporary activity so visible in the harbor, on the wharf, or just offshore, even if more and more frequently activity meant only small boating or the occasional airplane flight. Slowly, painstakingly, American authors convinced readers, especially would-be vacationers, that the coastal realm boasted riches beyond anything accessible in mountains or forests.

Certainly, the authors argued, the ocean contained all manner of living things unlike those in mere freshwater ponds, lakes, and rivers, and the wise visitor ought to know something of the wildlife cast up on the beach or lurking in tide-pools. Being familiar if not intimate with such flora and fauna not only made a casual walk more interesting but shifted the walk, indeed the whole coastal realm visit, from the category of pure leisure to that of an educational outing, to something like the insect-collecting effort in which Poe's Legrand and Jupiter discover the scrap of parchment. While the body relaxed in the salt wind and the soothing sun, and even in the surf, the mind identified and classified wildlife. Moreover, insisted many writers, some of the wildlife ought to be *taken home.*

Nothing is more easily missed yet more bizarre than the continuing propensity of Americans for bringing home seashore animals and plants. Tourists who would never think of bringing home sun-bleached bones from the far Western quicksands described by Albee immediately and casually begin to collect shells strewn along sandy beaches. Others often collect marsh grasses and other plants like sea lavender and sea oats, and some even gather rockweed and seaweed, pressing wet pieces between the pages of some summer novel or wrapping them in local newspaper. And young children, pails in hand, gather rock crabs and minnows, sea worms and hermit crabs, and beg permission to bring them home. Permission or not, many ocean animals do ride far inland, and sometimes at motels and restaurants a day's ride from the sea the barefoot historian finds parents struggling in automobiles, removing carpeting and back seats in search of something escaped and very, very dead. The lone shell or the pail-ensconced clam traveling inland bespeaks a long American effort to bring inland something of the natural history of the coastal realm, an effort that blossomed in the middle of the nineteenth century.

The aquarium craze began about 1845 and peaked perhaps ten years later. Almost always an urban phenomenon, it nonetheless reached far inland. Railroad express agents learned that the heavy barrels they off-loaded from passenger trains in Buffalo, Detroit, Pittsburgh, and even Chicago contained ocean water, seaweed and rockweed, and a few crabs and anemones, prizes destined for upper-class households smitten with keeping what the British called "sea-gardens." Gardening somehow lay at the root of the craze, for urban women already fascinated with flower gardening and scientific botany discovered the marine aquarium as a first step toward mastering zoology.[10] But the women also understood the marine aquarium as a year-round reminder of a summer holiday at the shore, something requiring the most delicate care. While ordinary women kept freshwater aquariums containing pond lilies and frogs, venturesome women maintained marine aquariums like those described in a New York magazine in 1856. *Household Words* published two detailed articles offering proven advice to would-be keepers of "marine menageries," emphasizing not only that the craze had captivated upper-class urban Britain but that the British had solved such maintenance issues as aeration and temperature stabilization.[11] By the mid-1850s, in fact, an entire aquarium-supply industry had developed along the Atlantic coast, and would-be aquarium owners had only to contact the right person to receive everything in one shipment.

Everything always included a tank. Pioneers in the 1840s had tried large bottles, especially empty quinine bottles, but by the 1850s most entry-level guidebooks, like Henry D. Butler's *Family Aquarium*, suggested a slate-bottomed, glass-sided box, every seam made watertight and every surface, including the cement that sealed the joints, made chemically pure. In the years after the Civil War, diehard hobbyists shifted to the one-piece glass tanks that still gather dust in antique shops, for they knew by the early 1870s that any chemical contamination, including contaminants leached over time from joint-sealing cements, killed the frail animals confined in very little water.[12] Along with the aquarium came a cask filled with the "makings" of the aquarium. Early on, certainly by 1850, devotees knew that water collected near the shore might not only be brackish but might well be polluted, so they followed expert advice. "No cask that has been used for spirits, wine, acids, chemicals, etc., will answer," Butler warned. Only a new cask, with new bungs, will serve, and the cask should be filled at sea: "a trifling sum will tempt the cook or steward of any sea-going vessel to fill you a cask from the clear and open ocean." How exactly vacationers contacted such mariners Butler does not explain, but he makes clear the utter necessity of beginning with pure ocean water and maintaining its purity by instantly removing any dead rockweed or animal. Marine aquariums, argued experts like Butler and Shirley Hibberd, author of *The Book of the Marine Aquarium*, tend to go out of balance quickly, disappointing owners who discover that everything died overnight. "It must be frankly confessed that you can have your tank fitted up with still greater perfection (and at very little expense) by those who make it a specialty and a profession," Butler warned at the close of his book.[13] The right person almost insured success.

Robert Carter discovered one right person in 1864 while sailing with a marine zoologist and other friends. Carter's *Summer Cruise on the Coast of New England* emphasizes the emerging mixture of scientific inquiry and aesthetic exploration of historical, often quaint villages that became the standard summer-holiday mix within twenty years. He marveled at his colleague's fascination with collecting everything from jellyfish to periwinkles, and understood the difference between professorial research and the hobbyist inquiry that supported the alongshore aquarium-stockers. One stocker, a full-time shoemaker providing aquaria as a sideline, rowed Carter ashore to show him his shop and personal collections. By 1864, perhaps because of the Civil War, his business had slackened dramatically, but Carter glimpsed its outline. "To those who ordered from him the materials for stocking an aquarium, he sent a keg or barrel of sea-water, and a box of two compartments,—one containing the sea-weeds and some of the animals, the other containing the more delicate animals in a bottle or jar."[14] During the height of the craze, Carter discovered, the shoemaker had worked full-time at supplying his wares and had even sent specimens to the Smithsonian Institution.

What the alongshore shoemaker and his competitors knew, of course, was something of the ruggedness of individual species of plants and animals. Although sound tanks and pure ocean water shipped in new casks contributed to success, the real key was local knowledge. The professionals shipped only rockweeds and animals able to withstand temperature changes and salinity shocks, and so greatly pleased amateurs infuriated with their own collections. The professionals understood, moreover, that once a marine aquarium had been balanced and proven successful, its owner might well send for additional plants, and even less common animals.

Of course, many marine aquarium owners collected, too, devoting their summer holidays to collecting, pressing, and drying the seaweeds they classified, and preserving in alcohol the animals they collected at low tide, and bringing home living specimens as well. Butler and other writers knew by the 1850s that many well-educated women, having mastered the classification of terrestrial plants and inland birds, had determined to demonstrate their intellectual capabilities by making sense of the inshore ecosystems.[15] Moreover, women wanted to collect their own specimens not only because it was a remarkably intriguing activity but because it demanded physical exertion. Although Butler and others warned the ladies to hire assistants with crowbars to break off chunks of rock exposed at low tide and to carry shoreward baskets filled with glass jars, women apparently delighted in getting soaking wet and utterly exhausted by doing the collecting themselves. Searching for a particular uncommon mollusk provided an excuse to venture day after day into the tide-pools, among the rocks exposed at low tide, even into the edge of the surf.[16] Shirley Hibberd viewed such activity with concern, certain that beginners would drown while following the experienced gatherer far out at low tide or suffer broken legs while clambering down rockweed-covered boulders. "The naturalist must encounter a few perils," Hibberd declared after recounting one harrowing misadventure, "and he does encounter them boldly, and with little regard to such a trifling matter as a grazed shin, or an impromptu bath; but it is not to be expected that all who commence the study of nature at the sea-side, will care to encounter the perils which an experienced student thinks lightly of."[17] Aquarium-keeping might be conducted sedately, but personal collecting meant activity beyond the dry land, in the realm of the forces George Howard Darwin described in *The Tides* in 1898, in the limicole zone of dubious behavior, in the marginal space especially perilous for women not absolutely certain of their social position.

Although it ended abruptly in the 1860s, the aquarium craze had long-lasting impacts. Certainly it taught summer visitors that bits and pieces of the seashore might be taken home, even taken home and kept prisoner. As Emerson collected especially beautiful beach stones even as he understood that they lost their beauty when dry and away from their natural setting, other visitors brought home shells as souvenirs of brief stays in the coastal realm. Until at least the turn of the century, parlors far inland from the sea often displayed an especially large or beautiful shell or two on some shelf in the "what-not," or a dried starfish or piece of driftwood worn by swirling sand.[18] Moreover, the craze taught visitors that some living things could endure the railroad journey home, and could endure—and even thrive—in a glass box. Free for the catching, crabs and other animals could be permanently contained without being killed, could be kept in an utterly limited place as ornaments, exactly as the aquarium industry explained at the Centennial Exposition in Philadelphia in 1876.[19] A bit of ocean wildness could be perfectly confined, almost like a model ship built inside a bottle. But most important, the aquarium craze introduced women to a spectacularly successful reason for hiking out along the beach, for climbing over rocks, for wading and splashing through tide-pools. As an intellectual rationale for sustained physical activity, nothing surpassed alongshore botanical and zoological searching, the looking down into the shallows as closely as any horizon-watcher scanned the glim.

At the turn of the century, when Augusta Foote Arnold published her massive volume *The Sea-Beach at Ebb-Tide*, women understood not only the intellectual accomplishments of half a century of collecting and classifying but the far less obvious, almost secret joys of scientific activity. Arnold's tome remains in print, still used by amateurs and professionals in identifying a specimen washed up on the sand or dipped up from a pool. But its first pages offer clues to the incredible power women took from the earliest era of collecting. Arnold does not explain why collecting and identifying specimens are good activities. In fact, she pauses only to say that knowing something of the plants and animals of the littoral makes visits to the coastal realm more interesting. But she tells her readers to go adventuring, to plunge into all manner of shores. "On the piles of wharves and bridges may often be found beautiful tubularin hydroids in large tufts just below low-water mark, branched hydroids looking like little shrubs, polyzoans, sea-anemones, mollusks, and ascidians." The wise collector should seek out such wharves and bridges, for the boring animals are not only beautiful but "most destructive," and should be noticed for their economic impact. In 1901, then, Arnold tells her readers to splash about under wharves, to venture into the very haunts of the nook-and-corner men. Elsewhere she advises her readers to go collecting at the height of great storms to see what deep-water specimens are washing shoreward and to venture "as deep as one cares to wade" off sandy beaches before collecting.[20] Arnold writes for women not afraid to get their feet, their thighs, even their hair wet, for serious collectors out having some exercise, some fun in the name of intellectual effort.

Science alongshore meant by the turn of the century something not always what Herbert Quick meant in *Virginia of the Air Lanes* or the Wrights demonstrated at Kitty Hawk. Science meant something that called naturalists, especially female naturalists, into the water,

into exertion, sometimes into danger. The "new woman" of 1880 found the shore exceptionally exhilarating; it provided the theater in which she could use her body while using her mind. Not surprisingly, upper-class women wholly secure in their social rank were the first to take to marine aquariums. After all, they had the wealth and the leisure to summer in the coastal realm and, moreover, often the opportunity to collect while their husbands remained inland, tending to business. When they began to collect and classify littoral wildlife, they provided models for middle-class women, who in time did the same thing, using their growing knowledge of marine biology not only to demonstrate to men that their minds could handle exceedingly complex issues but to offer an explanation, indeed a justification, for living actively along the edge of the sea.[21] Just as farm women might use blueberrying as an irreproachable reason for taking an entire day to hike up hillsides, to enjoy cool air and long-distance views, so women visiting the shore used marine biology as an irreproachable reason for hiring or borrowing a skiff or other small boat and rowing out into estuaries and marsh creeks, for hiking through marshes as wet as any Jewett described in *A Marsh Island*, for climbing along the bases of cliffs exposed at low tide, for getting splashed, even soaked.

Even today, the barefoot historian notices that a child in search of a crab scuttling away in the shallows is often a child in need of a mother to rise from her blanket, shrug off her shirt, and stride into the water. And in a few minutes, not for reasons of marine zoology, not because she keeps an aquarium, but because it is a good thing for the child to see a rock crab, to see mommy pick it up without being devoured, without being afraid, the woman suddenly plunges underwater and emerges victorious a moment later, crab in hand. And if contributing to her daughter's education has required her to plunge into the waves, so what? After all, the reason is irreproachable. And the barefoot historian, musing on Augusta Foote Arnold, wonders how many women learned to swim while plunging after some momentarily nondescript crab or rockweed.

At least as early as the turn of the century, magazine writers guessed at the liberating impact of seaside collecting. Mothers determined to educate their young children led the way into tide-pools and mudflats, suddenly different, far more adventuresome than they were at home. In a curious, almost cryptic *Living Age* article of 1908 entitled "Children on the Sands," an anonymous essayist mused about the transforming nature of the beach. To a child, the essayist argues, the beach is a world "of new meanings and new possibilities," filled with strange creatures "of uncouth shapes that belong to fairy-tales, that vanish as elves and spirits vanish." A crab appears to a child as something like a large spider, but in time the child learns to handle it. In the fields rabbits and butterflies escape, but at the shore "the baby fish and shrimps and crabs are there to be caught and admired in pails."[22] Drifting nearly invisible in the article, however, are the mothers, the women who would never pick up a spider but who teach the children, who somehow become willing to romp and splash after the baby fish, who swing the net with abandon. At the seashore in 1908, young children might discover not only the strangeness of the shore but the strangeness of their mothers, the mothers suddenly unrestrained, suddenly secure in marine knowledge. While *Scribner's Monthly* published articles on starfish and lobsters aimed at middle- and upper-income families increasingly eager to know something of the coastal realm ecosystems in which they vacationed, or in which they intended to vacation, the collecting women of the seashore, at least in summer, became almost as daring as Virginia in Quick's adventure novel.[23]

But context, too, made the net-dragging, jar-carrying women seem adventurous, for they and their families explored a place already classified as quaint, backward, incredibly conservative. As early as the 1850s, about the time the aquarium craze spurted toward its crest, magazine authors and other writers had begun to scorn the backwardness of coastal regions away from great harbors, establishing the tradition in which Wasson, Jewett, and James wrote so comfortably at the turn of the century.

In 1853, for example, Nathaniel Parker Willis wrote at length about Cape Cod, "the earth's most unattractive region," a Saharan place seemingly unchanged from the days when Timothy Dwight rode across its sand-drifted paths. Willis condemned the built landscape in scathing terms, finally determining that "the houses and their surroundings seem of an unsuitable *inferiority* of style, to those who live in them," and marveling that many wealthy people chose to live in such tiny cottages. Moreover, the people seemed odd, not merely outlandish, but almost a subspecies, one speaking a dialect marked by a very broad yet soft "a" sound. About another characteristic, Willis waxed eloquent, determining that "flatness of chest in the forms of the feminine population of Cape Cod is curiously universal" and apparently stems from the hard work women do while their men remain at sea, work that makes Cape Cod men remarkably muscular and particularly dignified in old age. "There is one class of unusual personal beauty on Cape Cod," Willis determined. "I never saw so many handsome old men in any country in the world." Cape Codders seem inured to physical activity, often to activity that Willis condemns as wholly unnecessary, as when they drive passenger-carrying wagons into the sea and off-load people into rowboats rather than building proper wharves.[24] Though he is often sarcastic, however, Willis frequently muses on the admirable qualities of the seashore people—their thriftiness, their lack of ostentation, their good health—and he concludes his lengthy descriptions most definitely confused not only about the natural environment of the place but about its indigenous population, the "natives" of the sandy coast.

What Willis understood as thrift or comfort the locals knew as near poverty, a close-to-the-edge existence that became more pronounced with every decade of the nineteenth century. What Dwight noticed as spartan comfort at the beginning of the century had become the conservative, almost parsimonious style of living Thoreau and Willis remarked at mid-century, a style of living that kept even well-to-do Cape Cod families living in tiny houses. And by century's end, the lack of extra cash had made the whole New England coastal realm, away from major port cities, almost a de facto historical preserve, a place in which old habits and old

Cape Cod–style houses attracted tourists as early as the 1840s, when John Warner Barber described them in his Historical Collections; by 1930 the houses struck many as diminutive, cute, old, and quaint, never as the product of mid-nineteenth-century poverty. (author's collection)

things endured essentially because they could not be replaced. From the years of the Embargo Acts onward, most of the New England coastal realm had become poorer and poorer, its people determined to live simply and frugally, to go without rather than go into debt. By 1883, when one *Century Magazine* essayist published a lengthy description of Cape Cod, the whole cape had become almost antique, filled with "historical associations," essentially quaint. While F. Mitchell admitted that "no one who travels through Cape Cod and visits the people in their houses can fail to notice an almost universal thrift and comfort," the whole of the cape appears vaguely threadbare, carefully patched and repatched, incredibly aged. Moreover, since about 1870 a handful of city families had discovered the excellence of the beaches as summer resorts, had begun boarding in farmhouses and even building summer cottages, and had discovered the local people as worthy of study.[25]

At this time, roughly between 1880 and 1910, summer visitors began to collect coastal Yankees almost as the women of the 1850s and 1860s had begun to collect crabs and rockweed. In one way, of course, the local people were collected as summer help, as cooks, carpenters, gardeners, and above all navigators of catboats and other pleasure craft bound for far-off sandbars and marsh islands. But collecting proceeded in another, far less forthright way, a way focused on making "the acquaintance" of "true" locals, the "real natives." Summer visitors and casual tourists began to study Atlantic coast locals from Maine to Florida, and especially in New England, for the locals appeared to represent the "purest" survivors of original English colonial stock.[26] As native-born Americans became increasingly wary of the immigrant groups that poured into east coast cities, the natives of the coastal realm seemed somehow a rediscovered population descended directly and perfectly from the gunkholing Pilgrims who had settled the sandy region three centuries earlier. Moreover, because they had not only a long tradition of worldwide navigation, albeit navigation greatly diminished in the nineteenth century, and hence routine contact with foreigners, shipwrecked and otherwise, the coastal realm people were more at ease with newcomers, even curious newcomers, than other isolated populations, say the farmers along the New England–Canadian border or those of the Ozark hollows. Near the end of the nineteenth century, travelers and summer residents and—above all—magazine writers learned that the old man digging in his garden, or in the garden of a summering family, had a thousand stories, had been to China, had been master of a full-rigged ship. The old man, they discovered, was "an old salt."

Between 1880 and 1920, people resident along this stretch of coast became identified as "folk," and individuals became known as "characters," as men and women worth getting to

know, to see how they handled a catboat or a kitchen, to hear their stories, to hear their
speech, especially the dialect they took for granted as the only pure English. As late as 1881,
when Sally Pratt McLean published her *Cape Cod Folks*, many authors wrote in the vein of
Willis, condemning the coastal people as poverty-stricken, ignorant, backward, and stub-
born, people in desperate need of an up-to-date urban schoolteacher like the one McLean
provided.[27] In popular literature, at least, the more acute understanding of locals as badly
schooled poor people struggling hard against vastly changed economic circumstances, the
understanding expressed as early as 1868 by Charles Nordhoff in *Cape Cod and All Alongshore*, appeared rarely indeed. Nordhoff wrote about cape boys who "take to the water like
young ducks" and are "born with a hook and line in their fists, so to speak," and he painstakingly described aged fishermen, each "an old Banker, which signifies here, not a Wall Street
broker-man, but a Grand Bank fisherman." Nordhoff admitted that the few well-to-do tourists
who visited the cape found both the scenery and the people "picturesque," and agreed that
"a moderately flush Wall Street man might buy out half the Cape and not overdraw his bank
account," but he praised the honesty and "conscience" of the locals, much of whose behavior—say the absolute refusal to fish on Sundays, even when offshore in vessels surrounded by
biting fish—seemed archaic, if not plain stupid.[28] But gradually the image created by Willis,

McLean, and other popular writers disappeared under a rising tide of appreciative if saccharine "local color" writing that introduced the coastal realm—especially to readers of *New England Magazine*—as a time-forgotten, strange place inhabited by odd locals, by characters. By 1910 the coastal people had become specimens or characters, something they remain —in the popular inland imagination at least—to this day. And the transformation, though undoubtedly part of the larger turn-of-the-century search for traditional American values, landscapes, and occupations, owed much to the best-selling novels of one alongshore writer.[29]

Joseph C. Lincoln engineered what can only be called the "quaintifying" of the New England seacoast. Literary scholars still ignore him, and even local historians know little of his career and certainly few of his original inclinations. Not even a proper bibliography of his short stories and novels exists, nor does a sales analysis of the novels he produced year after year, usually just before Christmas. In fact, no one has traced the marketing of his novels, although memoirs and autobiographies of inland farmers and other early twentieth-century readers sometimes mention his books as favorite recreational reading.[30] Moreover, Lincoln wrote in a larger context, having both a British counterpart, W. W. Jacobs, whose *Salthaven*, *At Sunwich Port*, and similar tales focused on the economically stagnant shallow-water ports of England, and American imitators, especially James A. Cooper, author of *Cap'n Jonah's*

Sally Pratt McLean depicted alongshore folks as crude and decrepit, and not at all quaint. (author's collection)

Fortune: A Cape Cod Story and other novels.[31] Finally, just as no one scrutinizes pirate-fiction illustration, no one has yet traced the impact of the line drawings and even full-color frontispieces that illustrate Lincoln's novels, although they derive from firsthand observation of local small craft, local topography and plants, and, of course, local characters, often aged men stooped over canes but still staring seaward. Year after year Lincoln brought forth another Christmas novel, and year after year he published his short stories in *Everybody's, Collier's,* and other magazines. Bit by bit he created an image of the ordinary coastal realm inhabited by extraordinary people.

Most of his novels take their titles from lead characters around whom several stories revolve. *Cap'n Eri, Mr. Pratt, Keziah Coffin,* and *Kent Knowles: "Quahaug"* all focus on cantankerous, extremely old-fashioned men and women who eventually demonstrate immense breadth and depth of intellect and character, usually when confronted with some summer person enamored of engine-powered boats, sleazy love affairs, or get-rich-quick schemes. "When one of that kind is out sailing with me, and begins to lord it and show off afore the girls, the *Dora Bassett* is pretty apt to ship some spray over the bow," remarks the narrator of *Mr. Pratt,* an elderly mariner who makes his living chartering out his catboat. "A couple of gallons of salt water, sliced off a wave top and poured down the neck of one of them fellers is

CAP'N ERI

—Frontispiece.

the best reducer I know of; shrinks his importance like 'twas a flannel shirt." [32] Another set of novels, each with a title like *Cy Whittaker's Place* or *Rugged Water* or *Blair's Attic* evoking some piece of coastal realm scenery, presents similar plots but with slightly more attention to alongshore economics and landscape, particularly to landscape elements being discovered—and purchased—by summer visitors. "For years it stood empty," Lincoln remarks in *Cy Whittaker's Place* of an untenanted house owned by a far-off master mariner. "The weeds grew high about its foundations; the sparrows built nests behind such of its shutters as had not been ripped from their hinges by February no'theasters; its roof grew bald in spots as the shingles loosened and were blown away." [33] Books like *Thankful's Inheritance* and *The Portygee* introduce plots involving threats to coastal community stasis and comfort, and the uncanny way the communities—or the "salt air"—work to incorporate strangers into places that become home. "Albert looked about him over the rolling hills, the roofs of the little towns, the sea, the dunes, the pine groves, the scene which had grown so familiar to him and which had become in his eyes so precious. 'It is *my* country,' he declared, with emphasis." [34] Alberto Miguel Carlos Speranza, the waif-orphan-become-congressman of *The Portygee*, becomes a son, if not a native one, of the coastal realm and its representative in Washington. Nothing elevates any of Lincoln's fiction to great literature, and none of it remotely approaches the complexities so evident in Jewett's fiction. But Lincoln's prose reads easily enough, and it catches the Cape

Cod–Plymouth County coast accent as readily as any of Lincoln's locals catch fish. It correctly calls a northeast storm a "no'theaster," not what summer visitors—and television newscasters —call it, a "nor'easter." Moreover, and certainly most important in any analysis of American coastal realm landscape, Lincoln's fiction is visually accurate. "The brilliant yellow gleam a mile away is from the Orham lighthouse on the bluff," Lincoln writes in establishing the nighttime landscape of *Cap'n Eri*. "The smaller white dot marks the light on Baker's Beach. The tiny red speck in the distance, that goes and comes again, is the flash-light at Setuckit Point, and the twinkle on the horizon to the south is the beacon of the lightship on Sand Hill Shoal."[35] Whatever literary specialists might say about his prose style once they condescend to study it, Lincoln's descriptions of objects in space seem as realistic as any in contemporary nonfiction.

Nowadays Lincoln novels sell immediately at high prices in alongshore secondhand bookstores, and coastal public libraries bind and rebind their tattered copies. Tourists discover Lincoln books every summer and try to assemble complete sets, to read them in chronological order, to immerse themselves in innumerable related novels. In reading Lincoln novels, the summer visitors glimpse something even the locals call—in an echo of a near-forgotten Patti Page song—"old Cape Cod," or old Scituate or old Marshfield, or old Norwell or simply "the old beach." On the one hand, summer visitors and locals alike acknowledge Lincoln's precise description of everything from baiting fishing hooks to gunkholing in a catboat to

Here Captain Pratt, secure in the cockpit of his catboat, muses on chartering himself and his craft to summer people likely to get seasick. (author's collection)

In Cape Cod Stories *Lincoln focused on the rising summer-folk passion for carrying away pieces of along-shore quaintness, including sofas the locals found uncomfortable to sit on but too solid simply to throw away. (author's collection)*

splashing across a marsh to launching a boat through surf to protecting rosebushes in winter gales and winter freezes. His loving and long descriptions of decrepit structures in particular elevate their subjects from decrepitude to quaintness, making unpainted shingles weathered a whitish gray by sunshine and wind-borne salt more beautiful than chemically stained shingles. On the other hand, his descriptions of people at the beginning of the century not only remind contemporary locals of long-dead old-timers but sometimes illuminate contemporary mannerisms, attitudes, patterns of speech, even words. Devoted readers of Thoreau's *Cape Cod* know the alongshore meaning of *guzzle*, but readers of Lincoln, curled up against cockpit cushions, in porch chair or on beach blanket, learn such topographical terms, too, along with others, like *conniption fit*.

By 1910, perhaps a decade earlier, tourists began to seek out the "real" beach, the old one, the one that looked like those Lincoln described. Whereas Gilpin spoke to the most educated of British home-island tourists, Lincoln addressed more general readers, and in time the readers began to arrive along this stretch of coast determined to use Lincoln novels the way other visitors used Arnold's *Sea-Beach at Ebb-Tide*. Upper-class summer visitors were no longer the only out-of-towners enjoying the beach, the local shops, the local conversation, and in the 1920s many bought acreage to protect their cottages from the influx of middle-class families who boarded with locals or bought—or rented—small summer cottages. And while everyone arrived more or less determined to enjoy the fresh air, low-priced accommodations, local boats, relaxed pace of life, and even swimming, a surprising number of middle-class visitors arrived to see the place Lincoln described, to experience living in a Lincoln novel.

Ascertaining the impact of Lincoln's fiction in any quantifiable way is impossible now. Summer visitors no more figure in turn-of-the-century census records than their point of view

RESIDENCE OF JOSEPH C. LINCOLN, CHATHAM, CAPE COD, MASS.

So popular were Lincoln's novels that his house became a favorite tourist destination. (author's collection)

shows up in local newspapers from the same era. All the barefoot historian can do is inquire of old-timers about stories told by their grandparents and parents, the "summer visitor" stories that locals tell in the winter, out of earshot of people who pay well for simple things like marine-railway haulage or a new wooden catboat. Bits and pieces emerge, some almost quantifiable, say the lengthening summer-only home-delivery routes of milkmen and truck farmers after 1910, and some scarcely quantifiable, say the anecdotes involving summer people who told locals explicitly that they reminded them of Lincoln characters. Local men and women who analyzed the motives of early twentieth-century tourists, usually in order to profit from them, have left behind few diaries, few ledgers, only hand-me-down stories about summer people determined to have particular experiences. Summer visitors wanted desperately to ride in local boats, to rent local boats, to buy local boats, so much so that some boatbuilders kept building catboats and other local small craft long after they knew of improved versions.

When locals cannot dig their own clams, they eat the next best thing, the locally canned clam chowder whose brand name echoes a near-forgotten Patti Page song. (author's collection)

Others wanted to learn how to cook haddock and halibut, crab and lobster, and especially chowder, and local women learned to charge for cooking lessons, to organize dinners to raise money for charities, to publish cookbooks like *What We Cook on Cape Cod*, which featured an introduction by Joseph C. Lincoln himself.[36] Still other summer visitors wanted to learn the identities of lighthouses flashing in the dark or the names of salt-marsh wildflowers. Many locals learned to behave as the tourists expected, to emphasize the broad "a," to talk in the most taciturn manner possible, to spice one's answers with maritime vocabulary, to remember always to say "port" for "left" and "starboard" for "right," to predict the weather with much explanation, to mumble "holy mackerel" when startled, to behave as characters in an alongshore theater. And by the 1920s, as Lincoln glimpsed in 1929 in his novel *Blair's Attic*,

ANSON T. RICE,
Fine Shoes,
115 Broad St   Providence

the tourists wanted to carry away a piece of the shore, a piece of the experience, much as aquarium-keepers wanted to collect the best specimens and take them home.

Bringing home authentic coastal stuff at first meant carrying home shells, smooth stones, perhaps driftwood. In the 1920s, however, the growing nationwide interest in antique collecting had not only swept into the rural coastal regions but had gathered force as summer visitors determined to bring home maritime antiques. As early as 1907, in "The Antiquers," Lincoln had begun to fathom both the behavior of tourists desperate to buy all sorts of junk and the willingness of locals to sell it to them. Residents at first parted easily with all sorts of washed-up nautical wreckage, but by the close of the 1920s they knew that tourists would pay high prices for anything vaguely old, even for storm-worn oars, lobster buoys, and bronze hardware like porthole rims. Instead of auctioning estates in midwinter and whenever else death made necessary, heirs and auctioneers began to store "old stuff" for summertime auctions, and as Phoebe Atwood Taylor made clear in 1943 in her mystery *Going, Going, Gone*, auctioneers cleverly placed one antique item in every box of junk and sold each box as a lot.[37] Moreover, by the 1920s, perhaps earlier, locals had begun to manufacture antiques, carving a schooner nameboard or building a ship's wheel, sometimes exposing the new-made artifact to a winter's storms or sinking it for a few months in the harbor, where currents and worms made it look properly salty.

By the 1960s, antique dealers had to order items in bulk, especially glass balls from Portugal. The balls, used in times past to float the tops of great trawl nets, had once been a coveted find on a sandy beach, and by the 1920s they had become good items to offer for sale. As the Portuguese fishermen converted to cheaper, far less fragile foam floats, the balls became rarer and rarer, until some antique dealers hit upon importing them. Wooden lob-

Well into the 1920s, tourists now and then glimpsed a schooner or other commercial vessel working the quiet harbors, and knew they were watching "a piece of history." (author's collection)

Outward Bound. Scituate, Mass.

COPYRIGHTED
F. W. HILL, BOSTON, MASS.

SCRIBNER'S MONTHLY.

VOL. XVIII.　　　　SEPTEMBER, 1879.　　　　No. 5.

SANDY HOOK.

EVERY American knows Sandy Hook by name, and to the voyager from Europe it is the extended right arm, lamp in hand, which offers first greeting to the land of promise. Of itself, it is not particularly inviting. It consists of a long, low, sandy peninsula, of drift formation, the continuation of a sand-reef skirting the Jersey coast, which projects northward five miles into the lower bay of New York, and forms the eastern break-water of Sandy Hook Bay. In width it varies from fifty yards at the Neck, near Highlands Bridge, where jetties of brush-wood form but a frail protection against easterly storms, to a full mile at the point where the main light is located. Those who look upon it from excursion boats or incoming steamers see only a strip of white sand-beach and a thick growth of cedars, broken here and there by light-houses and low buildings; but closer inspection discloses many interesting details by which this outline is filled in.

The scenery of the Hook is not varied, but it is unique. Situated within twenty miles of America's metropolis, and threatened on every hand by advancing lines of hotels and summer boarding-houses, this isolated spot, owned and set apart by the government for certain special purposes, has resisted every attempted inroad of civilization, and in many places retains the same wildness that it had when its Indian possessors gazed upon their first pale-faced visitors.

A first glimpse of Sandy Hook, discerned dimly upon the horizon of American history, is found in the diary of Robert Juet, of Limehouse, the companion of Henry Hudson during his third

MAIN LIGHT, SANDY HOOK, AND KEEPERS' HEN-COOP.

VOL. XVIII.—46.

IMAGINARY CAPE COD

ster pots, especially the round-top ones featuring steam-bent wooden hoops, struck so many summer visitors as perfect coffee tables that some lobstermen discovered a profitable retail trade in slightly used pots, inverted, set on a boatyard-like cradle, and topped with a piece of plate glass, the last for display purposes only, since the glass shipped badly. Nautical antiques, especially ship's wheels, exploded in price, and sent many locals scouring industrial cities for things that looked nautical and might be sold as such, especially after a few months of immersion in harbor mud. But in the 1960s, the summer-long visitors who left for Ohio or Illinois or Colorado with a lobster pot strapped to their car roof or a 1930s lifeboat oar projecting from their station-wagon rear window began to covet something else, to covet it in an almost pirate-like way, to covet the certainty that the coastal landscape would be unchanged the following Memorial Day, when they topped the last hill and the dunes and harbor and village lay spread out before their eyes.

The desire to pickle the coastal landscape somehow, to salt it down or away in permanent stasis, had been building since the turn of the century when yachtsmen learned the joys of wandering about tiny harbor towns. Yachting magazines after 1900 record the passages of families bound from one quaint harbor town to another in a series of daylong, coast-hugging

By the 1920s, even parts
of wrecked ships had
become picturesque and
quaint. (author's
collection)

Tourists still bring inland
bits and pieces of the
beach, decorating their
garden sheds with
lobster buoys. (JRS)

Lobster pots, even
metal ones piled behind
fish houses, can be
quaint if the fish house
has weather-beaten
unpainted wooden
shingles. (JRS)

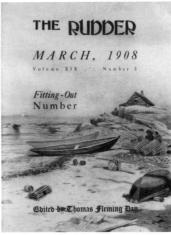

Part of the joy of yachting lay in the "romance" of anchoring night after night in one quaint harbor after another. (Harvard College Library)

sails beginning and ending in antique places. In 1908, for example, Winfield M. Thompson published a lengthy *Rudder* piece, illustrated with full-color plates, on Marblehead Harbor. "In its coloring and contour the stranger may find a resemblance to some of the ancient sea villages of Devonshire," he explains at the beginning, before cataloguing old houses, old streets, old tales, each suggesting "historical associations." Thompson details the illusion of antiquity that suffuses the reality of antiquity, and determines that Marblehead houses "might terminate in a ruined castle" were the air less clear. But the locals are specimens, too. "The speech of the people still savors of the nautical, notwithstanding changed social conditions," he writes of men and women who use oceangoing words with ease. "Even the youngsters express themselves in more or less salty phrases."[38] Other harbor villages appear in other articles, and sometimes authors like Edward H. Colton, who published "The Cruise of a Cape Cod Derelict" in 1910, focused on the quaintness of long-wrecked sailing ships washed ashore on sandy beaches.[39] Husband-and-wife teams often described their summer yachting vacations from two points of view, his being the care of the rigging and engine, hers being the quaintness of landfalls, the wonders of shops, the delights of restaurants tucked between ship chandleries and boatyards.[40] Finally, pen-and-ink illustrations of shallow-water harbors bordered by antique, sometimes ruinous structures, especially wharves, are everywhere in early twentieth-century yachting magazines, not so much as a record of the quiet, impoverished ports that welcomed yachtsmen, but because the yachtsmen savored the pleasant certainty that poverty kept the harbors unchanged year after year.[41] Gunkholing in yawls and catboats led yachtsmen and small boaters from one lost-in-time place to another, and boating magazines existed, in part at least, to guide novices into the hobby of collecting harbors, antiques, and old salts.

But the preservationist impulse had gathered force in the 1920s from others than boaters. Probably the rise in automobile touring after 1920 convinced many long-time tourists that the coastal realm had been spoiled somehow. Lots of shiny new cars eroded the image of antiquity, of stasis, of tradition. So, beginning in the 1920s, nonfiction writers began explaining how to drive into the true or "unspoiled" coastal realm, to avoid the crowds that ruined certain towns. In *The Old Coast Road from Boston to Plymouth*, a guide of 1920 to the region traversed by the easternmost north-south roads in Massachusetts, roads east of the easternmost north-south state highway, Agnes Edwards demonstrated that the landscapes of Joseph C. Lincoln existed in reality. The drawings by Louis H. Ruyl that accompany her text seem uncannily close to those that illustrate so many Lincoln novels, and they emphasize Edward's get-out-

140 — An Old Cape Cod Lane

DRIVE ALONG STATE ROAD. PROVINCETOWN. CAPE COD. MASS.

of-the-car point of view. "It is really a pity to see Scituate only from a motor," she counsels midway through the book. "There is real atmosphere to the place, which is worth breathing, but it takes more time to breathe in an atmosphere than merely to 'take the air.'" She advises the rambler to seek out the old lighthouse, to learn the story and route of Rebecca and Abigail Bates, to see the well, oaken bucket, and boyhood home of Samuel Woodworth. In time the wise walker discovers that Scituate is quaint. "Scituate has a quaintness, a casualness, the indescribable air of a land's-end spot," she asserts. "The fine houses in Scituate are refreshingly free from pretension; the winds that have twisted the trees into Rackham-like grotesques have blown away falsity and formality." Edwards insists on the depth of the past, and proves her points by enumerating details like the local boats afloat in the harbor and drawn up on the beaches. "It is the peculiar formation of the shore which has developed a small, clinker-built boat, and made the town famous for day fishing."[42] All one need do is walk about and breathe the historic atmosphere so evident away from motorcars, an atmosphere that might almost be bottled, canned, salted. A flood of books followed *The Old Coast Road*, all aimed at guiding the tourist not so much into the coastal realm but into antiquity.[43]

And by the late 1960s, town meetings in Scituate, Norwell, Cohassett, Hingham, and other towns had begun to shape historic preservation ordinances, establish historic districts, and discuss how best to preserve the historic authenticity of the local landscape. Since the early 1970s, especially after the energy crisis convinced many locals that well-to-do tourists were seeking out American places over European and deciding to spend three or four weeks in one place rather than drive from one motel to another, the towns that fronted the ordinary seacoast discovered the accidental advantage of having been poor and ordinary for generations.

By the late 1920s, state highways laced the dunes and salt marshes and motorists learned to look along the roads for quaintness and quiet. They expected to find "lanes," roads dead-ending at the ocean, everywhere in the coastal realm—and they expected to find them quaint, too. (author's collection)

So little coastal building had occurred that landscapes appeared only slightly changed from the 1950s, even the 1920s, and in some spots, say inland along the North River estuary, as the *Cruising Guide to the New England Coast* points out, from the 1870s. Moreover, the informal pressure after 1920 to build new structures in traditional forms had already guided much change along a gentle, almost indiscernible path. Elsewhere, resort areas had boomed, condominiums had choked off harborside marine industry, highway junctions had sprouted with shopping centers, and tourists had choked secondary roads. But along the ordinary coast, along coast neither spectacularly scenic nor spectacularly suited to yachting, beach-going, or partying, change had come slowly, so slowly that historical societies, garden clubs, preservation groups, and then town meetings determined to channel it, to make entire seafronts into aquariums of sorts, into salted set-pieces filled with local characters.

Light House, Scituate, Mass. built in 1811 - house of Rebecca + Abigail Bates _____ ___ __ __ war of 1812. They were my great great aunts. You have probably read about them in the school books. M. E. T. Sept. 07.

After the 1980s proved the worth of preservationist thinking, property owners began to backdate their structures. Aluminum and vinyl siding came off and white-cedar shingles went on. Plate-glass store windows vanished, replaced some early spring day with massive windows of tiny panes and thin mullions. Flat-roofed stores blossomed into gabled roofs. Such change rarely surprised summer visitors, and indeed tourists seem to have rarely noticed the grad-ual quaintifying of harborside structures. The few obviously modern buildings that seemed so modern, so future-oriented in the 1960s and 1970s suddenly seemed sleazy, almost grossly out of proportion and certainly built of out-of-context materials. When rebuilt in different massing and different surfaces, the buildings simply blended into a larger concatenation of space and structure that tourists sometimes say "seems so right," or "just what we expected." And they expect it because an entire industry has developed around it.

In Tennessee, changing planes in the Nashville airport, the historian stops short and stares at the American Airlines posters that invite Southerners to experience New England. The posters show wooden fishing boats and wooden small craft made fast along a ramshackle wooden wharf topped by unpainted, skewed, definitely old wooden shanties draped with nets and surrounded with wooden lobster traps. In the middle distance lies a harbor filled with

In November 1902, Rudder published this full-page photograph of the quaint—and dilapidated—alongshore built environment. (Harvard College Library)

A little wharf activity struck many 1900s artists as quaint, especially if the wharf stood nearly decrepit. (author's collection)

POST OFFICE, DENNISPORT, MASS.

What tourists found in many alongshore towns derived from economic stagnation, but postcards from the 1900s presented the stagnation as quietude in a friendly salt-air environment. (author's collection)

In the 1970s, National Park Service photographs enticed visitors to the Cape Cod National Seashore to examine quaint wrecks. (National Park Service)

In the Depression, canny locals began to build structures intended to look quaint, attracting tourists unable to afford European vacations. (author's collection)

110—Quaint Cape Cod, Mass.

COPYRIGHT BY E. D. WEST CO.

4A-H710

The picture postcard view of New England harbors always finds quaintness amid decrepitude, beauty in tottering wharves, peace next to boarded-up buildings. (Photograph by Gene Ahrens, author's collection)

NEW ENGLAND

sailboats and cabin cruisers, and beyond, at the harbor mouth, a lighthouse, sand dunes, and a vignette of salt marsh. No automobiles, no modern trucks, no neon-attired people interrupt the sunlit, warm, antique scene. The historian looks closer, squints, and realizes that in the posters the tide is high, masking the weed and mussels attached to the pilings. He knows that he looks at a picture of home, a picture intended to attract tourists, an image of a set-piece on display.

Local magazines, usually ones filled with advertisements for inns and rental condominiums, emphasize the salted coastal realm, teach far-off readers what to expect. The vacationing Tennesseeans walk into the airport, see the image, and hope that at the end of their long flight, at the end of their long rental-car drive along badly signed, badly paved roads, they will round a curve, crest a rise, and see what is in the poster, experience what they wish to experience. They have learned the lesson expressed in so many regional magazines, lessons like the one taught in the August 1990 *New England Monthly* cover story, "The Hidden Cape: There's Still a Cape Cod Unsullied by Crowds—But You Have to Know Where to Look." Here Barry Werth explains how the old cape faces a new century, a new millennium. "Virtually the entire sector is part of a historic preservation district, which means that everything erected north of the mid-Cape highway must conform to strict standards of taste and authenticity." And if the standards have subtle, unofficial results, so be it. "This has led, among other things, to a proliferation of small hand-carved signs, many of which, happily, are too miniscule to be read by passing motorists." Werth wanders further from the village, finding the great low-tide beaches almost empty of people, even at midsummer. "In the entire expanse I could see one other person—a man reading, naked."[44] At the end of the twentieth century, Werth finds the quiet, the diminutive structures, the antiquity, the empty beaches that figure in so many Lincoln novels.

Cape Cod–style houses, sandy roads, and wind-twisted trees all fitted into the larger coastal-realm image advanced by writers, photographers, advertisers, and postcard sellers. (author's collection)

WINDY WILLOWS, CAPE COD, MASS.

Updegraff and other early twentieth-century writers peopled the edge of the air with sirens as bold as those walking the edge of the sea over which airmen flew. (Harvard College Library)

Away from the crowded resorts, away from the traffic, the well-schooled vacationers find the "old coast," not the gold coast, and discover the uncrowded beaches, the dimly lit, packed-to-the-rafters antique shops, the salt-stained, wind-burned towns, the quaintness of everything.

And juxtaposed against the quaintness is the behavior at the marge, the marginal behavior of the new man and the new woman, the people sans restriction, sans inhibition, the people like Virginia of *Virginia of the Air Lanes*, the people sometimes sans clothes, the man reading, naked. Herbert Quick was no oddball, no isolato. Others glimpsed the emerging juxtaposition of quaintness and absolute modernity. In "The Siren of the Air," for example, Allan Updegraff told readers of the December 1912 issue of the *Century* an odd tale about women who somehow lurk in the sky, at the fringes of airplane range. In the clouds, at the edges of things, the young men in their flying machines glimpse a new woman, a woman like the sirens of old along the seashore.[45] Out of place, free of traditional restriction, the sirens sing songs of love and destruction, songs of something new, maybe feminism. In the clouds and in the surf frolic women anything but quaint.

11 BIKINIS

6436-10

In the summer of 1765, Charles Willson Peale accompanied local young men and women to Plumb Island, a sandy barrier island just off the Massachusetts coast, ostensibly to gather wild plums. Stocked with "stores of good things to eat and drink," the party spent little time plum-gathering. "It is diverting to see the young lads and lasses locked together and swiftly rolling down the high and steep sand hills," Peale recorded in a remarkable diary entry, "alternately being almost buried in sand and often showing a fine leg to their laughing companions." Only when he left the group, stripped, and plunged into the surf, however, did the young visitor discover how forward coastal girls could be. "Where he had been only a short time before some females of their company came down, and whether it was out of mischievous fun that they came, to prevent him from returning to put on his clothes, or whether the sight of a naked man was frequent to people inhabiting the seacoast, and therefore not so liable to produce a blush, the writer does not know," he continued, shifting his prose from first to third person, perhaps in memory of his embarrassment. After reassuring himself that the coastal women "have as much modesty as in other places of America," he notes that he "was too modest to expose himself naked before those ladies." Forced to choose between confronting the women and the dangers of the surf, he chose the less perilous option, "swam beyond the breakers," and waited, growing more chilled by the minute, until the women departed.[1] On the marge of mid-eighteenth-century Massachusetts, Peale discovered marginal women, women willing to lock bodies with men and roll down immense sand dunes, women perfectly able to discommode a naked man. On the summer marge, Peale encountered marginal behavior, and escaped only by swimming beyond the surf, away from wildness to wilderness.

Ocean swimming fascinates many onlookers, and remains a mystery to many Americans,

Hollywood popularized the bikini as the badge of belonging to a new generation, but its films also emphasized slim, slightly muscled young men with tans darker than those of any pirate. (Museum of Modern Art, Film Stills Archive)

Copyright 1905 by the Rotograph Co
A 7307   The Beach, Brant Rock, Mass.

perhaps to a growing number of Americans, if seacoast behavior offers any clue at all. Along this stretch of ordinary seacoast, locals still swim, often with the splashy, thrashing, seemingly inefficient stroke peculiar to coastal Massachusetts natives. The stroke, a kinetic mix of American crawl, breaststroke, and even dogpaddle, evolves first from a desperate desire to keep warm while swimming, to move every muscle to improve bloodstream circulation. And, too, it derives from lack of practice, for the ocean is rarely warm enough for prolonged swimming, or even for swimming much outside the months of July and August. Locals "splash" their way along, sometimes for pleasure, sometimes to reach a moored catboat or inspect a rotting wharf, sometimes—so say the tourists—to scourge their bodies in some grotesque post-Puritan sin-cleansing ritual. Tossed by waves, sometimes blinded by roiling sand, slapped by seaweed, and always suspended in a blackness that might hide anything from loose fishing nets and mooring lines to sharks or jellyfish, the locals enjoy—or endure—an experience utterly alien to those of tourists raised with still, clear, warm swimming-pool water, the water that reminds coastal people of bathtub water, the water that makes swimming simple.

On the beach the barefoot historian ponders the mystery of ocean swimming, watching the locals splash about, studying the tourists as they wade inch by inch into what they call frigid water, and above all wondering about the many people who do not venture into the water at all. Do the not-swimming-people not know how to swim? How does one find out, short of asking? And will nonswimmers answer truthfully? In a nation in which socially accomplished, socially astute people swim, what of the nonswimmer? Is it a faux pas not to know how to swim, and an even worse faux pas not to know how to swim when others are frolicking in the surf fifty feet away? Do the men and women who know nothing about swimming mutter about the possibilities of skin cancer and hospital-waste-bred disease to mask their inability to move through a medium others enjoy? Or do nonswimmers fear the sea so much that anticipating even a brief small-boat passage along a shallow salt creek overwhelms them with worry, of fantasies of shipwrecks worse than the Thachers' or the modern calamity of *Tyehee?* What of the men who explain, always so quickly, that cramps or old sports sprains keep them from swimming? What of the women who explain that they have on dry-clean-only swimsuits or who wear bikinis so unstable that walking, let alone swimming, is chancy? Why are so few people "at the beach" actually enjoying the water? Might they just as well enjoy a sandpit?

Of course, they enjoy the air. Breezes cooled by the sea, and ever so lightly salted, invigorate the visitor from hotter inland places, even rural inland places washed by fresh air. Many beachgoers luxuriate in salt air, running along the damp sand or doing chest-heaving deep-breathing exercises or just standing sniffing to windward, perhaps exclaiming that the "air smells so good." Salt air is different from fresh. Its moisture content is usually higher, and its salt content is measurable on windy days, as owners of automobiles discover when it settles in crevices and corrodes paint, underlayment, and metal. It smells of salt marshes and fish, mudflats and seaweed, dune grass and unknown sources, the last intriguing visitors who ask, "What is that smell, there, that one?" And according to tradition, salt air makes the sick healthy and keeps the well fit. Everyone ought to swim in salt air.

When John H. Packard published his *Sea-Air and Sea-Bathing* in 1880, he drew together all that his fellow physicians and surgeons knew about the beneficial effects of physically embracing salt air—and salt water. For decades, American physicians had urged the sick, the melancholy, and—after about 1850 or so—the urban dweller to visit the seashore for reasons of physical health, not merely for a change of scenery or activity. Given an onshore breeze, Packard explained, the visitor would scent the salt air a few miles inland and begin to grow more healthy. "Upon most persons the effect of breathing this air is tonic and invigorating, producing an immediate sense of exhilaration, improving the appetite, and promoting digestion," Packard insisted. The air at the coast is "a pure air, containing oxygen in the form of ozone, besides finely divided sea-salts, and water which is rendered stimulating by the presence of the same salts." While he admitted that "this whole subject of ozone is as yet very obscure," he insisted that salt air works more quickly on the ill and the tired than mountain air, which must be breathed for months in order to work what salt air accomplishes in a week or two.

Packard cautioned his readers to discriminate between the beneficial effects of salt air and the simple pleasantness of leaving a hot inland place for a merely cooler locale a few miles away—say a lakefront. Certainly any coolness relaxes, and any breeze stimulates the skin and so improves blood circulation. At the seacoast, however, salt air relaxes and stimulates even better, producing a "continual but not unpleasant sense of drowsiness" that leads to "profound" sleep every night, along with better digestion.[2] If an invalid or otherwise healthy but tired individual spent only a few weeks at the coast breathing salt air and never went into the sea, that person would still begin to improve—or at least slow any decline.

Packard and other physicians worked in a darkening time of tuberculosis and other seemingly city-spawned diseases, but they took courage from emerging patterns of medical research and inquiries into allied fields—such as the efficacy of ozone on human health. After the Civil War, more and more well-to-do urban Americans began to summer at the seacoast resorts Packard catalogued from a medical point of view. By century's end, middle-income people, especially executives, were struggling to spend a week or two at the coast or to send their wives and children away from the city to summer at the shore in places less expensive than Newport and Old Point Comfort. The middle-income families, the ordinary summer visitors Joseph C. Lincoln described in his Cape Cod novels, arrived seeking relaxation and good health, essentially free with room and board. Blue-collar urban families struggled to spend Sundays at urban beaches, or a weekend at Atlantic City perhaps, but they struggled for the same reasons.[3] Even one day of salt air seemed to improve one's chances of avoiding some dread respiratory disease. Who knew what a whiff of ozone might cure, or prevent?

Bathing in the sea worked wonders far beyond any wrought by salt air alone, however, and Packard insisted that the wonders amply repaid any attendant hassles. Mere "bathing in salt-water stimulates the skin, and renders the circulation more active." By "bathing," Packard meant a sort of haphazard playing in the surf and in the shallow water that laps the damp sand, exactly the activities enjoyed by children of up-to-date parents who themselves were collecting seaweeds or trying surf-fishing. Allowed to get wet, to play in the waves all morning, to experiment with a rowboat or a tiny sailboat, to wander along with collecting net and pail in search of crabs and other littoral creatures, children absolutely blossomed. Certainly continuous exposure to sunlight filtered through salt air worked much of the change. "The effect of exposure to sunlight upon animal life is directly invigorating," Packard insisted, basing his conclusions on "often-observed facts." Sunlight striking skin wetted with salt water works even more powerfully, however, and Packard offered the simplest of proofs. "Compare the pale, bleached, puny children, just out of the nursery and the school-room, who are taken down to the seashore in the early summer, with the same children returning to town in September,— tanned, ruddy, and hearty." Sun and exercise in the media of salt air and salt water produce children who are special, particularly robust and particularly invigorated. Children who summer on the seashore enjoy something of the environment that makes coastal children, indeed all coastal people, particularly healthy. "The robust health and long life so common among residents on the shore, afford proof of the salutary effect of the pure and stimulating air upon them," he insists in a passage emphasizing that coastal realm people enjoy perhaps the best health of all Americans.[4] Coastal realm folk are somehow so well salted that even the common cold troubles them less frequently and less harshly than it does their inland compatriots.

Sea-Air and Sea-Bathing is itself a liminal book. Packard wrote at an awkward moment, one of great actual and potential change. Medical science had begun to dismiss much folk advice as old-fashioned, even ridiculous, but physicians had also realized the worth of some traditional information. Urbanization spawned a wealth of new health issues, and the systematic collection of health statistics not only had confirmed issues of longevity in the coastal realm Thoreau had remarked in *Cape Cod* two decades earlier but had raised, albeit quietly, unsettling questions about spatial and social patterns of illness.[5] And perhaps most important, Packard wrote in a time of subtle experimentation in matters of health maintenance, child-rearing, and social behavior, but experimentation conducted chiefly by people extremely secure in their social positions.[6] *Sea-Air and Sea-Bathing* addresses middle-income, well-educated readers, advising them to grasp what the upper-class had had since 1860, perhaps earlier. Packard argues industriously to convince readers that tanned skin is healthy, that children need to play along the beach, that vigorous activity is neither unladylike nor unbefitting a businessman of worth, that middle-class social inhibitions inhibit good health. Nowhere does Packard work more diligently to overcome middle-class inhibitions than in his lengthy insistence that the health benefits of ocean swimming far outweigh any possible social improprieties. In 1880, Packard argued against a rising chorus of voices aimed at tightening social restrictions, even at the marge.

Sea-Air and Sea-Bathing becomes a navigational guide almost as cryptic as Poe's pirate cipher; it needs at least a second reading, if not a third. It grapples partly with class division and partly with geographic division, but it grapples subtly indeed. Packard certainly knew that the rich had discovered the nineteenth-century health secrets coastal people simply accepted, but he knew, too, that the rich and the locals feel free to behave as they choose. So when he addresses the subject of bathing suits, he does so circumspectly, arguing that "as a matter of course, the practice of general bathing at a certain hour of the day necessitates the wearing of bathing-suits," but implying that some swimmers swim another course.[7] Rather than weigh themselves down with wool or confine themselves with long sleeves and other drapery, they swim efficiently and safely. They swim in the traditional manner, in the nude.

Eighteenth-century swimmers swam as Charles Willson Peale swam, without clothes, but only rarely do diaries and other sources record the lack of attire swimmers simply accepted. "Living near the water, I was much in it and on it," notes Benjamin Franklin in his *Autobiography*. "I learned early to swim well and to manage boats." Growing up near a salt marsh and a harbor undoubtedly contributed to his proficiency, one that later made him a good swimming teacher and, while living in England, able to swim three miles on request, perform underwater stunts, and actually consider opening a swimming school. Of course, Franklin admitted that he had read M. Thevenot's *Art of Swimming*, a French manual translated into English in 1699, and that he had practiced "all the motions and positions," but he knew the first step in swimming. He "stripped" before entering the water.[8] Franklin also realized the "novelty" of his swimming ability, knowing that in England and in the colonies, only people raised alongshore knew how to swim well, especially over distances. Indeed, most drownings not caused by shipwreck or capsized small boats along the coast of Cape Cod Bay from the mid-seventeenth century until the late eighteenth century appear to have resulted from visiting inland farmers deciding to bathe or simply cool off by jumping into a salt creek or even into the surf—and sinking.[9] Mad thrashing coupled with waterlogged clothing drew them under, exactly as Peale, Franklin, and other expert eighteenth-century swimmers understood.

Until the 1930s, perhaps into the 1960s, coastal people frequently swam nude, and only in the 1920s did they begin to remark the end of natural swimming. Certainly salt-marsh farmers swam nude when cooling off at noontime, and in *A Marsh Island*, Jewett approached their behavior closely enough to signal her knowledge to knowing readers. But as Joseph E. C. Farnham remembered in 1915, boys swam nude in places less isolated than salt-marsh creeks and gutters. "There was a sartorial abandonment, and a free exercise of personal dress—most frequently necessarily due to parental financial circumstances—which gave us a boyhood as unique as it was hardy," he recalled of his boyhood in a local-minded place. Always barefoot in summer, often with legs "attired in 'birthday equipment,'" boys constantly went

swimming, never bathing, which Farnham dismissed as something peculiar to newfangled resorts, something that demanded suits. Neither parents nor other adults imposed restrictions, including clothing, and older boys apparently taught younger ones to swim without clothes. Boys graduated from swimming in shallow salt creeks and learning to dive into the deeper holes called "pots" to swimming in the surf off barrier beaches and finally to diving from harbor wharves.[10] Other early twentieth-century reminiscences second Farnham's report, often adding that swimmers hid their clothes under driftwood to forestall pranksters from tying sleeves in knots and wetting the knotted clothes to make untying them even more difficult. In a *Century* piece of 1921 entitled "A Boyhood Alongshore," L. Frank Tooker describes the knot-tying and implies that the prank forced naked and nearly naked boys to stand unclothed on the beach, in full view of adult passersby.[11] What the adults thought of naked boys, Tooker and other writers do not mention, perhaps because local adults accepted alongshore nakedness, at least until summer people popularized ocean bathing—not swimming—in suits.

Early nineteenth-century advocates of swimming accepted nudity without question and labored to overcome the reluctance of inland people to immerse themselves in water, even in

Locals traditionally swam in the nude, and well into the twentieth century, Vanity Fair and other periodicals accepted such behavior, as this image by Clarence White of February 1915 makes clear. (Harvard College Library)

a tub. As late as 1826, a writer in *New England Farmer* argued painstakingly that weekly or daily bathtub bathing promotes not only cleanliness but good health, and the combative tone of his article makes clear the stubbornness confronted by advocates of bathing in the sea.[12] Moreover, as the anonymous author of a *New England Magazine* article of 1832, curtly entitled "Swimming," recognized, many nonswimmers approached swimming as something as difficult as learning to play a musical instrument, something that takes years to accomplish. Yet the author of "Swimming" concurred with their view, concluding that swimming "is as much an art as reading or writing, or playing upon a musical instrument," but that facility results in "the pleasure one enjoys in breasting the waves and feeling at home in the water." Only rigorous instruction, much like that of the newly imported European gymnastic system, can make a man a competent swimmer, although he might make quicker progress learning to swim on his back.[13] "Swimming" purports to be about swimming, but its concluding remarks on the need to establish swimming schools everywhere in the Republic imply that its ulterior motive is simply to establish the need for formal instruction in an art very difficult to learn. In enclosed pools, single-sex nudity could be managed, and the class could pay the swimming master for effective lessons in the art.

Female nudity, as this carefully posed Vanity Fair image of May 1915 by Lillian Baynes Griffin suggests, posed difficult problems for periodicals willing to accept naked male swimmers. (Harvard College Library)

The manufacture of ocean swimming into a gymnastic-like activity for the middle-class required some sixty years to accomplish, and in time skewed all understanding of what became known as "bathing." Only two groups escaped the long reshaping of what Farnham remembered as simple swimming or "going-in-a-swimming" into something that required hotels, vacations, lessons, lifeguards, and expensive suits. The knowing rich embraced traditional swimming as early as the 1860s—perhaps, as Packard implies, ten or twenty years earlier—and continued to practice it until World War II at least. And the alongshore common people persisted in traditional swimming until about the same time, when the flood of urban middle-class summer visitors finally clamped down. To understand ocean swimming is to understand a split subject, one part scarcely mentioned in history books and "genteel society," the other quite well documented in period publications.

Alongshore and upper-class girls and women certainly swam nude, at least when by themselves. Oral tradition makes clear the innuendo of Packard and other late nineteenth-century bathing advocates who attempted to explain the usefulness of bathing suits. Women taught their daughters to swim, and their daughters learned in the nude. After gaining proficiency, the girls—and their mothers—might thereafter swim wearing a cotton shift but might occasionally swim nude in secluded salt creeks and coves. Whereas boys and men swam often, girls and women did not, but everywhere alongshore, local women knew that sooner or later girls would be afloat in boats, and that boats might founder. Girls needed to know how to swim, and they needed to know that once in the water, they would drown in their skirts and petticoats. Cartoons involving nude female swimmers accidentally discovered were frequent in *Life* and other turn-of-the-century magazines, offering some slight substantiation of the upper-class and alongshore family tradition of girls swimming in the buff, girls and mothers swimming nude, and sometimes brothers and sisters swimming naked.[14] Isolation permitted such behavior, of course, as Jane Hollister Wheelwright recalls of her girlhood in the late 1930s and early 1940s. "We would head straight for the waves, naked as jaybirds. I do not know how long it took me to learn to swim. Certainly no one taught me," she writes of swimming with her brothers and sisters. "Teeth chattering, we would race up to the hot, dry sand and burrow down into its delicious warmth. With no suits to get scratchy with sand, we could wriggle and dig our way down until finally we were warm." Because her parents condoned nudity, she and her siblings thought nothing of it, even when she waved at passing trains. "The passengers, waving back at us, seemed so friendly; but years later it occurred to me that they were fascinated by our nakedness. This was during World War II, in the early 1940s, and even children did not play on the beach without something on."[15] Of course, by the 1940s the very rich had learned to fear the censure already directed at the locals.

As late as the 1870s, a few diehard advocates of swimming persisted in educating middle-class people in traditional swimming, including the tradition of swimming in something practical. In *The Art of Swimming*, for example, Charles Weightman gingerly answered the vexing question "Why should not ladies learn to swim?" by focusing on "the real point of the question, which concerns the men more than the ladies." By the point, Weightman meant "the real or presumed difficulty of finding proper bathing places away from the observations of the other sex." But more pulses beneath his argument than the problems of finding secluded places in which women might learn to swim while wearing suits "of the latest French importation." Gathering force against women swimming is a new force, one opposed to "promiscuous bathing." Weightman understood the force and tried to counter it by insisting that "fathers and

brothers can, without any indelicacy, bathe with their families," and asking "Why should the ladies of this country be behind their continental cousins?" If "the Princess Royal of England and Prussia is a very accomplished swimmer," why should not American women swim as well?[16] But appealing to common sense and the example of royalty did little to stem the rising opposition to women swimming.

Within fifteen years or so, advocates of swimming had learned to call the sport "surf-bathing" and to focus their attention almost wholly on men. An article of July 1890 in *Scribner's Monthly* exemplifies the turn-of-the-century attitude, as well as the anger with which men like Duffield Osborne confronted the women-shall-not-swim crowd. "Now, it is more than possible that, being a good swimmer, and having first made personal trial of both beach and surf, you may desire to offer your escort to—well, to your sister; and right here let me note a few preliminary cautions," Osborne remarks near the close of his article. After warning against problems ranging from the woman stepping into an unexpectedly deep pool of water to the crisis of undertow, Osborne focuses on real trouble: "Never take a woman outside the life-lines, and never promise her, either expressly or by implication, that you will not let her hair get wet. Above all, impress it upon her that she must do exactly as you say." Beyond that, he dismisses the subject of teaching women, saying that "no skill is called for and no suggestions needed," unless the surf is high. If the surf threatens to knock down the woman, the stalwart male should hold her as though about to waltz across a dance floor, and then—

somehow, for Osborne does not say—continue the lesson.[17] Nothing could be more ridiculous, and less useful, than Osborne's tentative, reluctant advice about teaching a woman to swim, but clearly the foolishness results from his inability to write explicitly about how a man must hold a woman—or another man—when instructing. By 1890, "genteel" women did not accept such embraces, out of or in the ocean.

Self-appointed monitors of womanly virtue precipitated an alongshore revolution in manners between 1860 and 1900, a revolution that altered middle-class behavior, then filtered down to coerce most working-class people, too, a revolution that still haunts most Americans. Felicia Holt explained it all in a *Ladies' Home Journal* article of 1890 entitled "Promiscuous Bathing," a title that by then meant something instantly recognizable to a generation of middle-class readers anxious to behave in ways that announced their respectability. Unsure of their position on the rapidly shifting sands of late nineteenth-century society, Holt's readers took to heart her warnings about being in the same ocean at the same time as men. Holt inveighed against the brevity of swimming attire, against the man "looking like a harlequin in his red or white jersey and short blue trousers, legs and arms perfectly bare; the girl in her *costume de bain*, of fine, white serge, if she be very luxurious, made tight, showing every curve," and she positively railed at the latest innovation, open-toed swimming stockings for women. "I fear the girl will soon begin to calculate the effect of what someone late called 'artistic bareness' on the mind of masculinity, and the man to be too conscious of the value of muscle and calf which he exposes." Bare shoulders in the ballroom, Holt continued, offer no precedent for bare toes on the sand, and for that matter, bare shoulders are wicked, too. Men and women, apparently even men and women married to each other, should not parade themselves so, and certainly should not share the same sand, let alone the same sea. Who will protect the shop girl, newly arrived at the shore, or the parentless girl visiting the coast for the first time?[18] Holt answers emphatically and immediately. She will.

She did. Between about 1870 and 1900, middle- and working-class women learned to avoid promiscuous swimming in order to preserve their reputations. All sorts of corollary activities from promenading along boardwalks to gulping salt air to carefully collecting rockweed struck the monitors as acceptable, but even bathing at strictly monitored resorts—say those in which men cleared off the beach after ten each morning so that women could splash about without pollution—were suspect. Draped in woolen bathing dresses, complete with shoes and bonnets, middle-class women enjoyed or endured surf-bathing with scant thought of actually swimming someplace, of going beyond the lifelines rigged by prudent hotel keepers. In this period evolved such safety devices as the swimming corset, a corset worn over the swimming dress and securely lashed to a light manila line made fast to a small windlass ashore, which offered the promise of a quick winching-in should the breakers knock one down. Women, Holt and others maintained, did not belong in the water.

What explains Holt's energy and almost biting anger? One answer may lie in her awareness that behavior and attire peculiar to the beach could influence inland behavior and attire. The neat and tidy world of late nineteenth-century middle-class women might or might not stand rigorous questioning by younger women, and so younger women ought never to enter a place that might set them to questioning larger realms. Holt knew that an earlier generation of women did swim—she begins with a cryptic reference that "perhaps in 'ye olden times' or even thirty years ago, it was less objectionable than it is now, due partially to the fact that Puritan and Quaker simplicity tinctured both attire and manners in those days"—but she clearly no longer connected such behavior with 1890s women.[19] The beach threatened to wreck middle-class morality not by offering a glimpse of future mores, not even by providing a glimpse of different mores, but by offering a glimpse backward, a glimpse of time that in one respect at least might have been better, freer, for women. Holt knew, or surmised, that in "ye olden times" some women had it better, and that knowledge twisted her entire line of argument. What might happen if women learned to swim?

Kate Chopin answered the question in *The Awakening*, which even now borders on the pornographic. Edna Pontellier, vacationing at a Gulf Coast resort and enjoying staring at the sea-glim, breathing salt air, and relaxing among the local Creoles, slowly but surely learns to swim. Her Creole neighbors know the sea. "Most of them walked into the water as though into a native element." But for weeks, as she struggles to learn to swim while wearing her heavy woolen suit, Edna makes no progress despite her lessons. "A certain ungovernable dread hung about her when in the water, unless there was a hand nearby that might reach out and reassure her." Finally, in a mad spurt of discovery and mastery, she learns to swim, not simply splash about. "She could have shouted for joy. She did shout for joy, as with a sweeping stroke or two she lifted her body to the surface of the water." Suddenly independent of the male hand, she realizes that she can swim where she chooses. "A feeling of exultation overtook her, as if some power of significant importance had been given her to control the working of her body and her soul. She grew daring and reckless, overestimating her strength. She wanted to swim far out, where no woman had swum before." Edna's swimming becomes a metaphor for social experimentation, experimentation that ends in disgrace and disaster, in suicide. At the close of the novel, she puts on her bathing suit and walks down to the waves. "But when she was there beside the sea, absolutely alone, she cast the unpleasant, pricking garments from her, and for the first time in her life she stood naked in the open air, at the mercy of the sun, the breeze that beat upon her, and the waves that invited her. How strange and awful it seemed to stand naked under the sky!"[20] In a few moments she strikes out into the glim, abandoning husband, children, friends, and finally life. No story better illustrates Felicia Holt's fears about women learning to swim.

In spite of a hundred arguments that women would enjoy coastal realm recreations, particularly small boating, far more thoroughly if they knew they could save themselves if currents or waves swept them away, many women simply did not learn to swim. Not knowing how to swim maintained what many women wished to maintain, an image of themselves as utterly deserving the protection of men and as utterly proper. The old abandon-ship cry, "Women and children first," might change if women could swim well enough to save themselves—and children, too. Between 1880 and 1930, many female visitors to the seashore arrived ignorant of swimming and left equally ignorant, well aware that however they looked in their bathing suits, the very wearing of the suits was fraudulent indeed. And at least some women who learned to swim in the half-century of enforced propriety did so alone, out of public view. "With more misgivings than I would have admitted, I ducked into water six or seven feet in depth, and started swimming out even deeper," remembered Jane Douglas in a magazine article of 1934 entitled "Learn to Swim—Alone." "That was my emancipation."[21] But emancipation came only after years of solitary, slow-paced effort, filled with experiments like swimming with her eyes closed. Emancipation did come, however, and Douglas understood what other female swimmers knew, that they no longer needed a male hand nearby. That strong male hand reaching out to nonswimming women surfaces in hundreds of stories, often written by men, which emphasize the duty of men to frail, incompetent, drowning women. Stories like Richard G. Swaringen's story, "Fear," which appeared in *Comfort* in 1921, preyed on the fears of non-swimming women and weak-swimming women and bolstered the attitude of men that they should be women's protectors and saviors.[22] And so long as women appeared at the edge of the ocean attired in the voluminous bathing dresses Felicia Holt and other monitors of the status quo dictated, they could not swim well enough to discover any glimmer of emancipation.

Not until after the turn of the century did national magazines begin to address the long-standing but rarely remarked persistence of traditional swimming all alongshore. *Collier's* led the way. Beginning in 1901, Margaret E. Sangster used her column, "A Woman's Viewpoint," to insist that American women begin doing more on their own, not simply exercising and eating better but uniting in "girl colonies" and other associations to advance small-group goals rather than a broad political agenda. A champion of "open-air life for women," Sangster well understood the immediate advantages of swimming. "The woman who can float at pleasure and swim for a moderately long distance will not fear a sudden tip overboard if her yacht prove treacherous. And every woman should swim well enough to be at home in the surf." But swimming, she pointedly remarks, enables women not just to enjoy boating but to go boating alone or with other women. "A woman never looks so thoroughbred and so independent as on a boat, unless perhaps when mounted on the back of a good horse, a trotter or a pacer," she concludes in an argument influenced by the debate then raging over women's bicycling attire.[23] Swimming offers immediate independence from men, and it leads to greater and greater independence from men all alongshore. By donning a sensible suit and learning to swim, the suffragette might gain an immediate sense of emancipation, a salty taste of freedom.

Some women certainly listened to Sangster or followed the far older tradition of sensible swimming, but contemporary evidence is fitful. The bathing-suit articles featured every spring in women's magazines offer at best only a skewed introduction to turn-of-the-century women's swimming, for the articles almost invariably show fashionable suits, almost always remarkably decorous, and the suits depicted in general-interest magazines are rarely different. But other sources demonstrate beyond doubt that some turn-of-the-century women had abandoned skirts and other drapery for suits that permitted vigorous activity. In 1911, for example, the Boston *Sunday Post* featured a long article describing the swimming exploits of two twenty-one-year-old women from Brookline, each about five foot seven, weighing about 135 pounds, and able to lift 200 pounds. The two upper-class women, swimmers since the age of fourteen, are depicted in three photographs wearing skintight sleeveless body stockings, one ankle-length, the other thigh-length, the thigh-length suit lacing from armpit to midthigh, leaving a four-inch-wide band of flesh crisscrossed with lacings. Suits like these figure in no fashion magazine plates, not even in men's magazines cartoons. Were they risqué, or risqué only when worn by middle- or lower-class women? "To be a good swimmer and diver a girl must have courage," remarks one.[24] But clearly a solid social position, one so solid that a skintight, revealing suit is only proper, helped as much as the courage to wear the suit and swim well.

Women thrashed out the contours of the swimming and swimwear arguments in the half-century after 1870 in the context of men who agreed with Felicia Holt. In "The Beach at Rockaway," for example, William Dean Howells found himself disconcerted at the "promiscuous swimming," finally determining it to be "queer," and talking with another man about their mutual distrust and dislike of it. Much more suffuses the novelist's essay than description and argument, however, for between the lines surges Howells's embarrassment. "It was indeed like one of those uncomfortable dreams where you are not dressed sufficiently for company, or perhaps at all, and yet are making a very public appearance. This promiscuous bathing was not much in excess of the convention that governs the sea-bathing of the politest people; it could not be; and it was marked by no grave misconduct." Howells found little out of order at the working-class beach. "Here and there a gentleman was teaching a lady to swim, with his arms around her; here and there a wild nereid was splashing another." Elsewhere a young man "pursued a flight of naiads with a section of dead eel in his hand." [25] Nothing else specific seemed disorderly, but Howells did not like what he saw. After all, it made him feel naked, somehow. Perhaps a key to his unease lies in his use of words like *nereid* and *naiad* to distinguish the active women. Somehow the active women are different, almost nonhuman. Moreover, the women are working-class women doing something similar to what "the politest people" do—or what boys and men do.

Although middle-class American women and men debated the appropriateness of women's swimming and the correctness of women's swimming attire for decades, during the post–Civil War years there was a shift toward the nationwide conclusion that boys ought to learn to swim, at least well enough to keep from drowning in brooks, lakes, and estuaries. After 1900, as more and more educators realized that millions of boys lived in cities far removed from the swimming hole James Whitcomb Riley extolled in a nation-sweeping poem of 1882, some school systems experimented with swimming-pool—or "plunge-bath"—instruction much like what the *New England Magazine* essayist advocated back in 1832. But many communities found the cost of indoor pools prohibitive, and so they embarked on the "dry land" method of teaching, by suspending boys from swings and demonstrating proper strokes. [26] Eventually some YMCA and YWCA groups erected pools for year-round use, but many city and even large-town boys, let alone girls, had no place to learn, let alone learn well. And when they did learn, they learned in a complicated, very untraditional way, a way of complex strokes and breathing patterns unlike the splash-and-hurry old-fashioned way alongshore children taught younger children. Moreover, buried in the turn-of-the-century swimming-instruction literature lies a hint of the class division that now and then surfaces in articles about seashore behavior. The two girl divers described in the Boston *Sunday Post* article were from Brookline, the wealthiest suburb in Massachusetts, where they had the benefit of a mu-

nicipal gymnasium and a municipal "natatorium." Brookline emphasized physical education for boys and girls both, and by 1916 taught swimming not only to schoolchildren but to their teachers. As the director of the program explained in 1916, the town held separate classes for boys and girls, and never allowed schoolchildren to enter classes with teachers. "It is not good judgment to admit school pupils to this class, and many teachers will not join it if pupils are allowed in the pool at the same time." [27] Apparently, Samuel K. Mason had figured out that girls learned to swim better apart from boys and that teachers fretted about their dignity when children discovered them to be ignorant of a proper upper-class accomplishment. In such an upper-class town, girls became expert swimmers and divers and developed into women like those described in the *Sunday Post*. When they went to the seashore they swam and dove as wonderfully, perhaps even more expertly, than the locals. And they certainly far surpassed the women and men who knew only how to bathe in the surf. But they could enjoy their sports only when correctly attired, which meant that they could enjoy them only at beaches not held hostage by lower- and middle-class standards of bathing dress. The rich may not have wished to associate with the ordinary people at most beach resorts, but ordinary people didn't want

Established alongshore families swam in very little even in the 1880s, and even in mixed company. (Society for the Preservation of New England Antiquities)

the rich around either. No one wanted the embarrassment, the abasement of expert swimmers flaunting their expertise—and brief suits. It was no simple boy-girl double standard that split American society, therefore, but something far more complex, something hidden to this day, something concealed in the salt marshes, the lonely beaches, the salt creeks, something hinted at by a faded bikini tossed on a splintery landing staging.

All alongshore, local people, especially children, evaded the restrictions enforced by newly arrived "genteel" summer visitors, the people devoted to surf bathing and bathing suits, the people who by 1900 sometimes forbade girls to learn to swim. And where summer people arrived late, say well into the twentieth century, locals retained both their conviction that girls should learn to swim and their ease with what Farnham termed "sartorial abandonment." Until well into the 1960s, alongshore people retained an easy familiarity with the human body, with bodily functions, and with exposed skin. Even now, the barefoot historian realizes that locals share a mind-set still evidenced by farm families, though rarely made public. Just as many young adults from isolated farms feel uneasy about describing playing basketball in jockey shorts or bras and panties, let alone describing parents and children skinny-dipping in the back-pasture swimming hole, coastal realm people only rarely discuss some matters with inlanders.[28] Perhaps fortunately, few inlanders look carefully enough at the alongshore scene to notice the details that give rise to century-old questions.

Consider women's swimsuits. Along this stretch of seacoast, nonlocal women favor one-piece suits. Right now the suits are "in," as any fashion magazine explains, and most women, and most fashion designers, state that one-piece suits hide or display curves better than bikinis and are generally "more attractive" or "sexier" than bikinis. A general survey of July beach attire reveals a mix of one-piece suits and bikinis, and a more precise survey reveals that no age pattern explains the wearing of bikinis—pre-adolescent girls wear bikinis, and women in their sixties wear them, too. On beaches unknown to tourists and frequented almost entirely by locals, a far higher proportion of women wear bikinis. And on the small boats that prowl the inlets and salt creeks, and even move beyond the breakers, women wear bikinis under their shirts and sweatshirts. Occasionally a tourist, usually a woman, notices the pattern and dismisses it as evidence of the backwardness of alongshore women too weak-minded to read *Vogue* and *Cosmopolitan.* But the barefoot historian knows better, as do all the locals, male and female. A bikini not only enables a woman to remove sand caught between fabric and skin more easily but stays cooler and dries faster than a one-piece. And it has other advantages.

A bikini makes bathroom operations easy, especially in a small boat, but anywhere else alongshore, too. For men and women unversed in the necessary etiquette, a small-boat passage sometimes arouses fears not of capsize but of navigating far beyond public restrooms, even beyond scrub juniper and pine. People traditionally relieved themselves over the rail of small boats, an easy enough manuever—in calm water—for men and women alike. By the early 1980s, however, such behavior struck environmentalists as small-scale polluting of the worst sort, and uninitiated onlookers as unspeakably rude. They "take umbrage at the bayman's necessary habit of relieving himself (and herself, today) directly over the rail of the usual small skiff," mused one observer in 1979 of the growing rift between summer visitors with urban notions of propriety and locals living according to tradition.[29] As environmentalists and prudes collected their forces, small boaters learned to be more circumspect. "So I rediscovered the age-old process of drawing a pail of seawater, which sat safely amidships while

This Life *photograph explained to readers of the March 25, 1946, issue something of the alongshore setting of the Bikini Island nuclear tests. (Harvard College Library)*

TINY PACIFIC ISLAND OF BIKINI
CALMLY AWAITS THE ATOM BLAST

It Coaxes a New Skin

— CALIFORNIA PRODUCT —

ANITA CREAM

Makes a Dark Skin Lighter, Clearer, Purer.

Anita Cream removes **Tan, Freckles, Moth and Liver Patches, Pimples, Muddiness, and all Skin Discolorations in the** only practical way known to **Dermatologists,** and promotes a growth of new skin as clear, velvety and healthful as that of youth. Anita Cream is different from any and every other face preparation. It is a California medicinal luxury for the toilet. Anita Cream can be had of your druggist or direct from us. Full size for 50c, postpaid.

A California Booklet and Sample, 10c

A dainty booklet containing pictures of California Missions, photogravures of stage celebrities, interesting letters, complete information and directions about Anita Cream and other toilet helps, together with sample box of Anita Cream, will be sent, postpaid, upon receipt of your druggist's name and 10c in stamps or silver. The booklet alone will be sent upon receipt of your druggist's name and 2c to pay postage.

Anita Cream & Toilet Co. 147 N. Spring St., Los Angeles, California, U. S. A.

*This **Collier's** advertisement of August 24, 1901, seeks to reassure women who have spent too much time in the sun. (Harvard College Library)*

I squatted anxiously, hoping that the fog would not lift at that moment," remarks Henry W. Taft in "Rowing the Maine Coast," a report of 1982 on how he learned traditional methods of navigating a peapod (a local rowboat without an enclosed head).[30] Women discovered early, probably as early as the 1930s, that the two-piece swimsuit allowed them to keep their breasts covered while they squatted over rail or pail, something that middle-class-aimed boating-magazine advertisements subtly imply after 1945. The models wearing racy two-piece suits are models who can use the primitive bathroom arrangements small boaters accepted well into the 1970s.

But more than comfort and practicality keeps local women in bikinis, as they now and then admit. The bikini is a badge of freedom from fashion dictates, an advertisement that its wearer not only recalls long-past days of sartorial abandonment but remains convinced that urban, middle-class prudery cannot last. Contrary to what women visiting from cities think, the bikini is perfectly designed for swimming. After all, remarks the local woman, if it reveals more skin than a one-piece, so what?

For many tourists, it matters a lot. Arguments made by Felicia Holt and William Dean Howells endure at the turn of the twenty-first century, still bound up with notions of public propriety, class mores, and marginal behavior. Nothing quaint characterizes the long-running arguments concerning skin bared along the coastal realm. Indeed perhaps nowhere else than along the coast can contemporary Americans cruise so easily into emotion-charged arguments about the roots of national civilization or so quickly realize the inanity of so many "history of bathing suit" articles featuring movie stars and fashion models.[31] It is not what men and women, or boys and girls, wear on the beach so much as how they wear it, and what they do in it. The most revealing bikini sometimes reveals more than its wearer intends—or perhaps exactly what she intends. The tanned skin revealed by the momentarily askew bikini, the tanned skin quickly hidden by a readjusted scrap of nylon, indicates what? That the wearer of the bikini usually wears other, even briefer suits? Or that the wearer swims or sunbathes free of any suit at all? Whatever it indicates, it announces that the suit is not the wearer's only clothing, for beneath the bikini lies another layer of attire—tanned skin. And that tanned skin continues to raise eyebrows and arguments a century after its first exposure on beaches prompted an outpouring of impassioned argument.

No tanned woman could be wholly moral, wholly chaste, argued turn-of-the-century experts, for a tanned woman could not blush. Among maidens "no grace was rated as high as 'a complexion,'" remembered Edith Wharton in 1934 of upper-class New York society seventy years earlier. "It is hard to picture nowadays the shell-like transparence, the luminous red-and-white, of those young cheeks untouched by paint or powder, in which the blood came and went like the lights of an aurora. Beauty was unthinkable without 'a complexion,' and to defend that treasure against sun and wind, and the arch-enemy sea air, veils as thick as curtains were habitually worn." But "the almost pagan worship of physical beauty" among both men and women turned into something markedly different by the 1890s, something Wharton hinted at in her memories of girlhood boating and swimming.[32] Around 1890, tanned skin replaced "a complexion" as the indicator of good health and of sexual vigor, although only among the

Other women, as early as 1901, tried to get a tan. (Library of Congress)

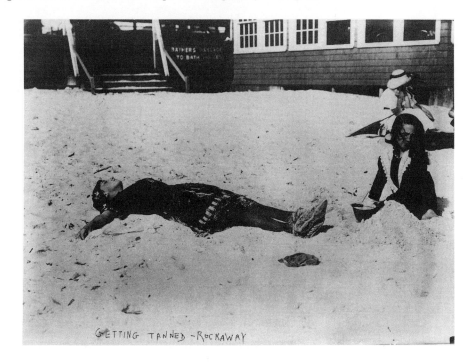

GETTING TANNED ~ ROCKAWAY

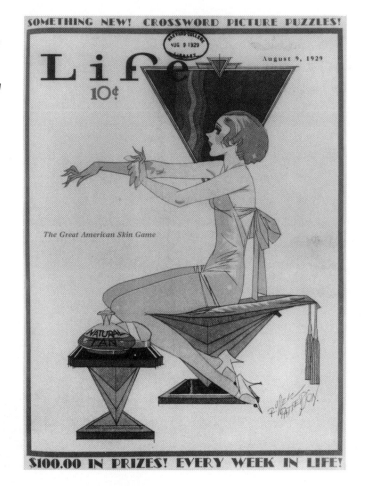

established upper class. While middle-class women worried about sunburn, freckles, and "tan" well into the 1920s, wealthy women abandoned their mothers' devotion to the pale skin that blushed so readily. For the established upper class—not the mere rich—perfectly secure in its social place, tanned skin evidenced a proper devotion to the emerging doctrine of fresh air, sunshine, and outdoor exercise, to the behavior only extremist middle-class women approached before being roundly attacked by authors like Felicia Holt or warned in the words of one *Harper's Weekly* advertisement of 1903 that "clear complexion indicates pure blood." [33]

Ocean swimming combined with suntanning to make men and women healthier, more resistant to tuberculosis and other "modern" diseases, and better able to sire and bear children. So ran the argument developed over almost a century, its logic based on the supposed superior vitality first of American blacks and Indians and subsequently of Polynesians, then even of pirates. Appearing in the pages of serious American periodicals as early as 1811, the argument gathered strength at the end of the century, as experts on "race deterioration" argued that American whites, and especially upper-class whites, had to become more fertile.[34] Beneath the gentlest suggestions that Caucasian women had lost their reproductive edge pulsed not only an admiration of dark-skinned people but a fear of dark-skinned vitality.[35] "Delicate persons, who seem too fragile for the boisterous play of the shore, take their daily dip to great physical advantage," announced one tourist brochure of 1896 aimed at attracting upper-class families to the Maine shore and into the sea. "Youngsters, after a season of such sports, look like young Indians; moreover they fare like Indians, at least three times a day."[36] To acquire a tan—not a mere sunburn—meant to acquire something of the exuberance of more virile "races," say the Creoles Kate Chopin describes in *The Awakening*. In 1899, her novel brushed up against the whole issue of race vitality; an important introductory scene reveals

By the 1930s alongshore undress was tied directly to tropical dark-skinned sexual wantonness. (author's collection)

C. G. 2 SPIRIT OF THE TROPICAL FLORIDA CYPRESS GARDENS

how distraught the middle-class husband becomes over finding his wife tanned, his first hard evidence that she is leaving behind her traditional roles and desires.[37] As Chopin implies, turn-of-the-century middle-class men had not readied themselves for the changes implicit in tanned skin. They valued the "complexion" so prized thirty years earlier, at least on women, and well into the 1930s, they worried that tanned wives had somehow "gone native" in more ways than one.

For boys and men, tanned skin had never been out of fashion, even among the urban rich. The mid-nineteenth-century farmwife may have hidden her face deep inside a sunbonnet—a hat that remarkably resembled the blinders worn by her horses—but her husband and sons exposed their faces to the sun. Throughout the decades of debate about swimming for women, abbreviated swimming costumes for women, and tanned skin for women, boys and men had enjoyed increasing freedom to swim, even briefer swimming costumes, and the right—and by 1900, almost the mandate—to tan skin. Perhaps because so many males swam nude and understood the efficient freedom of nude swimming, they made continuous, successful efforts to abbreviate bathing costumes, so that by the late 1930s all but the most conservative of men swam bare-chested even on public beaches. Certainly the example of competitive swimmers influenced the popular move toward reduced attire, as did the stunning example set by bare-chested matinee Tarzans Buster Crabbe and Johnny Weismuller, but perhaps tradition proved the most powerful of forces.[38] For almost all local men, and for those men who had summered on the coast as boys, "a boyhood alongshore" meant swimming nude or nearly so, and the memory of that freedom made any later suit-swimming less enjoyable.

So in 1880, in *Sea-Air and Sea-Bathing*, Packard offered some unsettling advice, advice that scandalized readers for decades. Very matter-of-factly, he explained that expecting women certainly ought to swim, "except after say the seventh month of pregnancy," that "sea-bathing is almost always of great benefit in those cases of debility from rapid growth, so frequent in either sex at or a few years after the age of puberty," and that sea bathing will help young girls discomforted "by delayed or difficult menstruation." The physician recognized that "exposure to the sun's rays seems to have a good deal to do with the result in all these cases," and that any exercise, not just swimming, seemed to help, but he dismissed all notions of stylish behavior as almost unworthy of comment.[39] Derived from common sense and clinical observation, his arguments later fitted into a new, far broader fascination with Darwinism, and especially with theories of the "survival of the fittest," and with the emergence of the "new woman," the tomboy.[40]

What alongshore behavior constituted romping or rowdiness, and made a woman not a lady? Readers of Whiz Bang found out. (author's collection)

By 1921 swimsuits had become briefer and more clinging, but not nearly streamlined enough for championship swimmers. (author's collection)

Along this stretch of coast, girls learn early the wisdom of Margaret Sangster, that to enjoy the coastal environment they must be in charge of their experiences. They must swim well, pilot a small boat well, find their way across the salt marshes as well as Jewett's *Marsh Island* heroine. Until post-1970s feminists altered terminology, such girls often understood themselves as "tomboys" in the eyes of girls visiting for the summer, girls unwilling to admit that physical strength and vigor contribute mightily to any coastal realm life. Beginning early in the twentieth century, the alongshore tomboy appears in novels influenced by Darwinism and "social Darwinism." The heroine of Jack London's novel *Adventure*, for example, wholly perplexes the hero. "She was certainly unlike any woman he had ever known or dreamed of. So far as he was concerned, she was not a woman at all. She neither languished nor blandished." Instead, "she persisted in swimming in deep water off the beach," and her "fingers were brown with tan and looked exceedingly boyish." In time, the hero tells her that "you are very primitive and equally super-modern," speaking in the encoded but blatant way Herbert Quick uses in his *Virginia of the Air Lanes* to describe the new woman. But although London struggles to describe the tomboy-woman who swims so far off the beach, he loses courage in the end. Try as he might, he fails to explain exactly her swimming costume—and implies that it is more or less only tanned skin.[41] The superb-swimming tomboy becomes the most feminine of women, the most enticing of women, the most challenging of women, the woman who swims better than men.

During the 1930s, upper-class alongshore attitudes slowly altered middle- and working-class thinking and behavior, as the most cursory survey of popular literature makes clear. For men and women both, bathing suits became briefer each year, and devotees of sunbathing—and those lucky enough to know secluded backyards and remote stretches of beach—sometimes wore homemade suits briefer than fashion magazines dictated.[42] Certainly the bikini, often dismissed as a postwar invention, existed as early as the 1930s in Europe and influenced backyard and rooftop behavior in America.[43] In the Depression, up-to-date Americans intended to arrive at beaches already sporting a tan, the emblem of leisure and restriction-free thinking, so backyard and rooftop sunbathing—and lying beneath sunlamps, too—became a popular spring pastime.[44] Tanned skin had become not so much the mark of a healthy body toned by vigorous alongshore exercise but merely something fashionable, and the bathing suit, by the 1960s routinely called a "swimsuit," became something stylish, too, and sexy. Briefer and thinner, changing contour from year to year, the swimsuit became the stuff of B-movies—in time the whole series of "Bikini Beach" films—and *Life* magazine photo spreads, of annual *Sports Illustrated* issues, and of fashion-magazine diatribes.[45] The hoopla masked—and still masks—the eerie unwillingness of many beach visitors to address what alongshore people so casually accept, the easygoing separation of near-nakedness from sexiness.

"That bikini-clad young woman has just come out of the family swimming pool," begins a telling *Cosmopolitan* article of 1960. "She not only swims, she golfs, gardens, hikes and sails. She is also the wife of a senior Pan American captain and the mother of a son at Yale and a daughter attending Northwestern." Betty Kinsley, comfortable in a bikini as brief or briefer than those featured in a *Cosmopolitan* summer-fashion story several months later, is fit, active, lithe, and certainly healthy, albeit nowadays an unsettling personification of the late Eisenhower era. Certainly no one at *Cosmopolitan* defined her as sexy or vulgar, as improperly showing off a finely toned body fit for anything, including entering the high-energy

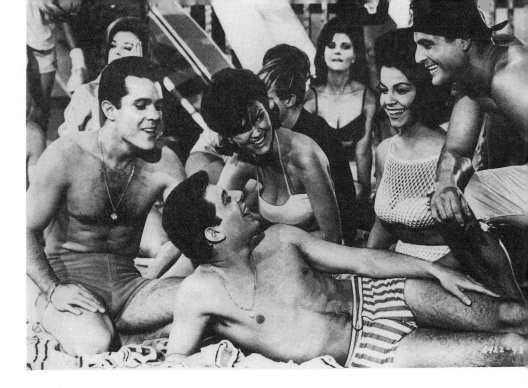

world of office work after her children had grown and left home. Instead, she is merely an ex-amplar of the lithe, energetic American woman Evelyn Archer Adams described in another article in the issue, "Are Woman's Troubles Out of Style?"[46] The magazine staffers understood Kinsley to be healthy and properly proud of herself, her whole self, and correctly comfortable in a scanty suit.

But for a century, some writers of fiction have accepted Felicia Holt's worst fears that brevity of attire leads directly to sexual misbehavior, at least alongshore. Surround a man with women in open-toed swimming stockings, and the man plunges into adultery; let a woman once realize that her brief attire embarrasses proper men, and she will play with that embar-rassment, growing bolder by the day. Male writers often accepted the argument as completely and as simplistically as Atkinson Kimball accepted it in an *Atlantic Monthly* piece of 1910 entitled "A Sea Change." A well-mannered, utterly faithful if dull fifty-year-old husband wor-ries about the briefly attired young men and women frolicking in the surf and on the sand. "Peckham viewed this scene with a certain disquiet; he knew there was no reason why he should disapprove of it, and yet he couldn't give it his approval." In time he blunders into the company of a single woman seeking "marine treasures," and finally puts his arm around her. He confesses his moral lapse to his wife, and they leave the shore at once, hastening inland

"Hello, Alice—er-r—are you sure it's all right for me to come in?"

In its issue of August 30, 1947, Collier's captured the difficulties of the prim and proper male confronting a mostly naked female. (Harvard College Library)

toward the restrictions and restraints that keep middle-class marriage whole.[47] The restraints still infuriate and titillate female authors and readers, too, as so much beach-set fiction read by sunscreen-slathered, beach-lying women makes clear. "She looked like a ballet dancer. The bikini that attempted to cover her was of some hot-pink shiny material that set off her tan, making it glow just a shade darker than honey." So begins the trouble Caroline Crane creates in her novel, *Summer Girl* (1979), trouble pulsating from a girl named Cinni, the personification of sin. The husband stares at the girl his wife has hired to look after their young children, watching her tanned body shift ever so gracefully inside the hot-pink bikini. "He marveled at her figure, full and womanly, barely restrained by the ridiculous inadequacy of the bikini bra."[48] Unlike his tired wife, a woman most unlike Betty Kinsley, the teenager possesses a vitality somehow suited to the beach, an energy unconsciously expressed in tanned skin and supple limbs. She belongs on the beach, belongs to something as elemental and as slightly restrained as her breasts—or the sea. For masses of Americans, the scantily clad woman or man at the seashore remains at the end of the twentieth century what she or he remained at the end of the nineteenth—an object of sexual attraction, sexual desire, sexual disaster, as dangerous as any Hollywood-imagined quicksand.

Only locals know what alongshore people—and the established upper class—have always known, that near-nakedness is neither disruptive nor sexy and that after one gets thoroughly used to it, it is utterly commonplace, simply sexless. But all alongshore, it remains oddly powerful, for in nakedness tanned skin becomes insignia, the badge of belonging to the sand, the rocks, the surf, the sun. Now and then a particularly honest writer probes the odd reluctance with which the contemporary feminist movement has dealt with alongshore nudity, and recently women belonging to the Coalition for a Topfree Equality have argued, unwittingly paraphrasing the long-term, matter-of-fact thinking of coastal realm people, that "if men saw women's breasts as often as they see men's there would no longer be this unnatural obsession with them."[49] But so long as powerful male attorneys, male newspaper columnists, and males everywhere raise the specter of the uncovered female breast as Felicia Holt raised the specter of uncovered female toes—"parents certainly have the right to decide whether their teenage boys should or should not be exposed to the naked breasts of protesting women," argued one law professor in 1992 in a newspaper attack on topfree women that reads uncannily like a diatribe from 1890 against open-toed swimming stockings—fears of beach nakedness will continue to twist all but those accustomed to such marginal appearance.[50] Like the paths that meander between sand dunes thick with razor grass, beach rose, and poison ivy, the path from Gloria Steinem's *Beach Book* (1963), perhaps the most conspicuously ignored feminist book in contemporary America, to late twentieth-century beach behavior is contorted and difficult to trace.[51] Never do historians of costume address what any thorough reading of American periodicals shows, the eerie, ahead-of-time brevity of swimsuits worn by Annette Kellerman—who announced in 1909 that she could not swim in "more stuff than you hang on a clothesline," sewed up a practical and modest suit, and found herself arrested in Boston—Gertrude Ederle and other champion female and male swimmers, even into the 1960s.[52] Never do they probe the feelings of light-skinned white Americans toward the dark-visaged, dark-tanned white man, the foreigner, the mariner, the pirate. "Think of going on board a coppered and copper-fastened brig, and taking passage for Bremen," Melville writes in the first pages of *Redburn* of a young man beset with an "inland imagination" that tricks and confuses. "And who could be going to Bremen? No one but foreigners, doubtless; men of dark complexions and jet-black whiskers, who talked French." The young man dreams of long voyages, and his seamanlike complexion afterward. "How dark and romantic my cheeks would look."[53] Women and men of the marge, of the sea, necessarily wear little, at least in warm weather, grow accustomed to their near-nakedness, and forget the impact of bare skin on those who know only the seediness of porn magazines and urban X-rated bars.

They glimpse what Anne Morrow Lindbergh glimpsed in 1955, that in shedding almost all attire and walking bare-skinned along the beach "one finds one is shedding not only clothes—but vanity."[54]

In **Self Portrait**, *Jane Tuckerman reveals the casual ease with which an artist views herself and her beach. (Courtesy of Jane Tuckerman)*

Something vaguely commonsensical endures along this stretch of not-particularly-fashion-able, not-particularly-trendy seacoast, something implicit in what local boys tell tourist boys who gape at the nearly naked girls. "What's with you, you never seen a girl before?" Middle-aged women in faded 1970s string bikinis climb sailboat rigging, old men in tattered shorts bleached white by salt and sun trudge across the clam flats, teenage boys in skin-tight, film-thin diving briefs revel in the warm sun after long dives, all go about their limicole business as locals properly attired in salted tan. While tourists gawk at them, define them as shame-less hussies or dubious men, the locals silently answer the question posed by Charles Willson Peale in 1765. The locals live familiar with skin, live comfortably in skin, go comfortably barefoot, go nearly naked into coastal realm places that reward fit bodies and local knowl-edge. They play in hazards.

12 RISK

Low tide poses all kinds of navigational problems, even to one man paddling a canoe. (JRS)

Estuaries mock inland understandings of rivers. Twice each day an estuary flows seaward and twice each day it flows inland. At full high tide and dead low tide, its water lies still, without current. Low tide brings to mind old jokes about tourists commenting on droughts and shrunken rivers, and high tide makes navigation seem so simple, so effortless. High tide masks all the hazards low tide reveals.

"The estuaries of rivers appeal strongly to an adventurous imagination," Joseph Conrad asserts in his *Mirror of the Sea.* "From the offing the open estuary promises every possible fruition to adventurous hopes." But always his view is that of the mariner closing the coast, gunkholing in, shallow-water lead line tossed and tossed again, and indeed Conrad imagines the thoughts and actions of the commander of the first Roman galley to explore the Thames, the "intense absorption" with which the officer watched the estuary unfold. In the end analysis, however, an estuary opening or mouth exists only to be entered, passed, and left behind. Scenery plays but the meagerest of roles. "A river whose estuary resembles a breach in a sand rampart may flow through a most fertile country."[1] The adventurer gunkholes inland, seeing what the freshwater river brings of inland mystery and treasure.

Conrad knew the other view, of course. His *Heart of Darkness* opens with a group of men yarning aboard a cruising yawl anchored in the Thames estuary. "The sea-reach of the Thames stretched before us like the beginning of an interminable waterway," the narrator begins. "In the offing the sea and the sky were welded together without a joint." The men, all mariners or sometime mariners, begin to muse on the estuary, how "the tidal current runs to and fro in its unceasing service," how many men had used the ebb to set out for adventure.[2] So much of English maritime literature begins with the Thames estuary falling astern, in the way that

H. M. Tomlinson describes it in "Outward Bound." "A great city is in the right geographical point, becomes bearable, and can be regarded with calm, only when the tide has turned seaward, the wide glare of neon and other lights sinks down in the night astern, and the only sound at last is the monody of unseen waters."[3] Airline travel all but eliminated the estuary exit. After World War II, ocean-liner passengers became fewer, then almost none. Today only working mariners putting out from New York or New Orleans, or from London, know much of great estuaries, and only small boaters, sometimes the smallest of small boaters, know much of many others.

Estuaries trouble many viewers, especially those unwilling to realize that some rivers do change direction. Locals, of course, simply accept the local estuary, often call it simply "the river," and find all other rivers, even great rivers like the Mississippi, really rather dull in their directional singleness. And even locals tend to think of entering or leaving the mouths of estuaries, not enjoying the bleak sandbars and dunes.

But one estuary along this stretch of coast offers more than perceptual difficulties, a twisted route inland, an opening on the sea. In a spectacular mouth-island beach, it offers an inkling of real trouble, of trouble striking near the foundation of inland thinking.

Trouble alongshore, real trouble, now and then comes stealthily, subtly, so menacingly that only vaticinal stillness heralds it. Such stillness gathers around coastal realm environmentalism now, like the stillness that follows an odd fluke of wind on a hidden summer beach visited only by a new elite, an elite physically fit for environmental exertion in the coastal realm, in the marginal zone.

Poverty stalks the halls of government. Gone is the nuts-and-bolts understanding of the Civilian Conservation Corps, gone the post-Sputnik federal Treasury largesse, gone the flitting good humor of the peace dividend. Whatever environmental catastrophes and opportunities lie ahead, coastal realm environmentalists now confront the "human-needs advocates," those who argue that medical care must precede wetlands reconstruction, that wilderness conservation ranks far below the needs of pregnant teens. How does one argue for plovers and herring and beach rose in a cacophony of voices raised in protest against more human services budget cuts? To strive for "politically correct" solutions to coastal realm environmental problems is to strive for wisdom beyond any that environmentalists now command or may ever command. But does everyone care to strive that way?

Perhaps not. A sort of tacit conspiracy now shows up everywhere in the Republic, a conspiracy of environmentally conscious, ever-so-quiet individuals who have made a separate

peace with conscience, politics, even friends, a conspiracy that enjoys the advantages of environmental decay, the delightful if perhaps short-term dividends of environmental funding cut recently or years past.

At the confluence of two estuaries along this stretch of coast, the conspiracy grows stronger each summer. Few visit the salt marshes, few explore the guzzles and gutters that snake seaward toward the estuaries' opening onto the sea. Simply put, the marshes no longer entice because the government-built gateways—the boat-launching ramps especially—have deteriorated almost beyond usefulness, and the government-maintained avenues, the dredged and buoyed channels, have silted in to become almost unnavigable, especially in bad weather. No longer can locals, let alone tourists, expect to launch large powerboats or even small sailboats and go adventuring or picnicking. The access ramps and inland channels, admits even the Division of Waterways, no longer entice, and grow less enticing with every tide. "In many harbors there has been a need to severely limit access to smaller, shallow-draft boats or to during mid- to high-tide periods," remarks the author of one recent report that catalogues "increasingly frequent calls for assistance to harbormasters, the Coast Guard, and the Environmental Police from recreational boaters grounded in or alongside shallow entrance channels," the "real potential for injury or loss of life due to collisions between boats negotiating narrow entrance channels," and the role weather can play "in the safe negotiation of shallow

The old boathouse, from which so many estuary excursions begin, sits precariously near the edge of destruction.
(JRS)

The hapless tourists who use this launching ramp at high tide learn that their trailer is snagged at the edge of the cement ramp. (JRS)

channels." In shallow channels, argues the author of this report, boats crowd together, and when the wind stiffens, "do strike the bottom while in troughs between the waves," or else crash into each other, leading to documented instances of damage, sinkings, and injuries.[5] Given these problems and dangers, the Division of Waterways concludes, people will congregate near the last deep-water inlets and harbors, and ever fewer people will try to navigate the dangerous shallows in small boats.

But some people still visit the marshes, not usually for the marshes themselves but because the marshland creeks and estuaries open on a distant, secret treasure—an isolated barrier beach of dunes and, at low tide, shining sand. Only since the *Portland* Gale of 1898, when the surf smashed a whole new inlet through a cobble beach and sealed forever the old estuary opening, has the sandy beach existed, and brutal storms and currents reshape it every winter. Essentially inaccessible from dry land, the beach enjoys an increasing solitude, for the Commonwealth of Massachusetts, too poor for years to dredge the channels around it, has unwittingly precipitated a giant silting-in. Now the owners of big engine boats and deep-draft

sailboats cannot even approach the beach except at the highest of high tides, and so treacherous is the tortuous channel that they cannot safely anchor and swim ashore. A few try to tow skiffs and other tenders to good anchorages, then use the small craft to reach the sand, and others try inflatable boats, often powered by mosquitolike outboard engines. But such secondary craft are hard to tow through the tide races that churn in the estuary mouth, and harder to stow on board during the deep-water passages to and from the beach.

So who comes?

The clerisy come, the cognoscenti, those active enough to paddle a seagoing canoe, pull a rowing boat, or, lately, manage a sea-kayak along a couple of miles of contorted, current-swept shallow channel. What an intriguing collection of people, gathered on the beach on a midsummer's Saturday, far beyond sight of the land-based crowds and, deliciously, away from the fiberglass-boat crowd, too. What a group, indeed. Perhaps it deserves more study than its chosen setting.

Visual analysis (what the barefoot historian calls lying raised on one elbow, staring from

behind sunglasses) produces discoveries about the beach visitors. In age they range from toddlers to the elderly, but nothing else makes them a random sample of Americans. They comprise a definable cohort. All are muscular, "fit" in contemporary American English, but scarcely fashionably beautiful. Many arrived under the "white-ash breeze," paddle- or oar-power, and most waded ashore through cold water and black, clinging mud, pulling and shoving boats and canoes the last hundred feet. All, in other words, exerted themselves to reach the beach—all were wet before they touched sand. What else? All tanned, all sporting sunglasses, all nearly naked, they confront the environment directly and actively. They walk, swim, visit, and explore, spreading out across the sands, knowing that the incoming tide will soon cover most of the vast beach. They leave no litter, and demonstrate perfect regard for not climbing among the dunes, not venturing into the zone marked by faded plover-nesting signs. No radios, no beach umbrellas, no aluminum chairs (but once a playpen) come ashore with them, and they play no games but casual Aerobie (and once, on the Fourth, volleyball). In the midst of them muses the barefoot historian, noting shell-seeking families following the wrack line, remarking the absence of cigarette smokers—and lately wondering less about

Even outboard-engine boats prove nearly impossible to launch through mud exposed by an outgoing tide. (JRS)

the beach environment than about a population of beach visitors distributed evenly along a beach, a population of active, almost restless visitors.

For the group is a definable one, day after day, almost always the same, relaxed and proud in its competency to reach the beach—and quietly smug in its understanding that governmental poverty sometimes works wonders. Hooray, then, for impoverished state government? Three cheers for the welfare advocates? May they carry the day, and keep the dredges far away? May government never again repair the launching ramp so that trailer-size motorboats can roar along the channels? Given a little more time, just a little, and soon engine boats of any size will stay away, even at high tide, and the mudflats will become a "natural" wetland, unalterable under current law. Only those who make effort will enjoy the beach beyond the marshes, one of the most dangerous beaches along the Atlantic coast.

At the nameless beach the ocean kills, actually rather rhythmically. In 1989, seven people drowned in the tremendous currents that snare swimmers, waders, sailboats, and even lobster boats (and on one stunning occasion, tossed two Coast Guard rescue vessels ashore on the sandbars). A year later, the current caught no one, because everyone went especially care-

fully, chastened by death, but in a few years carelessness will lead again to drowning, then care. Bereft of lifeguards, and with all the boats drawn up on the marsh side of the dunes and so useless in any oceanside rescue, the island beach is a rare place by late twentieth-century U.S. standards. It stands isolated by more than miles of salt marsh and twisting salt creeks. It is gruesomely far from paramedics. Suppose a foot sliced on a broken mussel shell. What then? Suppose a sudden coronary seizure? Suppose a falling tide and a rising onshore wind?

In 1992, a sorely tried and battered Coast Guard erected a wordy sign of warning about how quickly estuary currents and onshore winds can combine to make navigation difficult, if not deadly. "The channel is subject to change and should never be entered except by smallcraft with local knowledge," ran the caution in part. But its odd last sentence chilled the blood of locals and rare tourists alike, for it admitted intermittent defeat, impotency: "Modern rescue equipment limitations do not provide an effective rescue response under these conditions." Under some conditions, occurring quite frequently, even the Coast Guard could not help.

A few weeks later, vandals defaced the new sign. By dark they scratched away the long last sentence, the sentence rumored to be spreading fear among tourists everywhere along the coast, to the distress of innkeepers, marina owners, and common victuallers.

Physically fit people are capable of long passages across open water and can take their small boats far closer inshore than can the owners of yachts. (JRS)

But at the far-off spectacular beach the real danger lurks above the currents, almost always to the west. Let clouds appear on the forested horizon far west of the marsh, let breezes shift ever so slightly, let the sky go tyndall blue, let happen any number of atmospheric effects John Brocklesby delineated so long ago, and the locals fall still and watchful, judging sky and puffs of wind on cheek or thigh, gauging the tide. Everyone thinks suddenly of the long pull home, the risky gunkholing in shallows whipped by tidal currents and winds. Sliced feet and heart attacks are possibilities, but possibilities less remote than the drownings caused by overconfidence in a dangerous place.

But what of human threat, that other risk? Pirates have long gone, although their gold coins wash ashore, one or two every few winters. So have the U-boats, for all that the submarine watchtower still looms on a headland to the south, grossly unpicturesque in its concrete indestructibility, unsuitable for postcards or tourism brochures, a monolith that juts upward next to a wholly artificial hill, a grass- and tree-covered hollow mound once used for ammunition storage, now part of a U.S. Air Force reservation. No, pirates and U-boats have gone. And what about crime? Apparently none, not even profanity or petty theft. Now and then the harbormaster chugs past in the hazy distance, anxious to slow the large speedboats whose wakes erode the salt marsh, and rarely—recall the fiscal difficulties—the state environmental police, dapper in green uniforms, immaculate in deep-draft, high-speed fiberglass boats, ready to stop drunk motorboaters, ready to rescue those out of gas, ready for anything, except probing the extensive shallows that separate them from the little boats drawn up on the flats. Police seem unnecessary anyway, for a gentle camaraderie infuses the fortunate few enjoying the beach. No disputes break the peace, no bikini top drifted askew elicits a catcall, no drunks stagger along the sand. Beachgoers walk back through the dunes now and then to check on anchor lines and ebbing tide, not to see if some sneak thief has swiped water bottle or life jackets.

Here, then, is no human threat such as those described in books about urban beaches.[6] However dangerous the currents and waves and shallows, the beach visitors enjoy a crime-free place, and whatever risks they run originate in challenging the physical environment. Are they up to the challenge?

Exertion, more than risk-taking, distinguishes the beach visitors. All accept the risks posed by sudden squalls and turning tides, and even by the mind-boggling rapidity with which the incoming tide isolates, then submerges sandbars, forcing strollers to wade fast, then swim furiously. But exertion is the key to understanding the visitors as a specific type of people, for only exertion makes the beach accessible. Moreover, however they value the beach environment itself, many appear to enjoy at least as much the challenge of getting to the beach, the long row or paddle along the salt creeks threading the marsh, the half-mile effort to cross the rollers moving in from the sea, the careful navigating among sandbars, even the wading through thigh-deep mud. Their approach to the beach explains them not as intellectuals but as "physicals," as people of all ages confronting the alongshore physical environment with minimal assistance. All, for example, go barefoot, and do so efficiently, not hopping up and down over mussel shells and sharp stones, not complaining about wading in cold water and mud, not shrieking about hot sand on the ocean side, but moving effortlessly along a bit of littoral secreted away from tourists certainly, as well as from locals unwilling to exert themselves to reach it.

Trim, fit for the exertion of getting to the beach and then actively using it, the visitors move comfortably among themselves, relaxed in their tanned skins, their casual postures demonstrating familiarity with barefoot walking and casual, unthinking acceptance of physique. Simplicity, almost starkness, governs. Stylishness and ornament hardly shape attire, for clothing scarcely exists. Whatever magazine-dictated fashion elsewhere, even women well into their seventies wear bikinis, casually remarking that the scraps of cloth dry faster, stay cooler, make bathroom functions simpler in small boats, and trap less sand than maillots, and forthrightly admitting that the token suits flatter women rightfully proud of their bodies, "in shape" not for girlie posters but for activity that requires muscle, coordination, and endurance. Men wear remarkably little, too, eschewing the long "jam" swimsuits so common elsewhere, but which impede swimming, running, and drying off (and wading ashore through deep, sticky mud), preferring the briefest of shorts and racing suits, and dismissing flappy swimsuits as fit only for the beer-bellied. Husky or skinny, young and old, the men move as constantly and effortlessly as the women, clearly accustomed to walking barefoot, to moving around. Slender, active children and teenagers, wearing as little or less than adults, jump overboard to tow boats ashore, struggle with deep-mud, low-tide launching efforts, and race across massive shallows. Practical hats, sunglasses, cotton shirts, sunscreen, and little else come ashore from the small craft—except food.

Small boats carry small coolers. Meals are compact, simple, and usually washed down with ice water poured from insulated jugs. Soft drinks and beer come ashore, too, but not in "suitcase" quantities, and never in kegs. Eating often preoccupies visitors just arrived, ravenous from the exertion of paddling or rowing, or pushing the small, outboard-engine-equipped

The hiker who finds this view around a bend in the barrier beach feels somehow disappointed in the effort. (JRS)

skiffs off one sandbar after another. But the beach hosts few complicated picnics, and fewer cookouts (none with gas grills off-loaded from big boats), perhaps chiefly because the small craft have little room for fancy victuals and junk food but also because the visitors eat casually, sometimes while walking, often while sitting on damp sandbars near the waves. In the end, however, exertion explains the paucity of big coolers and fancy eating. Everything must be carried from the small boats up one beach, through the dunes, and out into the immensity of sand that lies exposed by low tide. Cuisine, like clothing, gets stripped down to absolute essentials.

Questionnaires have no proper place on the beach, and the barefoot historian gathers information about string bikinis and apples informally, catch as catch can. Everyone assumes a sort of homogeneity of interest and values, a dislike of crowded, "stylish" beaches jammed with people and equipment, of four-wheel-drive beaches, of noise and trouble and rudeness. In the final analysis, perhaps, the beachgoers value the beach not only for what it is, a bleak, dune-backed sandbar between sandhills, but because it lies at the end of adventure and exertion, and because adventure and exertion screen out the beachgoers who elsewhere censure the scantily attired, physically fit active users of the littoral. Tanned and sunscreened enough to be at ease in the sun, strong enough to take the ground and shove off through mud, in shape for sustained maritime effort, the visitors value exertion, and take for granted bodies accustomed to an old-style sort of exertion, not exercise.

Somehow the hidden beach smacks of the late 1950s, perhaps the 1960s, the moment when ordinary beaches not yet converted into resorts existed as areas of simplified, stripped-down recreation. Summertime alongshore life in the 1960s is hard to interpret to contemporary young people. Mosquitoes rarely annoyed anyone, even after dark—the DDT-dusting plane kept woods, salt marshes, and backyards free of them—and ticks produced only a rare momentary yank, not Lyme disease. Sunshine meant not cancer but vitamin D, healthful energy, and a sort of vitality diametrically opposed to atomic fallout. Parents traditionally emphasized outdoor play but strived as well to create a healthier, fitter generation—the goal of the newly formed President's Council on Youth Physical Fitness and Sports. Children, adolescents, and teenagers ought to be active, lithe, slender, and tanned. They ought to be outdoors and moving. So opined federal authority.

Nowadays such remarks elicit mingled disbelief and derision, and cracks about pesticide pollution and home-made human engineering. But the site-specific physical environment became the theater in which parents, and mothers especially, perhaps especially the already bikinied mothers like Betty Kinsley, experimented with changes in child-rearing, taking risks, gentle and otherwise. Mothers encouraged their children not only to play actively but to play hard, and to tan deeply and extensively, to armor their bare feet against injury and screen their skin against sunburn. An absence of mosquitoes and greenhead flies edged on

Body positions, especially the positions caused by bare feet, are best noted in sketches, not photographs, as this New Yorker *illustration of July 21, 1980, makes evident. (author's collection)*

the drift toward near nakedness, but many mothers provided ever briefer attire for their children, not only to facilitate exercise and tanning and keep perspiring bodies cool but to display the supple, toned, what-the-federal-government-wants bodies of their offspring, bodies to be proud of, bodies that earned the council's gold seals and cloth patches, bodies that fathers and mothers rewarded with rowboats and camping gear. Physical fitness meant fitness for exploring the coastal realm, for swimming in cold waves, for running along beaches, for hiking back of the dunes, all simple enough activities, perhaps, but all involving almost naked bodies connecting directly with land, air, and sea. Physical fitness meant fitness for outdoor exertion and challenge, for everything from climbing giant trees to racing the tide across the shallows to walking for miles over loose sand.[7] Few contemporary university undergraduates grasp the extraordinary importance of the 1960s physical-fitness movement, let alone its relation to the old conservation movement, to the beginnings of contemporary coastal realm—and maybe inland—environmentalism.

Going barefoot, for example, meant going barefoot across marshes and shell banks, through mud, and over barnacle-encrusted rocks, meant crossing broiling hot sand, running through dry, drifting sand, walking on splintery landing stages—in other words, going barefoot meant touching the natural and built environments, meant knowing the feel of them through calloused soles. Tanning meant freedom from shirts, pants, hats, and lotion (impedimenta of fair-skinned teenagers), freedom to swim to one sandbar, hike a little, then swim again, and freedom to be socially comfortable while exploring the coast, for tanned skin, skin by definition routinely exposed, became a sort of clothing itself. A few techniques helped barefoot alongshore progress, especially the clenched-toed walk that carried walkers through razor grass, and a few others—say that of scooping a shallow hole in the sand and sitting in it to prevent backache—distinguished kids experienced in beach-using from summer visitors equipped with towels and chairs and umbrellas—people who hated the touch of sand on their skin. Too poor for outboard engines, local kids simply rowed and walked, and after donning a shirt out of some vestigial rule of propriety, walked or bicycled home. Bodies, not equipment, did the job of taking children and teenagers everywhere in the coastal realm, and especially everywhere in the estuarine environment—and bringing them home again—and bodies, so long as they functioned well, had become something to be proud of, to use in intimate exploration of the outdoors.

Nothing idyllic or nostalgic shapes these words, simply individual and collective memory that makes slightly clearer reports like *Trends in American Living and Outdoor Recreation*, a document presented by an optimistic Outdoor Recreation Resources Review Commission to the president and Congress in 1962. Who reads such reports today? Who ponders the changing meaning of terms like *accessible*? Did *accessible* once connote "accessible to fit individuals exerting themselves"? Did recreation planners and landscape architects presume a physical, active response to the physical environment, one often direct, unmitigated by equipment like four-wheel-drive vehicles? Did they anticipate, even presume, a population of vigorously active people exploring wilderness and other outdoor areas?

Consider one especially pointed report in *Trends in American Living*, an essay by Lawrence K. Frank entitled "Outdoor Recreation in Relation to Physical and Mental Health." Frank identifies an incipient but disturbing trend in outdoor recreation, the propensity of more and more Americans to "put on sometimes elaborate and costly costumes for their outdoor activities" and to "take along a heavy load of equipment." He attributes such behavior to their attempt to impose "urban order" on environments in which they feel uncomfortable, and to the growing power of advertising, but he concludes that such adults function badly outdoors because their childhoods had been warped. "Overprotective parents" may have robbed them of "native courage," or perhaps frightening experiences had shaped them into adults crippled by "lack of self-confidence." Such people needed help, but Frank insists that more attention ought to be given to the nation's children, to make them different, better. Spontaneous childhood play, in whatever weather and natural environment available, struck him as the purest sort of outdoor recreation, the disorganized, relaxing, body-using experience for which everyone should strive in purely individual ways. But many children, especially urban ones, lacked the "muscular strengths and skills" necessary to play in the natural outdoor environment. After noting the low scores on physical fitness tests, Frank argues that something

The Pirates of Scituate Harbor, Scituate, Mass.

F. W. HILL, BOSTON,

must be done for "outdoor recreation to realize its full potential as a factor in developing a nation of reasonably vigorous and healthy young adults." Children must be encouraged to learn the value of outdoor exertion, but first they must stop being ashamed of their bodies.

Frank emphasizes the profound change in attitudes marked by the 1960s. "The human body frequently becomes a source of anxiety and often guilt to the individual who is taught to distrust and feel ashamed of his own body."[8] Eliminate shame, and children will take pride in developing their bodies just as they take pride in intellectual accomplishment. In acquiring through exertion the skills needed truly to enjoy the outdoors, they will achieve not only physical health but the emotional balance that comes through genuine re-creation. They will value their bodies not for how they look but for how they work, for what they can accomplish outdoors, in the natural theater of human exertion.

For a moment in history, say between 1958 and 1968, physical exertion—not exercise, not sports—seemed the gateway to making young people competent in their bodies, proud of their bodies, able to play and work hard, join the military, raise children, stay healthy—and explore the coastal realm and other outdoor places scarcely altered, if at all, to facilitate

In The Young
Shipbuilders, *Elijah
Kellogg emphasized the
risks of juvenile boat-
building and stressed the
necessity of physical
fitness and quick-
mindedness. (author's
collection)*

access. "Physical fitness," a term so worn now that almost no one remembers its meaning, connoted the physical competency that leads to active enjoyment of the outdoors. It meant being fit for activities that linked the new President's Council on Physical Fitness with the equally new Green Berets, whose military fitness program, in pamphlets assuring readers that "you will experience the marvelous overall sense of well-being that comes with having the human machine in order," reached out to the civilian population already headed outdoors.[9] What happened immediately after, transformations ranging from the sudden emphasis on organized, team sports to the ramifications of the earlier onset of puberty to the beginning of the illicit-drug era, make Frank's tautly reasoned essay a period piece, an anachronism out of touch with the new reality so evident on most beaches.

Consider an incident that prompted a graduate-level research seminar and much of this book, an incident that shaped several summers of sustained scrutiny, an incident that lingers engraved not only in a salt-stained notebook but in the memory of a historian struck by the future.

An eight-year-old boy sits aloof on the sandy beach one headland north of the secret one, whines he is "sea-sick," nauseated by the smell of sea and seaweed, worried that people pee in the ocean, and disturbed by the dirtiness, the darkness of the sea, so unlike a clean swimming pool. The historian overhears, and stops thinking about dune preservation in the 1820s. Who is this boy? What is he?

Everything has changed, has it not? Since the advent of the "Bikini Beach" movies of the late 1960s, the films that taught Massachusetts and South Carolina teenagers how California kids behaved and suggested that they behave similarly, beach-going has changed dramatically everywhere. Beach-going became a social rather than an environmental event, a "happening," perhaps a script, and if a script, one that required innumerable props from coolers to radios to ukuleles to chairs to surfboards. It became an essentially passive, essentially social undertaking, a sort of advanced-placement beach-blanket-bingo charged with sexual vibration, not physical exertion. Glamour and glitz, not the natural alongshore environment, became all-important, became the legacy contemporary children inherit through videos and advertising.[10]

Now the boy sitting on the beach, afraid to walk barefoot across three feet of dried seaweed, must concern every beachcomber, every environmentalist. He is the new American child, the indoor child, the child afraid to accept physical risk, the child unfit for risk. According to the president's council—still in business but fighting a rear-guard action now—this child is likely to be overweight, very likely by age sixteen. High blood pressure, the council warns readers of its reports, is now a pediatric disease.[11] Prolonged television watching, video-game playing, too little bicycle riding, and the wrong kind of food add up to something more precisely measured but far less frequently mentioned than the twenty-year-long decline in SAT and other test scores. They add up to what anyone sees on most beaches here and elsewhere in the Republic.

Kids are fat. Obesity is more than a national problem, it is now a national scandal, something Europeans notice immediately, especially at beaches. By any standard from 1960s schoolroom charts to modern insurance-company tables to contemporary fashion articles, Americans are flabby, and many are so ashamed of how they look in a swimsuit that they choose not to visit beaches at all. But nowadays any honest observer at most beaches notices what medical researchers have so precisely documented. *Kids* are fat. At the beach, shorn of their stylish baggy pants, one-size-fits-all sweatshirts, and flapping overblouses, kids display jiggling, bouncing rolls of fat. In articles entitled "Long-Term Morbidity and Mortality of Overweight Adolescents," "Inactivity, Diet, and the Fattening of America," "Increasing Pediatric Obesity in the United States," and "Do We Fatten Our Children at the Television Set?

As early as the 1930s, New Yorker artists began to depict the difference between slim, active beachgoers and heavy, inactive sunbathers. (Harvard College Library)

Obesity and Television Viewing in Children and Adolescents," readers of the *New England Journal of Medicine, American Journal of Diseases of Children, Perspectives in Practice*, and *Pediatrics* learn not only how well documented, how meticulously measured the pattern of obesity is but how clearly understood are the effects of obesity, especially childhood obesity.[12] Relatively few beachcombers routinely read journals like *Morbidity and Mortality Weekly Report*, but those who do encounter articles like one from 1992 entitled "Vigorous Physical Activity among High School Students—United States, 1992," that make clear the stunning decline in youth physical fitness since 1980—let alone since the 1960s.[13] Somehow the words of Thoreau return. Somehow the beach *is* a vast morgue, or morgue-to-be, peopled by overweight, inactive kids whose obesity and passivity presage illness, injury, and early death.

Worry about physical appearance preoccupies many teenagers about to visit the shore, prompting dieting and frantic buying of the "right" clothes. But if most teenagers on the beach are already overweight, such worries diminish, and new worries, mostly about having fashionable clothing, take their place. How the body-clothes ensemble *looks*, not how it functions, concerns not only teenagers but even adolescents and adults. In the summer of 1992 every clothing store in the nation sold surfing attire to young boys, most of whom will never surf, not even skimboard. The surfer look was "in," and the attire—especially the baggy jams—was momentarily correct, even on very flabby people who venture only short distances from beach-front parking lots and fear the barefoot crossing of dried or wet seaweed, let alone the exertion of a long hike toward sandy emptiness or the risk of a quick, wave-tossed swim. Overweight girls and women learn to wear a stylish, often expensive cover-up over their suits, and to sit rather than swim or hike. Or, if everyone on the beach is fat and more or less inactive, then overweight kids and adults move slowly up and down the sand, or wade in the surf.

"It is impossible to sit on a beach these days and not be struck by the overwhelming obesity of the American public," asserts fashion-critic Holly Brubach in "On the Beach," a *New Yorker* article of 1991 on swimsuits and exposed skin. "There are, apparently—astonishingly—very few people in between, few 'normal' people, for whom eating and exercising are natural functions rather than conscious decisions." Something has changed, and Brubach knows it, encountering one swimsuit designer who believes that the "old formula—skin equals sex appeal—is now defunct," that the thong bikini is no longer sexy but simply a " 'part of a life style.' "[14] But the obesity puzzles, then obsesses Brubach and other fashion writers, who see it as a force skewing clothing design and behavior in clothing, as something more important by far than any analysis, however perceptive, of fashion. For the overweight, for the unable-to-be-active, skin alone is no longer what it was even ten years ago. "While sun-bathing, the new Amazons are wrapped not only in suntans that are like a copper-colored body stocking, but also in an invisible aura of untouchable innocence, childlike self-absorption, and defiant lack of sexuality," argued Kennedy Fraser in 1981 in her book *The Fashionable Mind*. "The current wave of nudity follows the general interest in sporty naturalness and health. It is in most cases studiously unprovocative."[15] But ten years later, the wave has withdrawn from most American beaches, leaving behind not glistening bodies but a sort of human wrack. As more and more Europeans and South Americans remark to themselves, and sometimes to the historian who listens, no amount of clothing can mask American end-of-the-century obesity and laziness.[16]

Beach-going, to judge from newspaper articles and casual conversation, no longer satisfies many Americans. On the one hand, those ashamed of their half-clothed bodies sometimes find the beach a place of embarrassment, sometimes even when most other people are overweight, too.[17] On the other, the beach nowadays presents a physical challenge in ways devoted, all-season beachgoers simply miss. For the out-of-shape visitor, the sun is hot, the wind dry, the water cold and dark, the sand gritty and hard to walk on—and there are more

mosquitoes nowadays, at least at dusk. How much better it is to go to the air-conditioned mall on a warm day, to get indoors, to watch wilderness documentaries on cable television, to play adventure games on home computers.

The American public has become two publics, divided not by race or gender or economic position but by the insidiousness of physical condition. And nowhere else than in the coastal realm does the split become more obvious, more grating.

On so many beaches now, the lithe, slim, bikinied woman, the trim, fit bikinied man attract unfriendly stares, sometimes catcalls, and always muttered exclamations of dislike. Neither woman nor man is voluptuous or handsome in any sort of Hollywood sense, but both are something inactive people increasingly loathe. They are supple, active, strong, and they are moving, seemingly effortlessly, in an environment that demands and rewards exertion. They are fit for the beach environment. As *Vogue* notes, the bikinied woman is not merely bikinied, but bikinied in something called a "workout bikini," a very brief suit that stays on during any sort of activity.[18] Increasingly, women's magazines like *Shape* and *Self* emphasize the beauty of strength, arguing in cover-story-featured groups of articles that "strength is beauty."[19] Only recently have other magazines recognized the prescience of Elizabeth Kaufmann's conclusion to her *Self* article "Put Some Muscle into It" (1989): "Women who have successfully sculpted their bodies get a heady feeling, both from the physical results and [from] the sense of control they have over their bodies. What's more, the effort is worth it: while you're building a strong, shaped body, you're investing in lifelong physical independence and confidence."[20] Such women are independent on the beach, in the salt marshes, on the rocks, and in the surf. They are the awakened, empowered women Kate Chopin envisioned in 1899, the women on the beach beyond the marsh, beyond the dangerous estuary.

And the men fit for the beach are empowered, too, although as yet they receive little attention in any popular periodical genre except that devoted to shallow-water, muscle-powered boating. In "Cruising Maine's Non-Navigable Waters," a quintessential *Messing About in Boats* article of 1991, Bob Miller exhorts readers of a new sort of boating magazine to "forget about gunkholing in Penobscot Bay" and instead find "uncrowded, even undiscovered places" that "really challenge your small-boating skills."[21] But such shallow-water places are accessible only to those men and women whose muscles drive canoes, kayaks, and pulling boats through the shallowest of water and among the most daunting of hazards. Well-muscled and well-calloused, the new sort of small boater navigates for exercise, and proves his or her toughness not only by wielding a paddle or pulling an oar but by jumping overboard and dragging the craft along by skilled strength until deeper water or some rarely visited, almost secret place is reached. And the new small boater understands that arriving in some secret cove or inlet may not produce a warm welcome. "My waterfront was rotten with sea kayaks this summer, which is good in a way and bad in another," Peter Spectre reports in a *WoodenBoat* essay of 1992. "On the good side, it was nice to see people enjoying the water without dragging a cloud of purple smoke behind them. On the bad side, many of my secret coves had more visitors than I would have liked, especially during those moments when I wanted to throw off my clothes and do a buff-colored half gainer, so to speak." [22] For the old-timer cognoscenti devoted to gunkholing in dories, yacht-tenders, and other rowboats, the appearance of sea kayakers is indeed disturbing, but perhaps only temporarily. Sea kayakers are comfortable enough in their skins to be comfortable around skinny-dipping small boaters. For that matter, they do buff-colored half-gainers themselves.

In fact the "small craft revival" first analyzed in the January 1978 issue of the *Ash Breeze* has subtly altered not just the sport of small boating but the understanding of the accessibility of coastal realm areas impossible to reach by land and very difficult to reach in engine-powered craft. [23] As vessels for reaching rarely visited alongshore places, kayaks and peapods, skiffs and yacht-tenders do remarkably well only in *practiced* hands, and once the hands—and backs and legs and chests—become well-practiced, even distant places are suddenly accessible, as Peter Spectre and other savants began to notice around 1990. What the "local boat" enthusiasts of the Traditional Small Craft Association hoped somehow to preserve in 1978 has become not a pickled relic but something that is suddenly receiving sustained scrutiny in general-interest periodicals like *Country Journal*. "Rowing is not hard work. Nor is it tedious," advised Carl Kirkpatrick in 1990 in a *Country Journal* piece introducing fine pulling boats and rowing to readers familar only with engine boats. "It has a quiet rhythm." [24] But the rhythm comes only with practice, and for many onlookers anyone pulling a boat is merely exercising, building muscles for the sake of muscles, "working out." Surprisingly few onlookers realize that many rowers are headed someplace. The local boat and traditional yacht-tender—and the high-tech kayak—are remarkable vessels and are reshaping alongshore notions of waterfront risk and waterfront access.

Muscle-powered small craft moving close inshore, often nearly among breakers and sometimes through them, cruise into risk. Sometimes they cruise into greater risks, say the wake from engine-powered craft powered by careless or drunk or envious people. And always they cruise under the sun, without bimini tops or fiberglass roofs or any sort of cabins. Their pas-

sengers encounter the risk of sun, of skin cancer, just as their passengers disembarking at some marsh island encounter the risk, however slight, of invading the habitat of ticks that carry Rocky Mountain spotted fever. Equipped with life jackets and sunscreen, the local boats and yacht-tenders, seagoing canoes and kayaks carry people who have accepted the risks and tried to mitigate them, even to eliminate them. Gunkholing under muscle power advertises a willingness to accept some risk, and many contemporary onlooking Americans, particularly those struggling to help "at-risk" babies, unwed mothers, and other at-risk fellow citizens, consider the small boaters arrogant or stupid or smug in their tanned and muscled competence, in their desire for risk.

   In the wildest of American wilderness, the small boaters who accept some risks are a perplexing problem to those reformers who are attempting to eliminate all risks in life, physical and otherwise. Nowadays in fact, expressions like "children at risk" conjure up poverty-stricken inner-city children destitute of parental discipline, medical care, and good nutrition, not half a dozen boys and girls afloat in a leaky sixteen-foot rowing boat.[25] Exactly as the President's Council on Physical Fitness and Sports fears, people who lack physical strength are unable to evaluate the behavior of those who are strong and competent in a difficult but usually manageable environment. Moreover, many reformers lack the competence, experience, and even basic understanding to evaluate the degree of risk involved in shallow-water gunkholing.

What indeed of the risks of skin cancer? Already some medical researchers have noted the disconcerting prevalance of breast cancer among white women who live in northern, cloudy regions, and have begun to wonder if a lack of sunshine explains the geographic pattern that separates women in New York and Cleveland from those in Honolulu and Phoenix. Is breast cancer accelerating because so many women now live essentially indoor lives? Is colon cancer accelerating among men for the same reason? [26] So perhaps the women and men rowing and paddling out to the great unnamed beach—or running or walking along its sands—accept the slight risk of skin cancer in an attempt to avoid other sorts of cancer. And besides, high-quality sunscreen, especially waterproof sunscreen, offers remarkably good protection from the sun so long as gunkholers and other adventurers remember to put it on. But slathering on sunscreen offends some beach visitors, who find lugging umbrellas, tents, and other equipment preferable to watching nearly nude men, women, and children slather themselves and each other. Is the husband slathering his nearly nude wife engaged in something remotely prurient? Should the bikinied mother not be slapping sunscreen on the backs of her boys? Maybe everyone should wear the robes of the Syrtis desert, not nylon scraps and sunscreen.

The barefoot historian wonders about sunscreen as he slathers it on his wife and sons, and wonders more about the oddity of being on an immense sandy beach in which everyone seems fit, is nearly nude, and is very, very devoted to slathering sunscreen SPF 45 on themselves, their families, and their friends. What of risk in such an inaccessible place?

Accessibility is the rallying cry of many would-be beach users, but the cry is remarkably shrill and cracked. Increasingly, the most casual observation of beaches reveals not only the truth of Holly Brubach's comments about obesity but the effects of that obesity. Near the parking lot the typical beach is crowded, but a short distance away the sands lie nearly empty. So what of accessibility and its advocates? Is the hidden agenda to build a linear parking lot paralleling the beach and a thousand ramps through the dunes?

All along this stretch of coast public officials are embroiled in arguments with all sorts of special-interest groups, and they are arguing—often loudly—about accessibility.

Local officials find themselves squarely between the four-wheel-vehicle owners who want to drive at will along the sandy beaches and sometimes into and over the dunes and environmentalists who want to ban all vehicles from the littoral. Numerous and wealthy, the four-wheelers

not only acquire additional support from dealers who specialize in expensive four-wheel-drive trucks and cars but from the growing number of infirm middle-aged and elderly people who are no longer able to hike to favorite surf-fishing and picnic spots. The four-wheelers argue that they have enjoyed the right to drive on the beaches since the 1920s and that their "ancient right" must endure. In opposition to their driving anywhere, on dunes or even sand wet at low tide, the environmentalists argue from geology and ecology, demonstrating beyond measure that these vehicles destroy dune-grass rhizomes and cause massive—and measurable—erosion, not to mention that they force endangered species toward extinction. The tiny sand-colored piping plover, already endangered, faces extinction simply because its young are either run over by vehicles or trapped in their deep wheel ruts. But the environmentalists have new allies, too, the physically fit, vigorous men, women, and children who hike up a long barrier beach expecting to find solitude but find around a turn of dune dozens of vehicles disgorging coolers and gas grills, radios and tents and, lately, a growing number of vehicles parked facing the surf, with windows up and engines running, their occupants enjoying air conditioning. Few local officials find satisfaction in hearing the disputes, which spread to state and federal levels and spawn lawsuits. Few know how to address vehicular accessibility since they are already plunged into a similar dispute over accessibility.

How do disabled people "access" the beach? Since 1988 that question has perplexed and infuriated everyone from paraplegics to geologists, from harbormasters to judges. Is the Americans with Disabilities Act a landlubber's law, one written with scant understanding of the coastal realm? Already it creates difficulties that seem beyond political solution.

Typical wheelchairs move poorly through beach sand. About that, everyone all alongshore agrees. But what solution is appropriate? In town after town, often well before the passage of any federal or state legislation, taxpayers and groups of volunteers essayed the first "handicapped-access" ramps, essentially tightly fitted boardwalks running from asphalt-paved parking lots up among the dunes, and terminating beyond in flat, deck-like areas. And almost immediately thereafter came trouble. Sand drifted onto the boardwalks, making walking, especially for infirm people dependent on canes, difficult if not impossible. Almost everyone, including bicycle tourists eager for a view of the sea and anxious to keep sand away from fifteen-speed gears, began to use the ramps, rather than trudge through the sand. And many wheelchair-bound users, at first delighted with the ramps and the thoughtfulness the ramps represented, discovered that the drifting sand and the steep slopes made for awkward ascents and dangerous descents. Finally, in the summer of 1992, the project manager of the Massachusetts Office on Disability arrived in his wheelchair and discovered not only the blowing sand but the steep gradients. Exclude the sand and level the ramps, he announced.

Silence first greeted his dictate, then shocked disbelief, then anger.[27] To level any route through the dunes, for any reason, means to invite the winter surf to enter the dunes, to crash through them, erode them, and dump them in the parking lots and salt marshes inland. Geologists have long known what Dwight and Thoreau knew, what every alongshore child knows. The geomorphology of dunes and sandy beaches demonstrates the inland-moving propensity so evident on a winter day. Only the dunes protect, and only because their grassy slopes are built upward. Cutting paths through them is madness, and building solid vertical walls parallel to the ramps is worse, for the troughs will only fill solid with sand, and perhaps more quickly than anything William Hamilton found near the mouth of the Bannow, in the barony of Forth. Suddenly the immediate shore became a battleground between those who favored handicapped access and those who favored late twentieth-century beach-management techniques.

After all, does not every alongshore high-school student now learn that all experts favor building nothing along or on barrier beaches? And nothing surely includes boardwalks, for whatever purpose.

In December 1992, ocean itself entered the discussion, smashing inland during a great gale that flung boats, wharves, and summer cottages across the salt marshes. After the storm subsided, the first beachcombers saw what local officials, then state officials, then federal officials saw. Although the surf had come ashore in many, many places, it had sometimes apparently followed the ramps despite their slopes, and had tossed the ramps asunder, tipping them grotesquely so that they resembled U.S. Marine Corps agility-training tools.

How does one make a wilderness accessible to the disabled? Does wilderness, or at least the alongshore wilderness that drowned the Thacher family and wrecked *Tyehee*, somehow invite only the physically fit, or at least the physically able? Surely no one, not even the most strident of those demanding "accessible" beaches, expects the wet sand to be accessible, or the surf to be accessible. Or do they? Is the Americans with Disabilities Act one more landlubber's law that melts at the edge of the sea, or could technology perhaps offer a solution?

All but forgotten now, the mid-1960s national engineering search for a stair-climbing wheelchair raises questions about the breadth of the Kennedy-era emphasis on physical fitness. Clearly, in the mid-1960s wheelchairs were developed that heralded chairs that might follow paths meandering up and over dunes, that could move over sandy and rocky beaches, that could climb gentle flights of steps.[28] But between the late 1960s and the middle 1980s, the social and political climate changed dramatically, but essentially unnoticed. Rather than continue to develop wheelchairs that can climb stairs, let alone roll along soft sand and heaped-up cobbles, Americans began to devise wheelchair-accessible structures and spaces, all designed around the traditional urban wheelchair. So accepted is the change that no one thought to ask why the project manager attempted the dune ramp in an urban chair. What able-bodied man or woman attempts to walk a sandy beach in formal business shoes? Sensible people wear sandals or sneakers or hiking boots or go barefoot.

No one expects to drive a two-wheel-drive automobile more than a few feet onto a sandy beach. Why expect a wheelchair straight from a hospital or office to function any better?

And what of the pirate tradition? Were one-eyed pirates, peg-legged pirates, or hooked-hand pirates somehow "handicapped" in the new sense of the word, unable to swarm up rigging, leap onto docks, wound and kill the strongest of enemies? Even children playing with Lego toys note the eye patches, wooden legs, and hooks, and wonder, ask questions, note adult confusion.

So local officials struggle to achieve sensible solutions, knowing that for once compromise seems impossible. Perhaps only the infirm should be allowed to drive on the beach, or perhaps the number of vehicles can be limited, especially on sunny Saturdays. Perhaps a flexible steel mat might be run across the dunes to help wheelchair users and those who walk with canes, but what of problems posed by drifting sand, and the continued existence of a grade? Can the arguments be fairly argued, in the town halls so near the sea?

Out-of-shape people not only require very little from the outdoors, they require very little outdoors at all. Do they support "handicapped access" secretly knowing that once accessible to people in wheelchairs a place is accessible to the fat, the lazy, the out-of-shape? As Americans become less vigorous, will the entire outdoor experience, and especially the alongshore experience, change?

Has it changed already? No longer can a landscape architect specify bridging a salt creek with a simple deck. Absolutely not. A footbridge must be *safe*, essentially foolproof, risk-free, have handrails, and must be handicapped accessible, too, in many situations. No matter that such a bridge encourages bicyclists and motorcyclists to use it, and indeed must be built heavily enough for the motorcyclist who may use it even though the signs prohibit such use; it is the bridge necessary to a society that can no longer balance, that has lost its common-sense balance. Every new public structure built along the shore must be safe, risk-free. The simple, traditional boardwalk, snaking unobtrusively across a marsh or sandflat so unobtrusively, now sports massive handrails noticeable half a mile away, handrails that have no opening greater than four inches square, so that no crawling infant may get stuck.[29] Every new-built wharf, bizarre as it sounds, must have railings, in some cases railings strong enough to stop rampaging fish-company trucks. How does one unload a fishing boat when the wharf has waist-high or higher railings? Should the railings be demountable? And if so, who will hold the key? And who will pay for such complicated structures built in such difficult situations?

When society no longer maintains the narrow causeway, the boat-launching ramp, the boardwalk, the footbridge, when such public-access structures become unsafe or even collapse from assaults by boring mollusks or winter gales, then come the vigorous cognoscenti, and they balance across the stringers, or even wade, finding on the other side some Eden perhaps, or at least a less-frequented place.

And what of these physically fit, risk-taking, exertion-minded individuals? How numerous are they? How do they think about matters? Do they consider themselves normal or special, or at any rate somehow distinguishable from the physically unfit?

The barefoot historian notes the ever-stronger alliance between coastal realm wilderness conservationists and the physically fit, and explores its implications every summer, hearing in casual conversation something no longer marginally important. Already the alliance re-shapes beach use, arguing that four-wheel-drive vehicles damage dunes and frighten nesting birds—and never saying that when the vehicles are banned for ecological reasons, the outer reaches of the great barrier beaches will be accessible only to those who walk barefoot across the sand. As human-services needs consume ever great portions of total state and federal funding and structures from boat ramps to boardwalks decay, more and more outdoor areas become accessible only to those fit for them, those who read articles like Miller's "Cruising Maine's Non-Navigable Waters" and determine to carry a canoe a mile or so to a hidden salt creek, those who work up a sweat hiking far into rough country, those who munch trail mix rather than junk food, those who rejoice that the outdoors grow emptier by the year, those who quietly scorn the seasick boy.

Environmentalists are very often in shape enough to walk, row, paddle, and even splash through cold, clinging mud to gain the secret beach, let alone explore a modest woods or large park. But right now, any environmentalist who explores some hard-to-reach coastal place is likely to enjoy the relative solitude with little thought to its implications. Geology, estuarine ecology, climatology, and other sciences have much to offer, but perhaps some site-specific sociology, psychology, and nondisciplinary hard thinking might do more to locate the secret beach in end-of-the-century American culture, to explain why environmentalists of all ages are so frequently physically fit, enamored of muscle-powered local boats and ultra-light kayaks, and outfitted in sensible clothing that reveals firm flesh backed with muscle. Environmentalists did not predict the clashes between the active and the passive, say the clashes between hikers and all-terrain-vehicle-riders, skiers and snowmobilers, windsurfers and motorboaters, nor do they foresee much now, say extrapolate anything about the clash between those who row and paddle small craft and those who operate motorboats. Elitism suffuses the clash, but the elite is difficult, in fact impossible, to identify. Are the elite the fashionably dressed people in the twenty-thousand-dollar speedboat or those in faded clothes paddling the five-hundred-dollar sea kayak? Are the elite those almost run down or swamped in giant wakes, those constantly threatened, terrified of powerboats? Finger-pointing experts on political correctness shiver at deciding. But the people fit for the secret beach, if not now an elite, are certainly an elite of the late 1990s and beyond, and they have figured out the odd dividends of what might be called "rigorous accessibility," the delights implicit in a coastal realm area uncrowded and unspoiled simply because it is accessible only to those who exert themselves. They use ecological argument to control growth and curtail access, and already they explore the subtle dividends of governmental poverty and rising human-services budgets. The clash between fit and unfit, active and passive, healthy and infirm, whatever the designation, must trouble every beachcomber and everyone of goodwill, for ecological argument pulses as a foil for social engineering.

Just along the coast from the hidden beach, four-wheel-drive enthusiasts clash with ornithologists trying to foster the nesting of piping plovers and least terns. Prohibiting motor vehicles in nesting habitats sparks rage. So bitter has the clash become that it is front-page news, especially since the Commonwealth of Massachusetts has begun to consider banning all vehicles from the barrier beach on which these shorebirds nest. Public access and handicapped access are not issues: no one talks about banning walkers from strolling a few miles from the parking area, from hiking far out along the beach. The debate concerns vehicles only, and many drivers now hate the plovers, for plover-protection threatens air-conditioned vehicular access to the outer beach.

To walk far up that beach on a summer day is to walk into the social trouble that swirls around beach environmentalism, around coastal realm—management thinking. The far outer beach is the playground for a sort of people every bit as definable as those on the hidden beach, but different. Children surrounded by toys, senior citizens watching the waves from air-conditioned, engine-idling trucks, middle-aged men fishing with four surf rods, women delving into two or even three coolers at a time, and everyone sitting on aluminum-frame chairs—the people comprise a completely different population than that a few miles away. For the barefoot historian, the place offers immediate and jarring confirmation of reports from his beach-exploring friends. The people in the four-wheel-drive vehicles, even in the thirty-thousand-dollar vehicles, are fat.

And those people have begun to stare at the hikers who reach their hideaway not after driving a few miles but after hiking that distance. What lies behind their stares? In their eyes, is the lithe man who strides along the beach a dark-visaged pirate, some half-forgotten ghost come to stir up trouble, some reminder of sexual virility different from that implied in advertisements for V-8 engines and coolers of beer? Is the woman, mother of four, also striding along even worse trouble? "I'm still wearing the bathing suit I bought in 1967," reports the lithe forty-three-year-old mother of four of a suit she bought when she was eighteen and exploring the beach she now visits once each year on a sort of anthropological mission to study "mostly overweight bodies" and "bottlecaps and boomboxes." The barefoot historian understands her thinking about beach-walking attire easily enough, reading in her letter simply Yankee common sense. "It was a good, tough suit back then. I bought it to last and, by God, it has lasted. It fits fine, I'm used to it, and so I see no reason to replace it. It's been out of style for decades, but I wear a bathing suit to swim in, not for fashionable sun bathing."[30] But how do the four-wheel-drive people evaluate the woman and her four daughters? How, especially, does any fat woman evaluate another woman who fits so perfectly the suit she wore in high school? How many women and men her age fit into suits they wore as seniors in high school? Holly Brubach finds few on the beaches she visits for the *New Yorker*, and perhaps she concludes that few such Americans exist. But they do exist, they do delight in the coastal realm, and they do dislike the stares of people ill at ease in the marginal zone, people who bring a hundred symbols of status to the place Thoreau said "is a wild, rank place, and there is no flattery in it."[31] Those overweight adults—and perhaps their overweight children—half scorn and half envy the few lithe, nearly naked people who move effortlessly, tirelessly past the conglomeration of parked cars, moving fast not out of pride but to escape the stares of men and women who watch something that disturbs them.

Those walkers, those *chasse-marées*, barefoot, sans equipment, almost sans clothes, perhaps have a different impression of the piping plover, and perhaps wish the little bird better protected. For when it is, only they will find their way to the outer beach, and only then will they suffer no unfriendly stares. And only then will they be able to accept the beach for the risks it offers naturally to those essentially au naturel, to those more afraid of being run down by some racing four-wheel-drive drunk than by drowning in the undertow.

In the meantime, the vigorous, the physically fit, learn about the secret beach, and determine to purchase a used pulling boat or a canoe, perhaps a fiberglass kayak, and go adventuring into intimate contact with the natural realm, risking a little, exerting a lot, joining the like-minded visitors unafraid of deep black mud. As the average age of Americans increases, as infirmity and obesity work their will on children, teenagers, and adults, the link between coastal-realm environmentalism and physical fitness becomes stronger and starker, for the physically fit have already discovered the dividends implicit in the decrepit boat ramp and silted-in channel, the fallen wharf and grown-over path. Quiet and muscled, clad in the brief bikinis and the loose-fitting shirts one mail-order house calls "pirate shirts," the cognoscenti smile and turn away from the engine-boat aground in black mud, from the mired four-wheel-drive vehicle surrounded by a rising tide. Like Jewett's *Marsh Island* heroine, like Chopin's swimmer in *The Awakening*, like the boy boatbuilders of Elijah Kellogg's *Young Shipbuilders*, they move with power and certainty everywhere in the marginal zone. And always they push on, for right behind them puff crowds of flabby people demanding improved access, especially bridges, flat roads and paths, and paved parking lots.

In the shallows danger comes silently, usually with the turning tide, sometimes with changing wind. Canoes heel over as paddlers struggle to maintain course, kayaks nose into the channel and pick up speed, rowing boats move slowly into the current, long oars beating in the ancient, choppy stroke that builds speed amid current. Here and there tiny outboard engines start, forcing slightly bigger boats against the ebb. Only a little danger spices the exertion of leaving astern the hidden beach, unless big motorboats surge along the channel. Then the shaking heads, the shouted orders to slow down, the water flipped from the kayak paddle into the boater's face indicate something of what divides two groups afloat on the same estuary.

All alongshore, thoughtful observers ponder the growing numbers of individuals, couples, and families afloat in the tiny muscle- or sail-powered vessels that poke around hazards, gunkhole along salt creeks, probe sandbanks and surf. In some inchoate way, the thoughtful observers half know, half sense that the vigorous few rightfully know the coastal realm, the hazards, the marginal zone, for the realm rejects the structures so beloved by the flabby, the inactive, the visitors unfit for the marge almost as quickly as it rejects the flabby, the inactive, the lazy themselves. Only those who exert themselves, who accept the risks, reach the secret beach and enjoy the realm through which Conrad hoped only to pass on his way inland, or to the safe, open sea.

13 THE COASTAL REALM

Sundown signals a general exodus from the beaches. Only Sny, the author's long-distance research vessel, remains as day ends, waiting while its owner scrutinizes the emptying beach. (JRS)

Opening oneself to the sea, hearing the rote, watching the glim, feeling the sun and the wind in the end gets one only relaxed, tanned, salted, and perhaps thoughtful. To know the coastal realm is to dig, probe, slop about in the ooze, explore the marge, risk a soaking or worse. At least along this stretch of coast, *realm* adequately designates the vague area *seacoast, seascape,* and *marge* never precisely name.

Certainly *realm* denotes a place ruled by someone, usually a king, but the term has long designated a primary marine faunal division, too. Unlike *country* or *nation, realm* suggests that a place takes its character, its very identity, from its ruler, not from anything intrinsic. And because the sea rules the whole territory of this book, and touches even some distance inland where its gales and salt air and fogs reach routinely, *realm* is a remarkably workable term.

Ruling proves hard to accept, especially in a Republic, and perhaps especially in a Republic so enamored of technology and so worried about environmental degradation. Nothing people do has much effect on the sea. Salt-marsh dikes and granite seawalls crumble only slightly more slowly than wooden wharves, and the biggest ships blundering among inshore hazards last only moments longer than *Tyehee* and smaller craft. Beyond the inshore ledges left dry at high tide, every building technology fails so absolutely that no architect, no engineer considers building anything in deep water. Whatever builders may do to mountains and swamps, forests and deserts, they can do little to the sea. The sea mocks those who look at it with an eye toward real estate development. It is not real estate, but realm.

Nowadays far more tourists watch the coastal realm than watched just years ago. As always along this stretch of ordinary coast, no one has exact or even haphazard figures, but obser-

vant locals notice the growing numbers of spectators and the declining numbers of people who actively use the sand, the rocks, the surf, the marsh creeks, the mudflats. Whatever else environmentalists accomplished in the 1970s and 1980s, they slowed the "development" of the land touching the water. To be sure, they slowed or stopped the building of hotels that touched the surf, of condominium complexes that sprouted on filled-in salt marshes, of paved roads that led among the dunes. In so doing they reaffirmed older, sometimes forgotten understandings of the coastal realm as essential wilderness, as a place that defies construction, as a place fit for the fit only, for able-bodied seamen and able-bodied landsmen.

In countless scientific journals, geomorphologists, ecologists, and other experts argue that the "interface" of sea and land ought to remain "natural," unshaped, unbuilt, uncontrolled. "Rather than trying to hold back the sea, coastal management policies should aim to prevent development in threatened areas," assert Orrin H. Pilkey and William J. Neal in "Save Beaches, Not Buildings," an *Issues in Science and Technology* article of 1992 aimed at ending "beach degradation." Although some professions still strive to shape and build, and report their fleeting successes and failures in *Landscape Architect and Specifyer* and other journals,

Now and then ocean itself reaches far into the coastal realm, as it did one day in the late nineteenth century, destroying the new-built summer homes of visitors unlearned in invasion. (Society for the Preservation of New England Antiquities)

In the aftermath of a 1990s winter gale, not only houses but entire neighborhoods stood topsy-turvy and nearly buried in sand. (JRS)

scientists now have experience and studies on their side.[1] And suddenly, in a long end-of-century period of financial stringency, both private owners and local and state governments find themselves unable to bear the rising costs of maintaining structures built in the coastal realm, in places *threatened* by the sea. In many alongshore places, lack of private and public funding has created a vast experiment in minimal maintenance, or no maintenance at all, and houses, roads, handicapped-access ramps, boat moorings, and wharves are swept away by winter gales or buried under drifting sand.

The stretch of coast from Gurnet Light to Minot Ledge is a poor example of the dangers of "developing" land adjacent to the sea, or even adjacent to the dunes or cliffs.[2] After almost four centuries of colonization, much of the stretch is still scarcely developed. Long-standing poverty and sometimes a deep distrust of risking limited capital near guzzles and dunes keeps structures few. As a site for testing coastal-zone management methods, the stretch raises awkward issues. "Many more storms like the last gale," one lobsterman tells the historian, "and there won't be anything back of that beach to manage." For decades the locals have been careful where they build, and only a rash of winter no'theast storms has raised new worries that global warming, some tropic-spawned, piratelike sea monster, stirs off New England. Newcomers build million-dollar houses where no local would stable sheep, and suffer storm damage that boggles old-timers. But few millionaires arrive. The coast is too ordinary, still too much a "desert of sand-hillocks," to use Melville's phrase.[3]

Yet the wildness, the openness, the naturalness attracts tourists, but nowadays tourists who want to *watch*, not do. Every year, locals note that more and more tourists arrive knowing less and less about the sea, about the land that touches it, about the activities of lobstermen, marine-railway operators, swimmers. And rather than plunge in, try surf-fishing or beach running or seaweed collecting, they stand back, safe behind sunglasses and binoculars, content to watch. Unlike the local woman who stands in the spill of the surf, staring outward at some sloop, some freighter, some unknown vastness, they remain on boardwalks and parking lots, and watch the woman watching vastness. Less and less often do they move from pavement to sand and cobbles and marsh, less and less often do they wade and splash and swim, less and less often do they rent boats.

Boat liveries vanish, but no one mourns or tries to preserve them. Rising liability insurance premiums force many livery owners to close. Insurance firms charge too much to insure

Whatever else will happen in the next few years, fishermen will no longer scrape barnacles from the hulls of their boats. Federal regulations now ban such traditional alongshore activity, for scraping removes poisonous paint that pollutes harbors. (JRS)

owners against the dangers encountered by inlanders renting a rowboat, or even an outboard-powered boat, or even an aluminum canoe to explore shallow salt-marsh creeks. Locals sometimes wonder at the gradual extinction of liveries, but every summer the newspapers print stories of incredible foolhardiness. Inlanders rent tubby, slow, impossible-to-capsize rowboats equipped with foam flotation and twice the number of required life-saving devices, and row straight out to sea, to exhaustion. Others rent poky powerboats and ram sea-kayakers. Many simply get drunk and blunder into the surf, or take the ground on distant mudflats. At evening, the livery owner waits in the shack, all but one skiff made fast to the landing staging. He waits, wonders, radios the harbormaster, asks the locals returning from fishing. No one has seen the gaudily painted rental powerboat. Has it sunk? Rammed a moored yacht? Drifted off toward Portugal? And will the renter, enraged over defective oars, outboard engine, directions, sue?

All alongshore, locals ponder the odd unwillingness of tourists, and permanent-resident newcomers, to embrace the physical environment they have come to visit in the old sense of *visit*, to visualize, to see. Why not try swimming in the surf? Why not try hiking far up the barrier beach? Why not try baiting a fishhook with a worm that bites? Why not, when the local diffidently offers the watcher a chance at doing something, do? Why simply stand and look, as if the whole coastal realm concatenation existed as something inside a nineteenth-century aquarium?

Is the sea now merely a theatrical setting, a mere scene, something to be watched from afar, as Thackeray's painter watched?

In the year 1922, Lincoln Colcord addressed himself to a related, equally difficult issue, the definition of the term *sea-story*. In a penetrating introduction to his *Instrument of the Gods and Other Stories of the Sea*, the seaman-turned-writer struggled to define a literary genre known to two groups of readers, the landlubbers and the mariners. Moreover, he struggled in an era when writers like Joseph Conrad had simultaneously transformed the genre and almost destroyed it, at least for many well-educated mariner-readers.

Landlubbers, Colcord determined, accept the sea "as a source of romantic plot material" and expect "unusual and exciting happenings in its sea tales, scenes of elemental struggle, of broad comedy or tragedy." In contrast, "the sailor himself, the native of the wilderness of waters," thinks of the sea "as a place where life goes on," and understands the sea-story as "a story of life touched by the influence of the sea." In the end, Colcord determines that if the sea has a character, or indeed is a character in so much sea-focused fiction, it is a mild character indeed, "notwithstanding the countless disasters of the deep, the gruesome record of maritime adventure." For Colcord, for any offshore mariner, it is land that is dangerous. "I have been afraid of the land many times, but never of the sea," he concludes in a passage deliberately ambivalent, deliberately mingling the insidious moral and social dangers of land-based society with the ever-present dangers inshore hazards present to the long-passage mariner attempting to make land.[4] Danger begins in soundings, where the seas run steep in the shoaling water, and danger worsens near hazards, near the coast. Inshore everything goes wrong, gets mixed, gets dirtied by the land, by dirt itself.

Land as dirt, as dirtiness, as pollution incarnate suffuses much writing about the sea, and particularly that composed by mariners like Tomlinson. His "Outward Bound," written in 1948, emphasizes the filth of the urban harbor: "A space opens in the smoke and fudge to let out the stars."[5] Even after World War II, Tomlinson catches the cleanliness offshore, the cleanliness in which even steamship seamen live out their lives, however sordid and sooty their lives may be. Whereas Colcord, himself born aboard a square-rigger in mid-Pacific passage, knew firsthand the tension with which mariners aboard the difficult-to-maneuver sailing ships approached any land, Tomlinson went to sea in a far more certain, far more mechanical time, a time of coal smoke above and ashes below. But Tomlinson, too, understands the dirtiness of the land and the pollution that great rivers sweep far out to sea.

At the beginning of the twenty-first century, however, the sea-story has become something far less definable, and something far less common in bookstores. Perhaps the audience has shrunk. No longer do educated Americans have the routine experience of ocean-liner passages. Then, too, the supply of experienced authors has shrunk. Even the most desperate, most industrious of lads learns that running away to sea requires a union card and seniority, and the few Americans who officer and crew oil tankers and container ships write only very infrequently about their discoveries, adventures, and boredoms. Although sea novels sell spectacularly well as "serious" historical fiction, as the Jack Aubrey–Stephen Maturin series by Patrick O'Brian demonstrates year after year, contemporary stories about contemporary seafaring are few, as few as genuine sea-pieces, marines, or seascapes.[6] Increasingly, stories of the sea are stories of discovering the sea by landsmen like Harvey Oxenhorn, whose *Tuning the Rig: Journey to the Arctic* appeared in 1990. The title jars as a ship taking the ground jarred Conrad. Something sounds wrong. And of course it is. *Journey* is the wrong word, connoting a walk, a hike, a horseback ride, anything but sea passage. Perhaps "voyage" is a better term, for after all the *Regina Maris* is on a scientific expedition, not exactly making a passage between two ports.[7] But *journey* grates and grinds from the first, convincing any reader before Oxenhorn says it that Oxenhorn is making his first trip to sea, that he still has both feet planted on land.

Ocean grows more alien every year. Even locals begin to speak of it oddly, in ways that make other locals start, then listen closely. "Mother Nature is fighting back the Atlantic Ocean," one local, a harbormaster, tells a newspaper reporter inquiring about beach erosion. "The ocean has been winning, but last night was the first time in a long time Mother Nature turned the scales a little bit."[8] Is ocean no longer part of "Mother Nature"? Is the barrier beach laced with snow fence and planted dune grass more natural than the surf? Or is ocean only threat, enemy, alien ruler, a natural force that mocks anything people do to contain it, control it, even use it in regular fashion?

At the close of the twentieth century, the juxtaposition of sea and land grates on everyone unaccustomed to marginal existence. At the actual, immediate touching of sea and land, too many subjects become too messy, lost in the turbulence of surf and wind and tide. If a sea-story is a story influenced by the sea, what then is an alongshore story?

Perhaps it is purely marginal. To define *coast, seashore, alongshore,* even *seascape* demands a willingness to look out not so much toward the offing, into the offshore glim, but inland, too, toward that edge, however vaguely defined, that marks the inland boundary of alongshore setting and alongshore knowing. Inland of that edge, people think in terms of control. In the marginal zone, everywhere in the coastal realm, locals, or at least some locals, think otherwise, and accept the risks a handful of nineteenth-century writers like Elijah Kellogg detailed.

In *The Young Shipbuilders*, Kellogg told juvenile readers in 1870 that exploring the coastal realm means exertion, means acquiring local knowledge of seamarks and currents, landmarks and hazards, means not only building one's own boat but accepting the possibility of wrecking the boat—and living to exert and learn and build again. The whole of his book is about risk in the coastal realm, about taking one chance after another because only chance-taking works. No one in *The Young Shipbuilders*, and especially not the women, merely watches. Everyone actively engages the coastal realm.

Kellogg's novel gathers dust now in alongshore libraries, but his lesson endures, for all that some children and adults find it unsettling. Only those locals and visitors who actively engage the coastal realm develop any deep acquaintance with it, and with the realm itself. Only they get well and truly salted, get the taste of the realm in their mouths, the feel of it under their

The beachcomber finds all sorts of things along the wrack line, including this naked Barbie doll. The doll reminds the beachcomber that voluptuousness matters little in the coastal realm, that strength matters far more. (JRS)

bare soles, the swirl of it over their skin. Only they put themselves in the sway of the coastal realm, temporarily accepting their marginality in a realm of immense force. Only they know how puny, how marginal inland technology and inland thinking become when set down along-shore. Only they know something of gunkholing and guzzles, marsh islands and salt creeks, wharf pilings and watchtowers, pirates and breaststrokes. Only they know that all along the coastal realm, small boaters probe as explorers probed centuries ago, in a marginal place, in the limicole zone between sea and land, all alongshore.

NOTES

INTRODUCTION

1. Lowell, *Fireside Travels*, 155.

2. Dictionary definitions of alongshore terms demonstrate beyond doubt the long-standing vagueness of American lexicographical understanding of the coastal realm.

3. Nordhoff, *Cape Cod and All Alongshore*, 147. The dictionary is formally titled *An American Dictionary of the English Language*. It is useful to follow this term through other general and specialized dictionaries. It does not appear in the *American Encyclopedic Dictionary*, for example, but the *New Standard Dictionary* defines it as both a view of the sea (which usage it traces to 1890) and a sort of picture. As a sort of view, not a sort of painting, the term appears as early as 1890: "This broad seascape was broken exactly in the middle by the Bishop's Rock and its stately lighthouse rising tall and straight out of the water" Besant, *Amorel of Lynonesse*, 2. Besant clearly understood that a *little* land did not make a "landscape," and thereafter lexicographers cited his usage.

4. Thackeray, *Shabby Genteel Story*, 86–87. Discussions of the meaning of *sea-story* fail to illuminate the definitions of *seascape* and *seashore*; see, for ex-ample, Colcord, *Instrument*, vii–xvi, for an insightful analysis of the defining elements of sea fiction.

5. "Fascination of the Sea," 800–801.

6. The expression "all alongshore" once had national currency. See, for example, Nordhoff, *Cape Cod and All Along Shore*, and Lincoln, *All Alongshore*.

7. Spenser, *Poetical Works*, 255.

8. *Webster's Third New International Dictionary*. It is instructive to note that this edition is the first to attach the connotation to the term. *Webster's New International Dictionary* offers only the denotation.

9. Maugham, *Moon and Sixpence*, 253–254. For this quotation I am indebted to my wife, who has reason to muse on the character of wreckers. See also Wasson, *Cap'n Simeon's Store*, 159.

10. Stevenson, *Wrecker*, esp. 10–11.

11. Thoreau, *Cape Cod*, 68.

12. Conrad, *Mirror*, 100–101.

13. U.S. Department of Commerce, *United States Coast Pilot*, 3:68–70.

14. Lindbergh, *Gift*, 17.

I. GLIM

1. R. C. Carrington, "Table of Distances" *Marine Survey of India* 40 (London: Her Majesty's Government, 1886), n.p. See also Taylor, *Haven-Finding Art*, 63, and Thomas, *Last Navigator*, esp. 26–27.

2. Hart, *Romance of Yachting*, 305. From the masthead of a late eighteenth-century warship, the lookout could scan some 1,200 square miles of ocean surface; if he searched for a tall floating object, his range increased beyond that (see Lavery, *Nelson's Navy*, 300).

3. The United States abandoned the old Admiralty or nautical mile for the new on July 1, 1954. See *Webster's Third International Dictionary*, under "nautical mile." Locke, *Essay*, 1:32. See also Thoreau, *Journal*, 7:437. As early as 1705, George Berkeley wrote about such matters in his *New Theory of Vision*, 44–45.

4. Jefferson, *Notes*, 81–82. Shaw, *Travels*, 1:133. See also Olmsted, *Journey through Texas*, 167–166, and Zurcher, *Meteors*.

5. Melville, *Moby-Dick*, 12–13. Information concerning USS *Constitution*, a warship open to the public at Charlestown, Mass., from Officer of the Deck, USS *Constitution*.

6. Brocklesby, *Elements of Meteorology*, 176–191. See also Taylor, *Haven-Finding Art*, 79; Tyndall, *Forms of Water*; and "Sea."

7. Thoreau, *Cape Cod*, 47, 214–215.

8. Wasson, *Home*, 7. See also Ruskin, *Elements of Perspective*.

9. Thoreau, *Cape Cod*, 47. Thoreau, *Journal*, 7:441.

10. Freneau, *Poems*, 274. For information on calenture I am indebted to Gwen Lexow.

11. Josselyn, *Account*, 9. Some understanding of calenture infuses Amy Lowell's "The Note-Book in the Gate-Legged Table," in *East Wind*, 90–106.

12. Whittier, *Complete Poetical Works*, 286, 294.

13. Thoreau, *Cape Cod*, 76, 141; see also 45–46.

14. Chopin, *Awakening*, 25, 189, 29, 47. See also "Fascination of the Sea," *Outlook*.

15. Darwin, *Tides*, 5–6.

16. Thoreau, *Cape Cod*, 122. Phrases like "we were obliged frequently to empty our shoes of the sand" make one wonder how much time he spent sitting, emptying his shoes.

17. Thoreau, *Cape Cod*, 122–123.

18. Stevens, "The Snow-Man," in *Poems*, 23.

19. McClennen, "Finding Your Way," 102–105. See also Noble, *After Icebergs*.

20. Casey, *Spartina*, 57.

21. The painting is reproduced in color in Brewington, *Marine Paintings*, 73.

22. See, for example, Stebbins, *Martin Johnson Heade*.

23. Thoreau, *Cape Cod*, 137–138.

24. Thoreau, *Journal*, 7:171.

25. Brocklesby, *Elements*, 167–168.

26. See, for example, Fisher, *Color in Art*.

27. The sheet is available at almost all photographic-supply stores.

28. "Artists' Oil Colours" (Harrow: Winsor & Newton, 1989).

29. The firm is the Hancock Paint Company, with whose products I paint my house, barn, and boats. On the understanding of coastal-realm color, see Burwick, *Damnation of Newton*; Sloane, *Visual Nature of Color*; Hopfner, *Wissenschaft wider die Zeit*; and Northern Eye Institute, *Seeing Contour and Color*.

30. Thoreau, *Cape Cod*, 206. Thoreau wrote at the beginning of an era especially intrigued by chromatics and long-distance perspective; see, for example, Rood, *Modern Chromatics*; Chevreul, *Principles of Harmony*; and Ruskin, *Elements of Perspective*.

31. Langland, "Alongshore," 61–62. For general background concerning the context in which the photographers worked, see Stilgoe, "Popular Photography."

32. Hanson and Turner, "Provincetown Pilgrimage," 238. See also "Seascapes."

33. Neary, "Sand-Dunes," 27. See also "On the Beach."

34. Davis, "Vessels," 136, 138, 140.

35. See, for example, Davidson, "On the Beach"; Wohlrabe, "Catching the Breath of the Sea"; Bell, "Fog Photography"; Stebbins, "Photography"; Dugmore, "Landscape and Marine Work"; Kincheloe, "Landscape Photography"; Stiles, "Summer Work"; Riley, "Masterpieces"; W. Johnson, "Wonderful Marine Photography"; Lardner, "Beach Photography"; and Brownell, "Nature Photographer." Symonds, "Perils of Marine Photography," deals with naval photography.

31. Thoreau, *Cape Cod*, 227.

2. HAZARDS

1. Stevenson, *Treasure Island* [c. 1959], 26–27.

2. Bradford, *Mariner's Dictionary*, 76. See also Rogers, *Origins of Sea Terms*, 59, and Dana, *Seaman's Friend*, 105.

3. Francis Higginson, "To His Friends in England, July 24, 1629," *Letters from New England*, 20–21. See also Romans, *Chart of the Coast*.

4. On the longevity of the dipsey lead, see Clifford, *Charlie York*. This biography demonstrates beyond doubt that many seventeenth-century terms and activities survived into the era of electronic navigation.

5. The derivation is given in *Webster's New International Dictionary*.

6. Rogers, *Origins of Sea Terms*, for example, misses the term completely.

7. See, for example, Lenfesty and Lenfesty, *Gunkholer's Cruising Guide*, and Dana, "Business of Cricking," in Devine, ed., *Blow the Man Down*, 164–170. Shellenberger defines the term in *Cruising*, x–xi. See also Eyges, *Practical Pilot*, esp. 2–18.

8. Bradford, *Mariner's Dictionary*, gives *gurnet* as a synonym. His is the best definition of *gunkhole* I have found.

9. *Webster's Third New International Dictionary*. This definition is especially vague. Nowadays mechanics use *gunk* to designate *dirty* grease and oil. A tradename product, Gunk, removes grease and oil from engines and hands.

10. Champlain, *Voyages*, 2:118–126. See also Smith, *History of Chatham*, 7–12.

11. Bradford, *Of Plymouth Plantation*, 59–63, xxiv.

12. Smith, *Description of New England*, 22–23. See also Bolton, *Terra Nova*; DeCosta, *Place of Cape Cod*; and McManis, *European Impressions*.

13. Rosier, "True Relation," 31.

14. Spenser, *Poetical Works*, 5. Spenser published *The Faerie Queen* in 1596.

15. *Journal of the Voyage of the Sloop*

Mary, 27.

16. Southack, *New England Coasting Pilot,* plate 3.

17. Gulielmus Hack, "Map of New England" [1690], MS in Pilgrim Society Library, Plymouth, Mass. Wood, "The South Part of New England, as it is Planted This Year, 1634," *New-England's Prospect,* frontispiece.

18. "Description of Cape Cod," 151.

19. Bradford, *Of Plymouth Plantation,* 68. See also Nickerson, *Land Ho!* esp. 55–143.

20. *King Lear,* III, ii; *Troilus and Cressida,* V, ii.

21. *Vocabulario queechi-español,* under *hurakan.*

22. Winthrop, *Journal,* 2:128–132.

23. Thacher, "Letter," in *Letters from New England,* 168–174. For other "sea deliverances" of the New England colonial period, see Mather, *Magnalia,* 2:343–354.

24. Deodat Lawson, "Threnodia" [1693], *Genealogy of the Descendants,* 26–31.

25. Romans, *Chart of the Coast.*

26. Eliot, *Four Quartets,* 36–37. On the name of the hazards, see Reed, "Rockport Names."

27. Gilchrist, "Wreck at Cape Ann." This essay is a masterpiece of maritime prose and a heroic effort in self-examination.

28. Locke, *Essay,* 1:31.

29. On deep-water anchoring, see O'Brian, *H.M.S. Surprise,* 268–270.

30. Melville, *Redburn,* 296.

31. Thoreau, *Cape Cod,* 3–9.

3. GUZZLE

1. Dwight, *Travels,* 3:66–67, 2:139.
2. Bruce, *Travels,* 4:552–556. Other

explorers saw similar pillars, of course; see, for example, Browne, *Travels,* 282.

3. Shaw, *Travels,* 1:238.

4. Marcy, *Prairie Traveller,* 44–47. Miller, "Quicksand," offers a fine fictional treatment of river-crossing efforts stopped by quicksand.

5. Abbey, *Desert Solitaire,* 104–105, 138–143.

6. Camden, *Britain,* n.p.

7. Boccalini, *New-found Politicke,* 1:42.

8. Milton, *Paradise Lost,* 2.938–941. The Revised Standard Version of the Bible uses "Syrtis," not "quicksand." For the cultural significances of sand in European culture, see *Handworterbuch,* under "sand."

9. See Anthon, *New Classical Dictionary,* 848–849. On the voice telling the master of the African vessel to announce the death of Pan, see Conrad, *Mirror of the Sea,* 151.

10. Beechey, *Proceedings,* 113–280. See also Shaw, *Travels,* 1:122–123. The contemporaneous prevalence of quicksand in African and Near Eastern deserts is intriguing, for few contemporary travelers mention it.

11. Gilpin, *Observations on Several Parts of England,* 2:127–143; the quotation is on 2:134. Gilpin found the same sort of dry-crust surface and bottom-welling water that Beechey discovered later in Africa. Unlike so many observers of landscape, Gilpin wanted to know how a quicksand *works,* not merely what it looks like. See also Fontane, *Wanderungen,* 2:14–42.

12. Ive, *Practice of Fortification,* 16–17.

13. On the Tripoli pirates, see Allen, *Our Navy,* esp. 146–159. On purchasing local vessels, see *Naval Documents Related to the United States,* 6:132. On the

British view of the corsairs, see Lane-Poole, *Barbary Corsairs.*

14. Scott, *Bride of Lammermoor,* 135, 272. Collins, *Moonstone,* 28–29, 31, 138, 176, 338–340.

15. Each year I encounter at least one student who knows quicksand only from these films and who fears it tremendously.

16. My female students often report that when they are embarrassed on the beach they wish "the sand would swallow them," then suddenly hope it never does.

17. Thoreau, *Cape Cod,* 52.

18. Stilgoe, *Shallow-Water Dictionary,* 27–28.

19. Wasson, *Home from Sea,* 94, 168.

20. *Century Dictionary,* under "gutter." For an explanation of the connotations of secrecy, dating to Chaucer, see Stilgoe, *Shallow-Water Dictionary,* 29–30. Thoreau, *Journal,* 10:63.

21. Deane, *History of Scituate,* 23.

22. Geller, *Pilgrims in Eden,* 15–21. Chamberlain, *Documentary History,* 2:412–416.

23. Thoreau, in *Cape Cod,* describes the continual picking of the beach.

24. Winsor, *History . . . of Duxbury,* p. 23.

25. Samuel Davis, "Letter to John Davis" [April 3, 1829], MS in Samuel Davis Collection, Pilgrim Society Library, Plymouth, Mass.

26. Many seventeenth-century deeds recorded in Plymouth, Massachusetts, refer to land long lost to the sea; see deeds beginning in Book II of recorded deeds. On the geomorphological process, see Dean, "Managing," and Dolan and Lins, "Beaches."

27. Wright, "Curious and Exact Relation," esp. 723.

28. Hamilton, "Memoir." This source is excellent on tidal scouring, too.

29. See, for example, "Remains of a

Large Forest," and Taylor, *Geology*.

30. Lyell, *Principles*, 306–307, 718–729; Thoreau, *Cape Cod*, 238–239.

31. "On the Nature of Soils."

32. "Letter" [1637], *Letters from New England*, 218.

33. Thoreau, *Cape Cod*, 90–116. See also Melville, *Redburn*, 294.

34. Phinney, "Cultivation of the Pitch Pine," 208, 210.

35. Stevens, "Anecdote of the Jar," *Poems*, 21.

4. SALT MARSHES

1. Kodak Reference Sheet E-73.

2. Stilgoe, *Shallow-Water Dictionary*, 37–40.

3. Thoreau, *Cape Cod*, 226. Thoreau acquired his interest in coastal-zone colors, in part at least, from Gilpin; see 137.

4. The geologist is quoted in Townsend, *Sand Dunes*, 188.

5. Lowell, *Complete Poetical Works*, 70.

6. On photography and attitudes toward scenery, see Stilgoe, "Popular Photography."

7. Stebbins, *Martin Johnson Heade*, n.p. Clark, "Summer," 490.

8. Jewett, *Marsh Island*, 66–67, 79–82, 129–130, 183, 179, 266–273.

9. Jewett, *Marsh Island*, 79–82, 86–87, 167.

10. *Plymouth Colony Records*, 5:12; 7:56, 140, 148, 282.

11. Jewett, "Tidal Marshes." See also Townsend, *Sand Dunes*, 188–228. On upland meadow grasses, see Stilgoe, *Common Landscape*, 183–184. See also Redfield, "Development."

12. Jewett, "Tidal Marshes." See also Hale, *Christmas*, xi.

13. On Piscataqua River gundalows see

Works Progress Administration, *Hands*, 21–24.

14. Some of this is in Jewett, "Tidal Marshes." Much is local lore. See also Ordway, "Merrimac River Gundalow."

15. Salt hay is still harvested in marshes north of Boston; only a little is taken along this stretch of coast.

16. Tudor, *Letters*, 204–207.

17. On British efforts, see Darby, *Changing Fenland*, esp. 71–129.

18. I can find no evidence that anything other than haying occurred in the North River marshes until 1860.

19. "Salt Marshes," *Cultivator*, 303. On sea greens, see Stilgoe, *Shallow-Water Dictionary*, 42–43.

20. "Salt Marshes," *Cultivator*, 303. See also "Salt Marshes," *New England Farmer*, 185.

21. "Salt Marshes," *New England Farmer*, 185.

22. "Salting Hay."

23. Clift, "Salt Marshes," esp. 344–348.

24. Goessmann, "Best Mode," esp. 335–339.

25. Shaler, *Sea-Coast Swamps*, esp. 385–387.

26. Shaler, *Sea-Coast Swamps*, 373. Shaler did what he could to advance research: see his *Beaches and Tidal Marshes*. See also Niering, "Vegetation Patterns."

27. Shaler, *Sea-Coast Swamps*, 385–386.

28. Board of Harbor and Land Commissioners [hereafter BHLC], *Report for 1882*, 17.

29. BHLC, *Report for 1899*, 27–29.

30. BHLC, *Report for 1900*, 41–43.

31. BHLC, *Report for 1901*, 68–69.

32. Shaler, "Inundated Lands," 388–390.

33. Stilgoe, *Metropolitan Corridor*,

315–333.

34. Warren, *Tidal Marshes*, remains a useful introduction to the subject of marsh reclamation.

35. Lay, "Tidal Marshes," esp. 104.

36. "Rural Objects." For a modern example of the attitude, see Balliett, "Weeset Journal."

37. Schmitt, *Back to Nature*, esp. 3–19. See also Stilgoe, *Borderland*, 165–207.

38. On marshes as unhealthful, see "Marshes, and Their Effects."

39. Norris, "Tide-Marsh," esp. 753.

40. Farnham, *Brief Historical Data*, 80–82.

41. I have been unable to learn who invented the traps.

42. Stilgoe, *Shallow-Water Dictionary*, 38–39.

43. Tooker, *Middle Passage*, 4.

44. BHLC, *Report for 1886*, 14, 69–71.

5. SKIFFS

1. Duncan and Ware, *Cruising Guide*, 237–239.

2. USGS Chart #1233. See also "Ducking Out."

3. "Coast Guard Won't Remove Warning Buoys at North River," Quincy [Mass.] *Patriot Ledger* (June 18, 1992), 8C.

4. Hart, *Romance*, under *yacht*.

5. Bailey, *Dictionary*, under *yacht*.

6. *OED*, under *yacht*.

7. Evelyn, *Diary*, 3:296–297.

8. *OED*, under *yacht*. See also Smyth, *Sailor's Wordbook*.

9. Webster, *American Dictionary*, under *yacht*.

10. Worcester, *Dictionary*, under *yacht*. See also Gerr, *Nature of Boats*.

11. Dana, *Dictionary*, under *yacht*.

12. *American Encyclopedic Dictionary, Century Dictionary,* and *New Standard Dictionary,* all under *yacht.*

13. Bradford, *Mariner's Dictionary,* under *yacht.* See also Morris and Howland, *Yachting.*

14. Middleton, "Yachting," esp. 181–183.

15. Todd, "Voyage," 53; on "boat niggers," see 43. See also Howland, "Neponset Estuary."

16. Bailey, *Dictionary,* under "tender." Falconer, *Universal Dictionary,* under "tender." Street, "Improving the Classic Dinghy." See also Burke, "Eddie Crosby's Last Boat"; Grayson, *Dinghy Book;* Coleman, *Dinghies;* and "Twelve-Foot Sailing Dingey."

17. Brown, *Watershots,* 94; see also 92–93.

18. Prout, *Studies of Boats,* esp. plates iv, vi, ix, xii; Prout, *Easy Lessons,* esp. plates xiii and xv.

19. Prout, *Hints,* plates xvi, xi. For other early nineteenth-century examples, see Cooke, *Sixty-five Plates.*

20. Parramon, *Painting Seascapes,* 4–11. See also Anson, *How to Draw Ships;* Aylward, *Ships and How to Draw Them;* and Dade, *Sail and Oar.*

21. See, for example, Howlett, *William Partridge Burpee,* esp. 24–30.

22. "Where the Cape Cats Breed," 481. See also Brooks, "Boats," esp. 307–314.

23. I suspect that the same might be true of breeds of horses, say, or even types of wagons.

24. "Where the Cape Cats Breed," 482–483. Leavens, *Catboat Book,* esp. 6–13. See also "Racing Catboat."

25. Leavens, *Catboat Book,* 7–9. Dunlop, "Catboats," 40.

26. Leavens, *Catboat Book,* 7–16. See also Case, *Joy's Pier.*

27. Day, "Catboat," esp. 6. See also Muller, "A Little Talk on Catboats."

28. Thompson, "Cats," 87–88. See also Thomas, *Building.*

29. Thompson, "Catboat Sailor's Yarn," esp. 257–258, 263.

30. "Crosby Cat."

31. Thompson, "Sea Wolf," esp. 127–128. See also Quamino, "Handy Craft," and Mower, "How to Build."

32. "First and Last Sea-Fishing," esp. 538.

33. Dwight, *Travels,* 2:114–115.

34. Bartlett, *Water Tramps,* esp. 194–203, 251–271.

35. Dunne, "Irish Success." See also Gardner, *Building Classic Small Craft,* 2:131–146.

36. Reiger, *Profiles,* 2–72. See also Cabot, "New England Double Enders."

37. Ingersoll, *Oyster Industry;* Massachusetts Department of Fisheries, *Report Upon the Mollusk Fisheries.*

38. Dunlop, "Catboats," 42.

39. "Building a Flat-Bottom Skiff," 71. See also Thompson, "Successful Small Tender," and Alvord, *Beachcruising,* 1–45.

40. Cooper, *Yacht Sailor,* 150–151.

41. How much this was known at first, I cannot determine, but early advertising for outboard engines shows them mounted on the transoms of yacht tenders.

42. I can find no evidence that liveries rented round-bottomed boats after 1925.

43. "Electric Launches"; Calahan, "New York Yacht Club"; Clark, *History;* "Rheclair."

44. "Power and the Editor."

45. "Power-Boat Legislation." See also Delorie, "Engineer."

46. Bieling, "Mishandling Boats," 269.

47. Hancock, *Motor Boat Club and the Wireless,* 37.

48. Winfield, *Rover Boys on the Ocean,*

esp. 12–13.

49. Graef, "Superstition" and "Safe and Dangerous."

50. Martyn, "Era," 187. See also "Fast Runabout" and "Lure of Shining Water."

51. Stein, "What's the Hurry?"

52. Chapman, *Practical Motor Boat Handling;* Waters, *Outboard Cruising;* Taylor, "Motor Sailors." See also Coast Guard Auxiliary, *Preview and Prospectus,* i.

53. Hunn, *Old Outboard,* 17–19. See also Miller, *Small Boat Engines,* and Schult, *Curious Boating Inventions,* esp. 66–86.

54. Lincoln, *Rise,* 70–83.

55. Damkoehler, "Vanishing Marina," 69.

6. HARBORS

1. Spectre, "Lyman," 61–63. The newsletter of the Lyman Boat Owners' Association is also an excellent source of information.

2. Spectre, "Lyman," 64–67; Brown, "Lyman," 54–55.

3. Brown, "Lyman," 54–56.

4. Venk, *Complete Outboard Boating,* esp. 15–17.

5. "Will There Be a Horsepower Race?" 160. See also *Outboard Motor Boat Book.*

6. Gardner, *Building Classic Small Craft,* 1:33, 175, 260–261, 278.

7. Hanson, "Kit Boats," esp. 84.

8. Kenealy, *Boating,* 13–14.

9. Fair, "Coasting." See also Kaiser, *Built on Honor.*

10. Davis, *Harper's,* 65, 75, 77–81.

11. Gardner, *Dory Book,* 11–15, 82–117.

12. Bechdolt, *Handy Book,* 105, 106–107. See also "Handy Rule." On contem-

poraneous professional boatbuilding see Wasson, *Sailing Days*, 155–167; Howland, *Sou'West*, 81–96; and Mendlowitz, "Pete Culler's Workshop."

13. *Boy Mechanic*, esp. 76–90. See also *Build a Boat*; Modern Mechanix, *How to Build Twenty Boats; How to Build Twenty Boats (No. 362)*; and Fawcett, *How to Build Twenty Boats.*

14. Howland, *Sou'West*, 82–83. "Amateur Building," 688. See also "On Boats" and Roberts, "Old Steam Shed." The author's father, John F. Stilgoe, routinely built one ten-foot skiff a day, using an incredible variety of jigs, patterns, and moulds.

15. Parker, "Yachts," 10. See also Eno, "In Winter."

16. Bradford, "Building," 25–28. See also "What Are the Young Men of the Cities Coming To?" People still build boats in difficult places, including Manhattan apartments: see, for example, Peck, *Boat.*

17. See, for example, "How to Build a Launch for $100"; "We Must Have a Boat"; "How to Build a Small Catboat"; and Thompson, "Successful Small Tender." See also "Twelve-foot Sailing Dingey"; "Blocks"; and Schock, *How to Build a Rowboat.*

18. Davis, "Yacht-Building," 694. See also Constant, "Buying a Sailboat for $300." The story concerning Butler's difficulties is pasted into a manuscript notebook that belongs to Judy Genthner.

19. "Fire of Spring," 133–134. See also "Double Garage."

20. Taber, *My Own Cape Cod*, 100.

21. *Appleton's Dictionary*, 1:344–349.

22. Spon, *Dictionary*, 434–438.

23. Wasson, *Cap'n Simeon's Store*, 260.

24. Howland, *Sou'West*, 88–91.

25. *Railway Dry Docks*, esp. 3–7. See also Sullivan, *Description of the American Marine Rail-Way.*

26. Davis, "How to Build a Small Marine Railway," esp. 533–535.

27. Calahan, *Ship's Husband*, 259–263.

28. *Report on a National Marina Survey*, esp. 2–7.

29. Damkoehler, "Vanishing Marina," esp. 69, 70.

30. Lee, "High and Dry."

7. WHARVES

1. Conrad, *Mirror of the Sea*, 66, 13. See also Minnoch, *Aground.*

2. *Black's Law Dictionary*, under *pier* and *wharf*. Sturgis, *Dictionary*, under *pier.*

3. *Century Dictionary*, under *pier* and *wharf.*

4. Amusement piers seem to have been British in origin, and very rare north of New York City.

5. Hawthorne, *House of the Seven Gables*, 294.

6. Melville, *Moby-Dick*, 12–13. I have seen no evidence of this argument in Melville criticism. For a nonfiction description, see Commissioners of the Sinking Fund, *Wharves.*

7. Farnham, "Day on the Docks," esp. 33–34.

8. Falt, *Wharf and Fleet*, 39–41.

9. Falt, *Wharf and Fleet*, 23–28.

10. National Research Council, *Marine Structures*, 21–71.

11. Amos and Amos, *Atlantic and Gulf Coasts*, published in 1985, is actually less useful than Arnold, *Sea-Beach at Ebb-Tide*, published in 1901, and focuses on beaches, not "coasts."

12. "Marine Borers in San Francisco Bay," esp. 446–447; Weiss, "Wooden Ships and Ship Worms," esp. 592.

13. "Protecting Piles from the Teredo"; "Damage Done by Marine Boring Animals"; Bartsch, "Status of *Teredo beachi*"; Kofoid and Miller, "Unusual Occurrence of Rock-Boring Mollusks."

14. National Research Council, *Marine Structures*, esp. 2–17.

15. "Wooden Ships Not Back Numbers"; "Return of the Wooden Ship." See also Goddard, "Passing of the Five-Masters."

16. Leavitt, *Wake of the Coasters*, esp. 183–196. See also Trott, "Down East Merchant Fleet." Also see "Unusual Use of Concrete," and Foulke, "Life in the Dying World of Sail."

17. National Research Council, *Marine Structures*, esp. 108–109, 355, 384–391, 150. See also "Damage Done by Marine Boring Animals."

18. National Research Council, *Marine Structures*, 165–220, esp. 174–175.

19. Coker, "Perpetual Submarine War." See also "Chemical Warfare Against Shipworms."

20. National Research Council, *Marine Structure*, 87–150.

21. "Protecting Piles from the Teredo" describes one scraping device that falls and rises with the tides. See also "Paraffin and Poison."

22. National Research Council, *Marine Structures*, 238–264.

23. "Sowing Rice: Reaping Ship-Worms," esp. 113.

24. Brewer, "Six Hundred Tons of Barnacles." This article mentions the difficulties faced by "composite warships" (wooden hulls sheathed with iron) when cruising in waters infested with *Teredo* worms.

25. "How Ships Spread Species."

26. National Research Council, *Marine Structures*, 523, 265–287.

27. National Research Council, *Marine*

Structures, 265–266.

28. "Regulations for Construction." See also Greene, *Wharves*.

29. Condit, *Port of New York*.

30. Averill, "Maine Boy," esp. 217–219. See also James, *American Scene*, 70–71, and Commission on Metropolitan Improvements, *Public Improvements*, 34–35.

31. Jewett, *Marsh Island*, 135–136.

8. SMUDGE

1. Military historians routinely ignore the towers.

2. See Hoyt, *U-Boats Offshore*.

3. Bronner, *Atlantic Ocean Environment*.

4. See Tetlock, ed., *Behavior, Society, and Nuclear War*, and Shaheen, *Nuclear War Films*.

5. Allen, *Naval History*.

6. For the local context in which the Bates girls acted, see Deane, *History*, 141–142. See also Napier, *New England Blockaded*.

7. On New Orleans, see Paris, *Four Principal Battles*. See also Marine, *British Invasion of Maryland*, and Ingraham, *Sketch of the Events*. On Jackson, see Eaton, *Life*.

8. Woodworth, *Poetical Works*, 2:7–10. See also his *The War*.

9. *Journal de département des Deux-Nethes*, nos. 573–638.

10. Barker, *Vainglorious War*.

11. Thoreau, *Cape Cod*, 141. See also Mahan, *From Sail to Steam*, 25–44, and Leggett, *Naval Stories*.

12. "Our Sea-Coast Defense," esp. 314–316.

13. Thoreau, *Yankee in Canada*, esp. 81–82.

14. "Our Sea-Coast Defense," 318–320.

15. "Our Sea-Coast Defense," 319–325.

16. Adams, *Crisis of Foreign Intervention*; on the Russian fleets, see Jordan and Pratt, *Europe*, 199–201. On Civil War coastal actions, see Mahan, *From Sail to Steam*, 156–197, and Bailey, *Gun-boat Service*. Civil War actions profoundly influenced German coast defense thinking; see, for example, Von Scheliha, *Treatise on Coast Defence*.

17. Griffin, *Our Sea-Coast Defences*, 12–15. See also Southwick, "Our Defenseless Coasts."

18. Maguire, *Attack and Defence*, 119–122. See also Abbot, *Course of Lectures upon the Defence of the Sea-Coast*.

19. Board on Fortifications, *Report*.

20. Griffin, *Our Sea-Coast Defences*, 13.

21. Spears, *Our Navy*. See also Gomez, *Spanish-American War . . . Blockades and Coast Defense*, and Young, *History of Our War with Spain*, esp. 477–486.

22. Wisser, *Tactics*, esp. 54–75. See also Young, *History of Our War with Spain*, 151–164.

23. Kipling, *American Notes*, 12.

24. Walsh, *Take Your Choice*, 25.

25. Wisser, *Tactics*, 176.

26. Hajewski, "With U-53 to America," is an excellent example of the resurgence of scholarly interest in World War I submarine activity immediately off the American coast. See also Domville-Fife, *Submarines and Sea-Power*, and Cremer, *U-Boat Commander*.

27. On the potential gassing of New York and other coastal cities, and especially the impact of the magazine articles on women, see Stilgoe, *Borderland*, 279–282. See also Fyfe, *Submarine Warfare*.

28. I have been unable to learn the title of the Three Stooges short film: academic libraries are remarkably weak on the sub-ject of Hollywood B-movies.

29. By far the best book on this neglected element of naval warfare is Treadwell, *Submarines with Wings*, esp. 57–74. See also Whitehouse, *Subs*, and Watts, *Japanese Warships*. For general background, see Lewis, *Fight for the Sea*, and Herzog, *60 Jahre Deutsche UBoote*. Even the passage of half a century does not diminish the unease with which contemporary British and United States military historians confront the technological superiority of Axis forces in 1939. On aircraft-carrying dirigibles, see Collier, *Airship*, 209.

30. The film "The Russians Are Coming, The Russians Are Coming," remains deeply moored in the local imagination along this stretch of coast.

9. TREASURE

1. Allen, *Our Navy*, 72–99.

2. Lego toys are extremely popular, and still unstudied. For examples of modern and contemporary views of pirates in juvenile fiction, see Orton, *Mystery in the Pirate Oak*, and Sobol, *Encyclopedia Brown Saves the Day*.

3. Stevenson, *Treasure Island*, [c. 1956], 1–5, 38, 56.

4. Gosse, *History of Piracy*; Woodbury, *Great Days of Piracy*; Gosse, *Pirate's Who's Who*. An especially fine introduction is Sauer, *Early Spanish Main*. See also Carey, *History of the Pirates*. For the connections between piracy and slaving, see Brooke, *Book of Pirates*.

5. Masefield's *Jim Davis*, little read nowadays in the United States, remains an important British boys' book. My British male undergraduates have almost always read it, usually at about age twelve.

6. Stoddard's book is critically important in any understanding of early twentieth-century American ethnocentrism. One finds traces of the dark-visaged man in such 1950s seacoast novels as Brooks, *Shining Tides*.

7. Thomas et al., *Films of Errol Flynn*, 91–97.

8. See Cawelti, *Six-Gun Mystique*, and Slotkin, *Regeneration Through Violence*.

9. Mason, *Golden Admiral*, in 1953 completed the image-making of Drake. See also Esquemeling, *Buccaneers of America*, esp. 100–222.

10. See Stevenson, *Treasure Island* (1895), *Treasure Island* (1902), and *Treasure Island* (McLoughlin Brothers edition, 1915). See also Forbes, *Doubloons—And a Girl*.

11. On tanning, see chapter 11.

12. Burg, *Sodomy and the Perception of Evil*, 43–104. See also *Trials of Eight Persons*.

13. Snow's *True Tales* and *Mysterious Tales* offer a good introduction to the notion of treasure trove and its relation to piracy. Snow lived in Marshfield, Massachusetts, and his books echo local oral lore.

14. "Antiquarians," esp. 307–313.

15. The search for the landscape of American romance deserves a book in itself, for American authors routinely tried to find the place of romantic associations, not simply the time.

16. "Antiquarians," 310–315.

17. Stilgoe, *Common Landscape*, 334–335.

18. Poe, *Works*, 1:146, 147, 153, 173, 163. On Stevenson drawing his map first, see Greenhood, *Mapping*, x.

19. Masterson, *Jurisdiction*, xiv. See also Daniel, *Sovereignty and Ownership*.

20. Dixon, *House*, 176, 79–80, 84, 86, 91–94, 97, 142, 161, 157, 5. Stratemeyer wrote hundreds of juvenile adventure books; on his career, see Billman, *Secret*.

21. Dixon, *House*, 142. See also Warner, *Secret of the Marsh*, and London, *Adventure*, 2–8. The Glencannon stories by Guy Gilpatric are collected in his *Glencannon Omnibus*.

22. For examples of the tropical connection, see Young, *Ship's Surgeon's Yarns*; Pease, *Night Boat*; and Allison, *Secret of the Sea*. On that connection and enthnocentrism, see Rohmer, *Yellow Claw* and *Golden Scorpion*.

23. On smuggling liquor from New England into Canada as the constant background operation of the Gloucester fishing fleet, see the turn-of-the-century fiction of James Brendan Connolly, especially his *Open Water*, *Gloucestermen*, and *Out of Gloucester*, all of which mention gunkholing into Canada or losing British revenue vessels in American shoal water. On rumrunning into the United States, see Allen, *Black Ships*, and Carse, *Rum Row*. Rumrunning has resurfaced as background to contemporary smuggling: see Wilcox, *St. Lawrence Run*.

24. Allen, *Black Ships*. See also Leeds, *Phantom of the Shore*.

25. Casey, *Spartina*, 104–105.

26. See, for example, her *Cape Cod Mystery*.

27. For a Florida-focused book, see Blair, *Scuba*. On New England, see Langton, *Dark Nantucket Moon*; Boyer, *Billingsgate Shoal*; Kiker, *Death Below Deck*; and Page, *Body in the Kelp*.

28. Borthwick, *Bodies*, 30–32. For a nonfiction study of contemporary piracy, see Villar, *Piracy Today*.

29. An example of news stories about local efforts to police the coast is Mallia, "Weymouth Harbormaster."

10. QUAINTNESS

1. Quick, *Virginia*, esp. 4–42 and 134–139.

2. Fear of air assault remains a curiously unexamined feature of early twentieth-century American social and intellectual history, in part because so few contemporary historians are trained in military history or are even interested in it.

3. James, *American Scene*, 32–37.

4. Gilpin, *Observations on the Coasts*, 11–13, 21–22.

5. Turner, *Picturesque Views*, n.p. This sort of book gave way to a more sentimental viewpoint later in the century, at least in Britain; see, for example, "Sea-Side Churchyard."

6. Local-color writing apparently evolved from the earlier search for a romantic American past.

7. Morison, *Maritime History*, 33–357.

8. James, *American Scene*, 36.

9. On tourism, see Palmer, *New England Tourist Camps*.

10. On women botanizing, see Stilgoe, *Borderland*, 34–37.

11. "Sea Views," esp. 507, and "Sea-Gardens."

12. Butler, *Family Aquarium*, 86–87; see also Wood, *Common Objects*.

13. Butler, *Family Aquarium*, 86. On water collection, see also "Sea Views" and "Sea-Gardens."

14. Carter, *Summer Cruise*, 21–22.

15. Thornton, *Cultivating Gentlemen*, 147–172. See also Agassiz, *Sea-Side Studies*, and *Domestic History of the Learned Seals*.

16. Butler, *Family Aquarium*, 82. See also Lewes, *Sea-Side Studies*.

17. Hibberd, *Book of the Marine Aquarium*, 39–42. Hibberd's book was reprinted many times, last in 1875.

18. Emerson, "Each and All," *Poems*, 14–15. See also Moulton, "Now We Hunt Shells."

19. Ingram, *Centennial Exposition*, 176–178.

20. Arnold, *Sea-Beach*, 13–16. See also "All Along the Shore" and "Summer on the Sands."

21. On this period, see Mayer, *Sea-Shore Life*; Shannon, "Spring Awakening"; Dexter, "Two Centuries of Naturalists"; Fenton, "Let's Visit the Sea Shore."

22. "Children on the Sands," 614.

23. Bishop, "Lobster at Home," and Ingersoll, "Stars of the Sea."

24. Willis, *Hurry-Graphs*, 40–71, esp. 56, 66–67.

25. Mitchell, "Cape Cod," 643–658; for the quotation, see 654.

26. Houghton, "Sandy Hook." See also Ralph, "Old Monmouth," and Bishop, "Fish and Men."

27. McLean, *Cape Cod Folks*. I purchased this book at a rummage sale held approximately a mile in from the sea; the book, smothered in dust and apparently long unopened, had been defaced.

28. Nordhoff, *All Alongshore*, 122–123.

29. On the turn-of-the-century search, see Stilgoe, *Borderland*, 283–300. The *New England Magazine* deserves a monograph analyzing not only its contents but its editors' motives: more than any other periodical except boating magazines, it fostered the new image of the coastal realm and its inhabitants.

30. See Hall, *String Too Short to Be Saved*, 71.

31. Conrad admired Jacobs; see *Mirror*, 45–46.

32. Lincoln, *Mr. Pratt*, 18.

33. Lincoln, *Cy Whittaker's Place*, 9.

34. Lincoln, *Portygee*, 361.

35. Lincoln, *Cap'n Eri*, 23.

36. Handy, *What We Cook on Cape Cod*.

37. Lincoln, *Cape Cod Stories*, esp. 218–219. Taylor, *Going*, 13–30.

38. Thompson, "Marblehead," 115, 120. See also his "In the Isles of Shoals." See also Carpenter, "Provincetown."

39. Colton, "Cruise." This sort of article is utterly removed from those on the heartbreak of pleasure-boat shipwreck.

40. See, for example, Hine and Hine, "Going Down."

41. *Port Jefferson* is a useful example of this sort of illustration; the small steamships are far off, beyond the old schooner. These drawings might illustrate the argument in Wasson, *Cap'n Simeon's Store*, 75, that the prosperous times are gone from small ports.

42. Edwards, *Old Coast Road*, 111, 116–122. See also Johnson, *Highways*, 185–198; Ingersoll, *Down East Latch Strings*; and Sylvester, *Maine Coast Romance*.

43. See, for example, Federal Writers' Project, *Massachusetts*, 621–626, 494–507.

44. Werth, "Hidden Cape," 27, 32.

45. Updegraff, "Siren," esp. 284–285.

II. BIKINIS

1. Peale, *Life*, 34–35.

2. Packard, *Sea-Air*, 19, 20, 21–22, 20.

3. On upper-class resorts, see Bremer, *Homes*, 1:561–562, and Grant, "North Shore."

4. Packard, *Sea-Air*, 22–23. See also Winthrop, *Life in the Open Air*.

5. On patterns of illness, see Stilgoe, *Borderland*, 38–48.

6. See, for example, Cremin, *Transformation*.

7. Packard, *Sea-Air*, 49.

8. Franklin, *Autobiography*, 7, 44–45.

9. "Plymouth County Coroner's Reports," MS filed in Probate Court, Plymouth, Mass.

10. Farnham, *Brief Historical Data*, 80–82. "I didn't know what a bathing suit was until I was fifteen," one eighty-eight-year-old Norwell man commented in 1988. Quoted in the Quincy [Mass.] *Patriot Ledger* (June 30, 1988).

11. Tooker, "Boyhood Alongshore," 500–501.

12. "Soliloquy on Bathing," 69. See also "Bathing."

13. "Swimming," 507–512. See also "Drowned."

14. Occasionally my students mention that their great-grandparents swam nude but that such behavior was supposed to be kept secret.

15. Wheelwright and Schmidt, *Long Shore*, 38–39.

16. Weightman, *Art of Swimming*, 70–72.

17. Osborne, "Surf and Surf-Bathing," 106–107.

18. Holt, "Promiscuous Bathing," 6. See also McCabe, "Modesty."

19. Holt, "Promiscuous Bathing," 6. See also Ladova, "About Bathing Suits."

20. Chopin, *Awakening*, 47, 189.

21. Douglas, "Learn to Swim—Alone," 43.

22. Swaringen, "Fear."

23. Sangster, "Swimming." See also "Out of Doors Woman"; "What Will She Wear?"; "An All-the-Year Garment"; Duryea, "What Shall Our Women Wear?" and "Bathing Suit."

24. "Popular Girl Must Learn Fancy Strokes."

25. Howells, *Literature*, 170–173. See also Pyke, "Summer Types."

26. Wingert, "Swimming on Dry Land." See also Wisby, "Learning to Swim," and Martyn, "Outdoor Swimming Pools."

27. Mason, "Importance," 316.

28. My students now and then remark about the intense conservatism that they define as both eastern and urban, and sometimes remark, too, on what they see as the impossibility of innocence in the urban east. On nudity at beaches frequented only by the established aristocracy of other nations, see Erwitt, *On the Beach*, esp. 18–28. On the vagaries of 1930s inland, rural skinny-dipping, see Smart, *R. F. D.*, 191–195.

29. Rattray, *South Fork*, 131–132.

30. Taft, "Rowing the Maine Coast," 54.

31. See, for example, Hollander, "Swimsuits Illustrated," and "Happy Birthday, Bikini." See also "Swimsuits through the Ages" and Smith, "Monokini Beaches."

32. Wharton, *Backward Glance*, 46.

33. Follett-Stevens, "Treatment for Tan"; see also "Wind . . . Sun . . . and Complexions." The warning comes from "Abbott's."

34. The argument also fostered suburbanization: see Stilgoe, *Borderland*, 38–48. By far the best explication of the argument is Schmitt, *Back to Nature*. For the argument in connection with women, see Leder, "American Dilemma."

35. See, for example, Melville, *Typee*. See also "Essay on the Causes."

36. Maine Central Railroad, *Maine Central Christmas*, 90.

37. Chopin, *Awakening*, 7.

38. Again, analysis of Tarzan movies is complicated by the unwillingness of film historians to focus on B-movies and to admit that box-office receipts say some-thing about how a film is received by society.

39. Packard, *Sea-Air and Sea-Bathing*, 74–75.

40. Rostnor, "Darwinism and Swimming." See also Sumner, *Social Darwinism*.

41. London, *Adventure*, 90, 134, 178–179. See also Freedman, "New Woman."

42. This information comes from alongshore sources determined to remain anonymous. See also "Trouble with Bikinis."

43. Hill, "Roof." Varron, *Professions of a Lucky Jew*, 81–82. See also "Yes, Amy!"

44. "Turn on the Sun." See also "Sun-Suit."

45. See especially Lewis, "Swinging Beach Movies." Students can write paper after paper on these subjects.

46. "What Female Complaints?" See also Adams, "Are Woman's Troubles Out of Style?" The follow-up article on bikinis is Germond, "Bikinis Are Here!"

47. Kimball, "Sea Change," 266.

48. Crane, *Summer Girl*, 44–45. Little has changed since 1956: see Williams, *Diamond Bikini*, and Wakeman, *Fabulous Train*, 224. Women have traditionally viewed male nudity with a casualness that informs women's fiction; see, for example, Welty, "Moon Lake," in *Thirteen Stories*, 212. Rarely does anyone examine what anyone, especially women, reads on beaches. See, however, "Novel Pleasures," for an introduction.

49. Muschamp, "Don't Look Now." For the quotation, see Martin, "Porn-Free," 19. A Gallup poll conducted in 1983 found that 15 percent of adult Americans had swum nude in mixed-sex groups and that 72 percent favored nude swimming and sunbathing on designated beaches, so attitudes may be changing (quoted in the Quincy [Mass.] *Patriot Ledger*, June 30, 1983).

50. Alan Dershowitz, "New York Nudity Rule Goes Too Far," quoted in the Boston *Herald* (July 27, 1992), 21. For a fine analysis on the "morality" of women in very brief attire, see Barrett, "Provocative Dress, Reasonable Doubt."

51. Steinem's *Beach Book*, now in danger of becoming unknown, is an extraordinary window on the difficulties that confronted 1960s feminists in addressing issues of women's physical strength. Few students of women's studies know the book, and few of their professors ever mention it in lecture, even when they do know it, or know of it.

52. On Kellerman, see Germond, "Bikinis Are Here!" See also "We Nominate for the Hall of Fame."

53. Melville, *Redburn*, 4–5.

54. Lindbergh, *Gift*, 31.

12. RISK

1. Conrad, *Mirror*, 100–101.

2. Conrad, *Heart of Darkness*, 3–4.

3. Tomlinson, *Mingled Yarn*, 31.

4. No one has yet analyzed the grassroots implications of ecological awareness, but such analysis is long overdue. An excellent introductory exploration is Ehrenfeld, *Beginning Again*.

5. Lewis, *State of Our Harbors*, esp. 12–13.

6. Edgerton, *Alone Together*, remains the best study of urban beaches.

7. On physical fitness after 1960, see runs of any "woman's magazine," especially *Ladies Home Journal* and *Cosmopolitan*.

8. Frank, "Outdoor Recreation," in

Outdoor Recreation Resources Review Commission, *Trends*, 220–224.

9. U.S. Army, *Green Beret Fitness Program*.

10. See, for example, Lewis, "Those Swinging Beach Movies."

11. President's Council on Physical Fitness and Sports, *Youth Physical Fitness in 1985*, 9–10.

12. See Must et al., "Long-term Morbidity"; Gortmaker et al., "Inactivity"; Gortmaker et al., "Increasing Pediatric Obesity"; and Dietz et al., "Do We Fatten Our Children?" Only rarely do journalists bring such articles to public attention (see Jane E. Brody, "Obesity in Teenagers Tied to Ailments in Adulthood," *New York Times*, November 5, 1991), and never do they make the connection between pediatric obesity and the national health-care controversy.

13. "Vigorous Physical Activity." This is an important, indeed critical, article.

14. Brubach, "On the Beach," 74.

15. Fraser, *Fashionable Mind*, 190–191.

16. My foreign students routinely remark on the obesity and inactivity of American high-school and college students.

17. Every springtime local papers run columns written by women about the terrors of shopping for swimsuits. The columns emphasize the concerns the women have about their overweight bodies. About

two months later, the local newspapers run articles by the same journalists about women who adore the most revealing of swimsuits.

18. Kase, "From Aqua Aerobics to Workout Bikinis."

19. See "Strength Is Beauty," a collection of features in the February 1991 issue of *Shape*.

20. Kaufman, "Put Some Muscle into It," 128.

21. Miller, "Cruising," 8. See also Robberson, "Where There's a Wind."

22. Spectre, "On the Waterfront," 15.

23. Gardner, "Small Craft Revival."

24. Kirkpatrick, "Fine Craft."

25. See, for example, Viscusi, *Risk by Choice*, and Zeckhauser and Viscusi, "Risk within Reason."

26. Cowley, "Can Sunshine Save Your Life?" This is a particularly important article.

27. On the visit of the project manager, see "No Day at the Beach," quoted in the Quincy [Mass.] *Patriot Ledger* (July 15, 1992).

28. On off-road wheelchairs, see "Curb-Climbing Wheel Chair."

29. On designing, I am indebted to my landscape architect colleagues, especially Roger Courtenay and George Hargreaves.

30. My correspondent is Patricia Foley, of Alexander, Maine.

31. Thoreau, *Cape Cod*, 217.

13. THE COASTAL REALM

1. Pilkey and Neal, "Save Beaches," 36. See also Taylor, "Waterfront Development."

2. See, for example, Clark, *Coastal Ecosystem Management*.

3. Melville, *Redburn*, 294.

4. Colcord, *Instrument*, xv–xvi.

5. Tomlinson, *Mingled Yarn*, 31.

6. O'Brian, *Master and Commander*, is the first of a series of best-selling novels set in the British navy during the time of Horatio Nelson.

7. Oxenhorn, *Tuning the Rig*, esp. 4–7. See also Buckley, *Steaming*.

8. Quoted in "A Beach on the Ropes Takes a Glancing Blow," Quincy [Mass.] *Patriot Ledger* (March 15, 1993).

BIBLIOGRAPHY

Abbey, Edward. *Desert Solitaire: A Season in the Wilderness.* New York: Ballantine, 1968.

"Abbott's." *Harper's Weekly* 47 (April 11, 1903), 591.

Abot, Henry Larcom. *Course of Lectures upon the Defence of the Sea-Coast.* New York: Van Nostrand, 1887.

Adams, Charles Francis. *The Crisis of Foreign Intervention in the War of Secession.* Boston: Massachusetts Historical Society, 1914.

Adams, Evelyn Archer. "Are Woman's Troubles Out of Style?" *Cosmopolitan* 148 (April 1960), 32–38.

Adeline's Art Dictionary. New York: Appleton, 1891.

Agassiz, Alexander. *Sea-Side Studies in Natural History.* Boston: n.p., 1865.

"All Along the Shore." *Munsey's Magazine* 15 (July 1896), 387.

"An All-the-Year Garment." *League of American Wheelmen Bulletin* 27 (January 7, 1898), 22–23.

Allen, E. S. *A Wind to Shake the World: The Story of the 1938 Hurricane.* Boston: Little, Brown, 1976.

Allen, Everett S. *The Black Ships.* Boston: Little, Brown, 1979.

Allen, Gardner Weld. *A Naval History of the American Revolution.* Boston: Houghton, Mifflin, 1913.

———. *Our Navy and the Barbary Corsairs.* Boston: Houghton, Mifflin, 1905.

Allison, William. *A Secret of the Sea.* Garden City, N.Y.: Doubleday, 1920.

Alvord, Douglas. *Beachcruising: An Illustrated Guide to the Boats, Gear, Navigation Techniques, Cuisine, and Comforts of Small Boat Cruising.* Camden, Maine: International Marine, 1992.

"Amateur Building." *Rudder* 18 (August 1907), 687–689.

American Encyclopedic Dictionary. Chicago: Conkey, 1896.

Amos, William H., and Stephen H. Amos. *Atlantic and Gulf Coasts.* New York: Knopf, 1985.

Anson, Peter F. *How to Draw Ships.* London: Studio, 1943.

Anthon, Charles. *A New Classical Dictionary of Greek and Roman Biography, Mythology and Geography.* New York: Harper, 1856.

"The Antiquarians." *Knickerbocker* 20 (October 1842), 305–320.

Appleton's Dictionary of Machines, Mechanics, Engine-Work, and Engineering. 3 vols. New York: Appleton, 1852.

Arnold, Augusta Foote. *The Sea-Beach at Ebb-Tide: A Guide to the Study of the Seaweeds and the Lower Animal Life found between Tide-marks.* New York: Century, 1901.

Averill, Albert E. "A Maine Boy at Sea in the Eighties." *American Neptune* 10 (July 1950), 203–219.

Aylward, W. J. *Ships and How to Draw Them.* New York: Pitman, 1950.

Bailey, Avery William. *Gun-boat Service in the James River.* Providence: Rhode Island Soldiers and Sailors Historical Society, 1884.

Bailey, Nathan. *An Universal Etymological English Dictionary.* London: Midwinter, 1737.

Balliett, Whitney. "Weeset Journal." *New Yorker* 59 (October 3, 1983), 109–120.

Bancroft, Hubert Howe. *History of the Northwest Coast.* San Francisco: History, 1886.

Barber, John Warner. *Historical Collections of . . . Every Town in Massachusetts.* Worcester, Mass.: Warren Lazell, 1848.

Barker, Arthur J. *The Vainglorious War, 1854–56.* London: Weidenfeld, 1970.

Barrett, Sarah Curran. "Provocative Dress, Reasonable Doubt." *Smith Alumnae Quarterly* 84 (Spring 1993), 19–21.

Bartlett, George Herbert. *Water Tramps; Or, The Cruise of the Sea Bird.* New York: Putnam's, 1895.

Bartsch, P. "Status of *teredo beachi* and *teredo navalis.*" *Science,* n.s., 57 (June 15, 1923), 692.

"Bathing." *New England Farmer* 10 (July 11, 1832), 413.

"Bathing Suit." *Collier's* 27 (July 27, 1901), 17.

Bechdolt, Jack. *Handy Book for Boys.* New York: Greenberg, 1933.

Beechey, Frederick William. *Proceedings of the Expedition to Explore the Northern Coast of Algeria from Tripoly Eastward.* London: Murray, 1828.

Bell, B. B. "Fog Photography." *Photo-Era* 43 (July 1919), 20–23.

Berkeley, George. *New Theory of Vision.* [1705] London: Dent, 1929.

Besant, Walter. *Amorel of Lyonesse.* New York: Harper, 1890.

Beston, Henry. *The Outermost House: A Year of Life on the Great Beach of Cape Cod.* [1928] New York: Ballantine, 1956.

Bieling, Walter M. "Mishandling Boats." *Rudder* 20 (November 1908), 268–269.

Billman, Carol. *The Secret of the Stratemeyer Syndicate: Nancy Drew, the Hardy Boys, and Million-Dollar Fiction Factory.* New York: Ungar, 1986.

Bishop, W. H. "Fish and Men in the Maine Islands." *Harper's Monthly* 61 (August 1880), 336–352.

———. "The Lobster at Home." *Scribner's Monthly* 22 (June 1881), 209–217.

Black, Henry Campbell. *Black's Law Dictionary.* 3d ed. St. Paul, Minn.: West, 1933.

Blair, Joan. *Scuba!* New York: Bantam, 1977.

"Blocks." *Rudder* 8 (May 1897), 161.

Board of Harbor and Land Commissioners. *Annual Report.* Boston: State Printers, 1875–1905.

Board on Fortifications or Other Defenses. *Report of the Committee to Collect Information and Report on Ships of War, Their Armor, Armament, and Draught of Water, also the Navigable Draught of Entrances of the Ports of the Country.* Washington, D.C.: Government Printing Office, 1885.

Boccalini, Traiano. *New-found Politicke.* London: Francis Williams, 1626.

Bolton, Charles Knowles. *Terra Nova: The Northeast Coast of America Before 1602.* Boston: Faxon, 1935.

Borthwick, J. S. *Bodies of Water.* New York: St. Martin's, 1990.

The Boy Mechanic. New York: Popular Mechanics, 1952.

Boyer, Rick. *Billingsgate Shoal.* New York: Ballantine, 1982.

Bradford, Gershom. *The Mariner's Dictionary.* New York: Weathervane, 1952.

Bradford, W. R. "The Building of Dubbalong." *Rudder* 23 (January 1910), 24–30.

Bradford, William. *Of Plymouth Plantation, 1620–1647.* Edited by Samuel Eliot Morison. New York: Modern Library, 1952.

Bremer, Frederika. *Homes of the New World: Impressions of America.* Translated by Mary Howitt. New York: Harper, 1853.

Brewer, Charles B. "Six Hundred Tons of Barnacles." *Harper's Weekly* 54 (October 29, 1910), 24.

Brewington, Dorothy E. R. *Marine Paintings and Drawings in Mystic Seaport Museum.* Mystic, Conn.: Mystic Seaport Museum, 1982.

Brocklesby, John. *Elements of Meteorology.* New York: Pratt, Woodford, 1853.

Bronner, Finn E. *The Atlantic Ocean Environment in Future Warfare.* Santa Barbara, Calif.: General Electric, 1958.

Brooke, Henry K. *Book of Pirates . . . With an Account of the Capture of the Amistad.* New York: Perry[?], 1845.

Brooks, Alfred A. "The Boats of Ash Point, Maine." *American Neptune* 2 (October 1942), 307–323.

Brooks, Win. *The Shining Tides.* New York: Morrow, 1952.

Brown, Bruce C. *Watershots: How to Take Better Photos on and around the Water.* Camden, Maine: International Marine, 1988.

Brown, David G. "Lyman Makes a Comeback." *Lakeland Boating* 12 (February 1990), 54–60.

Browne, William George. *Travels in Africa, Egypt, and Syria from the Year 1792 to 1798.* London: Cadell & Davies, 1799.

Brownell, L. W. "Nature Photographer at the Sea Beach." *American Photographer* 35 (August 1941), 572–576.

Brubach, Holly. "On the Beach." *New Yorker* 67 (September 2, 1991), 70–76.

Bruce, James. *Travels to Discover the Source of the Nile.* Edinburgh: Robinson, 1790.

Buckley, Christopher. *Steaming to Bamboola: The World of a Tramp Freighter.* New York: Congdon, 1982.

Build a Boat. New York: Motor Boating, 1927.

"Building a Flat Bottomed Skiff." *WoodenBoat* 31 (November-December 1979), 71–79.

Burg, B. R. *Sodomy and the Perception of Evil: English Sea Rovers in the Seventeenth-Century Caribbean.* New York: New York University Press, 1983.

Burke, John. "Eddie Crosby's Last Boat." *WoodenBoat* 34 (May–June 1980), 82–84.

Burwick, Frederick. *Damnation of Newton: Goethe's Color Theory and Romantic Perception.* New York: DeGruyter, 1986.

Butler, Henry D. *The Family Aquarium.* New York: Dick & Fitzgerald, 1858.

Cable, George Washington. *The Creoles of*

Louisiana. New York: Scribner's, 1884.

Cabot, David. "New England Double Enders." *American Neptune* 12 (April 1952), 123–141.

Calahan, H. A. "The New York Yacht Club: A Brief History." *Sportsman* 18 (July 1935), 45–60.

———. *The Ship's Husband: A Guide to Yachtsmen in the Care of Their Craft.* New York: Macmillan, 1937.

Camden, William. *Britain, or a Chorographicall Description of . . . England Scotland, and Ireland.* Translated by Philemon Holland. London: Bishop and Norton, 1610.

Carey, Thomas. *History of the Pirates.* Hartford, Conn.: Williams, 1829.

Carpenter, Edmund J. "Provincetown: The Tip of the Cape." *New England Magazine* 22 (July 1900), 533–548.

Carpenter, James M. *Color in Art: A Tribute to Arthur Pope.* Cambridge, Mass.: Fogg Art Museum, 1974.

Carse, Robert. *Rum Row.* New York: Rinehart, 1959.

Carter, Robert. *A Summer Cruise on the Coast of New England.* [1864] Boston: Cupples, 1888.

Case, Philip N. *Joy's Pier.* New York: Dorrance, 1951.

Casey, John. *Spartina.* New York: Knopf, 1989.

Cawelti, John. *Six-Gun Mystique.* Bowling Green, Ohio: Bowling Green University Popular Press, 1971.

Century Dictionary. New York: Century, 1911.

Chamberlain, Mellen. *A Documentary History of Chelsea . . . 1624 to 1824.* 2 vols. Boston: Massachusetts Historical Society, 1908.

Champlain, Samuel. *Voyages.* Translated by Charles Pomeroy Otis. 2 vols. Boston: Prince Society, 1878.

Chapelle, Howard. *American Small Sailing Craft: Their Design, Development, Construction.* New York: Norton, 1951.

Chapman, Charles F. *Practical Motor Boat Handling, Seamanship and Piloting.* New York: Motor Boating, 1917.

"Chemical Warfare Against Shipworms." *Science*, n.s., 57 (February 9, 1923), supp. x–xii.

Chevreul, Michael Eugene. *Principles of Harmony and Contrast of Colors and Their Application to the Arts.* London: Bell, 1890.

"Children on the Sands." *Living Age* 258 (September 5, 1908), 612–614.

Chopin, Kate. *The Awakening.* [1899] New York: Avon, 1972.

Clark, Arthur H. *History of Yachting.* New York: Putnam's, 1904.

Clark, John. *Coastal Ecosystem Management.* New York: Wiley, 1977.

Clark, Sarah. "A Summer at York." *Harper's Monthly* 65 (September 1882), 487–491.

Clifford, Harold B. *Charlie York: Maine Coast Fisherman.* Camden, Maine: International Marine, 1974.

Clift, William. "Salt Marshes: The Mode of Reclaiming Them and Their Value." Commissioner of Patents. *Report for the Year 1861.* Washington, D.C.: Government Printing Office, 1862. Pp. 343–358.

Coker, R. E. "Perpetual Submarine War." *Scientific Monthly* 14 (April 1922), 345–351.

Colcord, Lincoln. *An Instrument of the Gods and Other Stories of the Sea.* New York: Macmillan, 1922.

Coleman, Eric. *Dinghies for All Waters.* Levittown, N.Y.: Transatlantic, 1989.

Collier, Basil. *The Airship: A History.* New York: Putnam's, 1974.

Collins, Wilkie. *The Moonstone.* [1868] New York: Century, 1903.

Colton, Edward H. "The Cruise of a Cape Cod Derelict." *Rudder* 23 (May 1910), 507–509.

Commissioners of the Sinking Fund of New York City. *Wharves, Piers and Slips of the East River.* New York: New York Printing, 1867.

Complete Book of Small Boats. New York: Maco Magazine, 1954.

Condit, Carl. *The Port of New York: A History of the Rail and Terminal System from the Beginnings to Pennsylvania Station.* Chicago: University of Chicago Press, 1980.

———. *The Port of New York: A History of the Rail and Terminal System from the Grand Central Electrification to the Present.* Chicago: University of Chicago Press, 1981.

Connolly, James B. *The Book of the Gloucester Fishermen.* New York: John Day, 1927.

———. *The Deep Sea's Toil.* New York: Scribner's, 1905.

———. *Gloucestermen: Stories of the Fishing Fleet.* New York: Scribner's, 1930.

———. *Open Water.* New York: Scribner's, 1910.

———. *Out of Gloucester.* New York: Scribner's, 1902.

Conrad, Joseph. *Heart of Darkness.* [1899] Edited by Robert Kimbrough. New York: Norton, 1963.

———. *The Mirror of the Sea: Memories and Impressions.* [1906] London: Dent, 1946.

Constant, William Sinclair. "Buying a Sailboat for $300." *Country Life* 8 (August 1905), 414–418.

Cooke, E. W. *Sixty-five Plates of Shipping and Craft.* London: Cooke, 1829.

Cooper, C. William. *The Yacht Sailor.* [1873] London: Hunt, 1888.

Cooper, James A. *Cap'n Jonah's Fortune: A Cape Cod Story.* New York: Burt, 1919.

Cowley, Geoffrey. "Can Sunshine Save Your Life?" *Newsweek* 89 (January 6, 1992), 53.

Crane, Caroline. *Summer Girl*. New York: Dodd, Mead, 1979.

Cremer, Peter. *U-Boat Commander: A Periscope View of the Battle of the Atlantic*. Annapolis, Md.: Naval Institute Press, 1984.

Cremin, Lawrence A. *The Transformation of the School: Progressivism in American Education, 1876–1957*. New York: Knopf, 1961.

"Crosby Cat." *Rudder* 13 (October 1902), 470.

"Curb-Climbing Wheel Chair." *Science Digest* 36 (September 1954), 51.

Dade, Ernest. *Sail and Oar: A Hundred Pictures*. London: Dent, 1933.

"Damage Done by Marine Boring Animals." *Science*, n.s., 60 (October 24, 1924), supp. xiv.

Damkoehler, David J. "The Vanishing Marina." *Offshore* 11 (February 1986), 68–75.

Dana, Richard Henry, Jr. *The Seaman's Friend . . . A Dictionary of Sea Terms*. Boston: Groom, 1851.

Daniel, Price. *Sovereignty and Ownership in the Marginal Sea*. Copenhagen: International Law Association, 1950.

Darby, H. C. *The Changing Fenland*. Cambridge: Cambridge University Press, 1983.

Darwin, George Howard. *The Tides and Kindred Phenomena in the Solar System*. [1898] San Francisco: Freeman, 1962.

Davidson, W. B. "On the Beach." *Photo-Era* 29 (July 1912), 5.

Davis, Charles G. *Harper's Boating Book for Boys*. New York: Harper, 1912.

———. "How to Build a Small Marine Railway." *Rudder* 17 (September 1906), 533–542.

———. "Yacht-Building." *Rudder* 17 (December 1906), 694–700.

Davis, William S. "Vessels as Pictorial Subjects." *Photo-Era* 43 (September 1919), 136–141.

Day, Thomas Fleming. "The Catboat." *Rudder* 7 (November 1896), 560–572; (December 1896), 634–659; 8 (January 1897), 1–6.

Dean, Robert G. "Managing Sand and Preserving Shorelines." *Oceanus* 31 (Fall 1988), 49–55.

Deane, Samuel. *History of Scituate, Massachusetts, from Its First Settlement to 1831*. Boston: Loring, 1831.

DeCosta, B. F. *The Place of Cape Cod in the Old Cartology*. New York: Wittaker, 1881.

Delorie, A. J. "The Engineer on Power Boats." *Rudder* 13 (September 1902), 423.

"Description of Cape Cod and the County of Barnstable." *Massachusetts Magazine* 3 (February–March 1791), 73–76, 149–152.

Devine, Eric, ed. *Blow the Man Down*. Garden City, N.Y.: Doubleday, 1937.

Dexter, Ralph W. "Two Centuries of Naturalists on Cape Ann, Massachusetts." *Historical Collections*, Essex Institute, 122 (July 1986), 246–258.

Dietz, William H., et al. "Do We Fatten Our Children at the Television Set? Obesity and Television Viewing in Children and Adolescents." *Pediatrics* 75 (May 1985), 807–812.

Dixon, Franklin W. [pseud. Edward Stratemeyer]. *The House on the Cliff*. [1927] New York: Grosset & Dunlap, 1959.

"Dock Timbers and Sea Water Tests." *Science*, n.s., 57 (February 16, 1923), supp. ix.

Dolan, Robert, and Harry Lins. "Beaches and Barrier Islands." *Scientific Ameri-can* 257 (July 1987), 68–77.

Domestic History of the Learned Seals. New York: Whitehouse, 1860.

Domville-Fife, Charles William. *Submarines and Sea-Power*. London: Bell, 1919.

"Double Garage Makes It Easy to Park a Boat." *Popular Science* 166 (March 1955), 166–167.

Douglas, Jane. "Learn to Swim—Alone." *Leisure* 1 (August 1934), 43.

Dow, George Francis, and John Henry Edmonds. *The Pirates of the New England Coast, 1630–1730*. Salem, Mass.: Marine Research Society, 1923.

"Drowned." *Hours at Home* 1 (July 1865), 259.

"Ducking Out." *Offshore* 15 (March 1990), 101.

Dugmore, A. R. "Landscape and Marine Work." *Country Life* 12 (June 1907), 240.

Duncan, Roger F. *Coastal Maine: A Maritime History*. New York: Norton, 1992.

Duncan, Roger F., and John P. Ware. *A Cruising Guide to the New England Coast*. New York: Dodd, Mead, 1968.

Dunlop, Tom. "Catboats and Carping." *Martha's Vineyard Magazine* 13 (May–June 1992), 40–42.

Dunne, W. M. P. "An Irish Immigrant Success Story." *New England Quarterly* 65 (June 1992), 284–289.

Duryea, Charles E. "What Shall Our Women Wear?" *League of American Wheelmen Bulletin* 27 (April 15, 1898), 375–376.

Dwight, Timothy. *Travels in New England and New York*. [1822] Edited by Barbara Miller Solomon. 4 vols. Cambridge: Harvard University Press, 1969.

Eaton, John Henry. *The Life of Andrew Jackson*. Philadelphia: Bradford, 1824.

Edgerton, Robert. *Alone Together: Social*

Order on an Urban Beach. Berkeley: University of California Press, 1979.

Edwards, Agnes. *The Old Coast Road from Boston to Plymouth.* Boston: Houghton, Mifflin, 1920.

Ehrenfeld, David. *Beginning Again: People and Nature in the New Millennium.* New York: Oxford University Press, 1993.

"Electric Launches." *Country Life* 12 (June 1907), 241.

Eliot, T. S. *Four Quartets.* New York: Harcourt, 1943.

Emerson, Ralph Waldo. *Poems.* Boston: Houghton, Mifflin, 1897.

Eno, F. L. "In Winter." *Rudder* 13 (April 1902), 212–213.

Erwitt, Elliott. *On the Beach.* New York: Norton, 1991.

Esquemeling, John. *The Buccaneers of America.* [1893] New York: Dover, 1967.

"Essay on the Causes of the Variety of Complexion and Figure in the Human Species." *American Review* 2 (July 1811), 128–166.

Evelyn, John. *Diary.* Edited by E. S. deBeer. 3 vols. Oxford: Clarendon Press, 1955.

Eyges, Leonard. *The Practical Pilot: Coastal Navigation by Eye, Intuition, and Common Sense.* Camden, Maine: International Marine, 1989.

Fair, H. L. "Coasting." *St. Nicholas* 37 (February 1910), 322–323.

Falconer, William. *An Universal Dictionary of the Marine.* London: T. Cadell, 1769.

Falt, Clarence Manning. *Wharf and Fleet: Ballads of the Fishermen of Gloucester.* Boston: Little, Brown, 1902.

Farnham, Charles H. "A Day on the Docks." *Scribner's Monthly* 18 (May 1879), 32–47.

Farnham, Joseph E. C. *Brief Historical Data and Memories of My Boyhood Days in Nantucket.* Providence, R.I.: Author, 1915.

Farnol, Jeffery. *Black Bartlemy's Treasure.* Boston: Little, Brown, 1920.

"Fascination of the Sea." *Century* 58 (September 1899), 800–801.

"Fascination of the Sea." *Outlook* 75 (December 12, 1903), 883–884.

"Fast Runabout." *Rudder* 23 (March 1910), 190.

Fawcett Editors. *How to Build Twenty Boats.* New York: Fawcett, 1942.

Federal Writers' Project. *Massachusetts: A Guide to Its Places and People.* Boston: Houghton, Mifflin, 1937.

Fenton, Carroll Lane. "Let's Visit the Sea Shore." *Leisure* 3 (July 1936), 10–12.

"Fire of Spring." *Rudder* 19 (March 1908), 133–136.

Fisher, Howard T. *Color in Art.* Cambridge: Fogg Art Museum.

"Foiling the Woodpeckers of the Sea." *Illustrated World* 36 (December 1921), 563–564.

Follett-Stevens, Helen. "Treatment for Tan, Sunburn, and Freckles." *Conkey's Home Journal* 9 (August 1901), 21.

Fontane, Theodor. *Wanderungen durch die Mark Brandenburg.* 2 vols. Stuttgart: Cotta'sche, 1907.

Forbes, John Maxwell. *Doubloons—And a Girl.* New York: International, 1917.

Foulke, Robert. "Life in the Dying World of Sail." *Literature and Lore of the Sea.* Edited by Patricia Ann Carlson. Amsterdam: Rodop, 1986.

Franklin, Benjamin. *Autobiography.* Edited by Russel B. Nye. Boston: Houghton, 1958.

Fraser, Kennedy. *The Fashionable Mind: Reflections on Fashion, 1970 to 1982.* Boston: Godine, 1985.

Freedman, Estelle B. "The New Woman: Changing Views of Women in the 1920s." *Journal of American History* 61 (September 1974), 372–393.

Freneau, Philip. *Poems.* Edited by Harry Hayden Clark. New York: Hafner, 1929.

Funnell, Charles E. *By the Beautiful Sea: The Rise and High Times of That Great American Resort, Atlantic City.* [1975] New Brunswick, N.J.: Rutgers University Press, 1983.

Fyfe, Herbert C. *Submarine Warfare Past, Present, and Future.* London: Richards, 1902.

Gardner, John. *Building Classic Small Craft.* 2 vols. Camden, Maine: International Marine, 1977–1984.

———. *The Dory Book.* Mystic, Conn.: Mystic Seaport Museum, 1987.

———. "The Small Craft Revival." *Ash Breeze* 1 (January 1978), 1–2.

Geller, Lawrence D. *Pilgrims in Eden: Conservation Policies at New Plymouth.* Plymouth, Mass.: Pilgrim Society, 1974.

Germond, Barbara W. "The Bikinis Are Here!" *Cosmopolitan* 148 (June 1960), 78–82.

Gerr, Dave. *The Nature of Boats: Insights and Esoterica for the Nautically Obsessed.* Camden, Maine: International Marine, 1992.

Gilchrist, John H. "Wreck at Cape Ann." *WoodenBoat* 88 (July–August 1988), 74–87.

Gilpatric, Guy. *The Glencannon Omnibus.* New York: Dodd, Mead, 1939.

Gilpin, William. *Observations on the Coasts of Hampshire, Sussex, and Kent, Relative Chiefly to Picturesque Beauty.* [1775] London: Cadell, 1804.

———. *Observations on Several Parts of England, Particularly, the Mountains and Lakes . . . Relative to Picturesque Beauty.* 2 vols. [1772] London: Cadell, 1808.

Glasspool, John. *Open-Boat Cruising.* Lymington, U.K.: Nautical Publishing, 1990.

Goddard, Robert H. I. "The Passing of the Five-Masters." *American Neptune* 4 (January 1944), 58–67.

Goedde, Lawrence Otto. *Tempest and Shipwreck in Dutch and Flemish Art.* University Park: Pennsylvania State University Press, 1989.

Goessmann, Charles A. "On the Best Mode of Subduing and Utilizing for Tillage the Salt Marshes in This State, After They Are Drained." Massachusetts Board of Agriculture, *Report.* Boston: State Printer, 1875. Pp. 328–342.

Gomez, Severo Nunez. *The Spanish-American War . . . Blockades and Coast Defense.* Washington, D.C.: Government Printing Office, 1899.

Gortmaker, Steven L., et al. "Inactivity, Diet, and the Fattening of America." *Perspectives in Practice* 90 (September 1990), 1247–1255.

———. "Increasing Pediatric Obesity in the United States." *American Journal of the Diseases of Children* 141 (May 1987), 535–540.

Gosse, Philip. *History of Piracy.* [1926?] New York: Tudor, 1946.

———. *Pirate's Who's Who: Giving Particulars of the Lives and Deaths of the Pirates and Buccaneers.* Boston: Lauriat, 1924.

Graef, Ernest W. "Getting the Engine in Commission." *Rudder* 13 (March 1902), 120–122.

———. "Safe and Dangerous Gasoline Tank." *Rudder* 14 (September 1903), 494–495.

———. "Superstition on the Decline." *Rudder* 13 (November 1902), 515–517.

Grant, Robert. "The North Shore of Massachusetts." *Scribner's* 16 (July 1894), 3–20.

Grayson, Stan. *The Dinghy Book.* Camden, Maine: International Marine, 1989.

Greene, Carleton. *Wharves and Piers.* New York: McGraw-Hill, 1917.

Greenhood, David. *Mapping.* Chicago: University of Chicago Press, 1964.

Griffin, Eugene. *Our Sea-Coast Defences.* New York: Putnam's, 1885.

Hajewski, Thomas J. "With U-53 to America." *Sea History* 56 (Winter 1990), 44–45.

Hale, Edward Everett. *Christmas in Narragansett.* New York: Funk & Wagnalls, 1884.

Hall, Donald. *String Too Short to Be Saved: Recollections of Summers on a New England Farm.* [1960] Boston: Godine, 1979.

Hamilton, William. "Memoir on the Climate of Ireland." *Transactions*, Royal Irish Academy, 6 (1797), 27–55.

Hancock, H. Irving. *Motor Boat Club in Florida.* Philadelphia: Altemus, 1909.

———. *Motor Boat Club at the Golden Gate.* Philadelphia: Altemus, 1909.

———. *Motor Boat Club on the Great Lakes.* Philadelphia: Altemus, 1912.

———. *Motor Boat Club at the Kennebec.* Philadelphia: Altemus, 1909.

———. *Motor Boat Club off Long Island.* Philadelphia: Altemus, 1909.

———. *Motor Boat Club at Nantucket.* Philadelphia: Altemus, 1909.

———. *Motor Boat Club and the Wireless.* Philadelphia: Altemus, 1909.

Handwörterbuch des deutchen Aberglaubens. Edited by E. Hoffmann-Krayer. 14 vols. Berlin: De Gruyter, 1927–1942.

Handy, Amy L. *What We Cook on Cape Cod.* Barnstable, Mass.: Barnstable Improvement Society, 1911.

"Handy Rule for Spiling Plank." *Rudder* 17 (January 1906), 34.

Hanson, John K. "Kit Boats for the Eighties." *WoodenBoat* 38 (January–February 1981), 84–87.

Hanson, Raymond E., and Herbert B. Turner. "A Provincetown Pilgrimage." *Photo-Era* 44 (May 1920), 236–248.

"Happy Birthday, Bikini." *People* 26 (July 28, 1986), 47–49.

Hart, Joseph C. *The Romance of Yachting.* New York: Harper, 1848.

Haupt, Lewis M. "Effective Beach Protection." *Cassier's Magazine* 35 (October 1908), 22–29.

Hawthorne, Nathaniel. *The House of the Seven Gables.* [1851] Boston: Houghton, 1883.

Herzog, Bodo. *60 Jahre Deutsche UBoote, 1906–1966.* Munich: Lehmanns, 1968.

Hibberd, Shirley. *The Book of the Marine Aquarium.* London: Groombridge, 1860.

Hichborn, Philip. *Report on European Dockyards.* Washington, D.C.: Government Printing Office, 1886.

Hill, Ellen. "The Roof." *Leisure* 1 (July 1934), 13–14.

Hine, T. A., and C. G. Hine. "Going Down to the Sea in Totem, 1909." *Rudder* 23 (March 1910), 142–150.

Hockett, Francis. "Wooden Ships." *New Republic* 11 (July 14, 1917), 302–304.

Hollander, Anne. "Swimsuits Illustrated." *American Heritage* 41 (July–August 1990), 58–68.

Holt, Felicia. "Promiscuous Bathing." *Ladies' Home Journal* 7 (August 1890), 6.

Hopfner, Felix. *Wissenschaft wider die Zeit: Goethes Farbenlehre ans rezeptions geschichtlicher Sicht.* Heidelberg: Winter, 1990.

Houghton, George. "Sandy Hook." *Scribner's Monthly* 18 (September 1879), 641–653.

"How Ships Spread Species." *Harper's*

Weekly 54 (June 4, 1910), 32.

"How to Build a Launch for $100." *Rudder* 19 (May 1908), 402–406.

"How to Build a Small Catboat." *Rudder* 18 (December 1907), 887–892.

How to Build Twenty Boats. New York: Rudder, 1958.

Howells, William Dean. *Literature and Life*. New York: Harper, 1902.

Howland, Llewellyn. "Neponset Estuary—1898." *American Neptune* 11 (April 1951), 83–94.

———. *Sou'West and By West of Cape Cod*. Cambridge, Mass.: Harvard University Press, 1948.

Howlett, D. Roger. *William Partridge Burpee: American Marine Impressionist*. Boston: Copley Square Press, 1991.

Hoyt, E. P. *U-Boats Offshore: When Hitler Struck America*. New York: Stein & Day, 1978.

Hunn, Peter. *Old Outboard Book*. Camden, Maine: International Marine, 1991.

Hurd, Archibald S. "The Race for Naval Power." *Cassier's* 34 (August 1908), 355–370.

Ingersoll, Ernest. *Down East Latch Strings: Or Seashore, Lakes, and Mountains by the Boston & Maine Railroad*. Boston: Boston & Maine Railroad, 1887.

———. *Oyster Industry*. Washington, D.C.: Government Printing Office, 1881.

———. "Stars of the Sea." *Scribner's Monthly* 22 (September 1881), 650–657.

Ingraham, Edward Duncan. *A Sketch of the Events which Preceded the Capture of Washington by the British*. Philadelphia: Carey & Hart, 1849.

Ingram, J. S. *The Centennial Exposition Described and Illustrated*. Philadelphia: Hubbard, 1876.

Irvine, Lucy. *Castaway*. New York: Random House, 1983.

Ive, Paul. *Practice of Fortification*. London: Tobie Cooke, 1589.

Jacobs, W. W. *Salthaven*. [1908] New York: Scribner's, 1909.

———. *At Sunwich Port*. [1901] New York: Scribner's, 1909.

James, Henry. *The American Scene*. New York: Harpers, 1907.

Jefferson, Thomas. *Notes on the State of Virginia*. [1785] New York: Harper, 1970.

Jewett, Amos Everett. "The Tidal Marshes of Rowley and Vicinity with an Account of the Old-Time Methods of 'Marshing.'" *Historical Collections, Essex Institute*, 85 (July 1949), 272–291.

Jewett, Sarah Orne. *A Marsh Island*. Boston: Houghton, 1885.

Johnson, Clifton. *Highways and Byways of New England*. New York: Macmillan, 1915.

Johnson, W. A. "Wonderful Marine Photography." *World's Work* 12 (August 1906), 7827–7843.

Jones, Thomas. "Pilot's Chart of Nantucket Shoals." Boston: Jones, 1786.

Jordan, Donaldson, and Edwin J. Pratt. *Europe and the American Civil War*. Boston: Houghton, 1931.

Josselyn, John. *Account of Two Voyages to New England*. Edited by Paul J. Lindholdt. Hanover, N.H.: University Press of New England, 1988.

Journal de département des Deux-Nethes. Anvers. Nos. 573–638 (February 7–May 6, 1814).

Journal of the Voyage of the Sloop Mary. [1701] Edited by E. B. O'Callaghan. Albany: J. Munsell, 1866.

Kaiser, Frederick F. *Built on Honor: Sailed with Skill—The American Coasting Schooner*. Ann Arbor, Mich.: Sarah Jennings Press, 1989.

Kase, Lori Miller. "From Aqua Aerobics to Workout Bikinis." *Vogue* 177 (July 1991), 92.

Kaufmann, Elizabeth. "Put Some Muscle into It." *Self* 9 (December 1989), 123–128.

Kellogg, Elijah. *The Young Shipbuilders*. Boston: Lee & Shepard, 1870.

Kenealy, James P. *Boating from Bow to Stern*. New York: Dodd, Mead, 1966.

Kiker, Douglas. *Death below Deck*. New York: Ballantine, 1991.

Kimball, Atkinson. "A Sea Change." *Atlantic Monthly* 106 (August 1910), 264–270.

Kincheloe, W. X. "Landscape Photography in Florida as a Summer Vacation." *Photo-Era* 49 (July 1922), 10–18.

Kipling, Rudyard. *American Notes*. Boston: Brown, 1899.

Kirkpatrick, Carl. "The Fine Craft of Rowboats." *Country Journal* 17 (May–June 1990), 50–57, 82.

Kofoid, Charles A., and Robert C. Miller. "Unusual Occurrence of Rock-Boring Mollusks in Concrete on the Pacific Coast." *Science*, n.s., 57 (March 30, 1923), 383–384.

Ladova, Rosalie M. "About Bathing Suits." *Harper's Weekly* 58 (September 13, 1913), 11.

Lane-Poole, Stanley. *The Barbary Corsairs*. London: Unwin, 1890.

Langland, B. F. "Alongshore with a Camera." *Photo-Era* 41 (August 1918), 58–62.

Langton, Jane. *Dark Nantucket Moon*. New York: Harper, 1975.

Lardner, F. "Beach Photography at Block Island." *Photo-Era* 39 (September 1917), 135.

Lavery, Brian. *Nelson's Navy: The Ships, Men, and Organisation, 1793–1815*.

London: Conway, 1989.

Lawson, Deodat. "Threnodia." [1693] *Genealogy of the Descendants of Anthony Collamore.* Compiled by Charles Hatch. Salem, Mass.: Newcomb & Green, 1915. Pp. 26–31.

Lay, Charles Downing. "Tidal Marshes." *Landscape Architecture* 2 (April 1912), 101–108.

Leavitt, John F. *Wake of the Coasters.* Middletown, Conn.: Wesleyan University Press, 1970.

Leder, Priscilla. "An American Dilemma: Cultural Conflict in Kate Chopin's *The Awakening." Southern Studies* 22 (Spring 1983), 97–104.

Lee, Robert. "High and Dry: Inland Boatyard's Performance Defies Recession." Quincy [Mass.] *Patriot Ledger* 112 (October 22, 1992), 34.

Leeds, Lawrence. *The Phantom of the Shore: A Folktale.* Philadelphia: Acorn, 1928.

Leggett, William. *Naval Stories.* New York: Carvill, 1834.

Lenfesty, Tom, and Hatty Lenfesty. *A Gunkholer's Cruising Guide to Florida's West Coast.* Camden, Maine: International Marine, 1990.

Letters from New England: The Massachusetts Bay Colony, 1629–1638. Edited by Everett Emerson. Amherst: University of Massachusetts Press, 1976.

Lewes, George Henry. *Sea-Side Studies at Ilfracombe, Tenby, the Scilly Isles and Jersey.* London: Blackwood, 1858.

Lewis, David D. *The Fight for the Sea: The Past, Present, and Future of Submarine Warfare in the Atlantic.* Cleveland: World, 1961.

Lewis, Leslie R. *The State of Our Harbors.* Hingham: Massachusetts Division of Waterways, 1990.

Lewis, Richard Warren. "Those Swinging Beach Movies." *Saturday Evening Post*
238 (July 31, 1965), 83–87.

Lincoln, Joseph C. *All Alongshore.* New York: Coward-McCann, 1931.

———. *Blair's Attic.* New York: Burt, 1929.

———. *The Bradshaws of Harniss.* New York: Appleton, 1943.

———. *Cape Cod Stories.* New York: Burt, 1907.

———. *Cap'n Eri: A Story of the Coast.* New York: Burt, 1904.

———. *Cy Whittaker's Place.* New York: Grosset & Dunlap, 1908.

———. *The Depot Master.* New York: Appleton, 1910.

———. *Kent Knowles: "Quahaug."* New York: Appleton, 1914.

———. *Keziah Coffin.* New York: Appleton, 1909.

———. *Mr. Pratt.* New York: Barnes, 1907.

———. *The Portygee.* New York: Burt, 1920.

———. *The Rise of Roscoe Paine.* New York: Burt, 1912.

———. *Rugged Water.* New York: Burt, 1924.

———. *Thankful's Inheritance.* New York: Burt, 1915.

Lindbergh, Anne Morrow. *Gift from the Sea.* [1955] New York: Random House, 1975.

Locke, John. *Essay Concerning Human Understanding.* Edited by Alexander Campbell Fraser. [1690] New York: Dover, 1959.

London, Jack. *Adventure.* New York: Macmillan, 1911.

Lowell, Amy. *East Wind.* Boston: Houghton, 1926.

Lowell, James Russell. *Complete Poetical Works.* Boston: Houghton, 1896.

———. *Fireside Travels.* Boston: Houghton, Mifflin, 1883.

"Lure of Shining Water." *National Geo-*
graphic 24 (December 1913), back cover.

Lyell, Charles. *Principles of Geology.* [1830–1833] Boston: Little, Brown, 1853.

McCabe, Francis R. "Modesty in Women's Clothes." *Harper's Weekly* 58 (August 30, 1913), 12.

McClennen, Douglas. "Finding Your Way." *WoodenBoat* 59 (July–August 1984), 102–105.

McLean, Sally Pratt. *Cape Cod Folks.* Boston: Williams, 1881.

McManis, Douglas R. *European Impressions of the New England Coast, 1497 to 1620.* Chicago: University of Chicago Department of Geography, 1972.

Maguire, Edward. *The Attack and Defence of Coast-Fortifications.* New York: VanNostrand, 1884.

Mahan, Alfred Thayer. *From Sail to Steam.* New York: Harper, 1907.

Maine Central Railroad Co. *Maine Central Christmas.* Portland: Maine Central, 1896.

Mallia, Joseph. "Weymouth Harbormaster Finds Adventure along 17.5-mile Coast." Quincy [Mass.] *Patriot Ledger* 156 (November 26, 1992), 29.

Marcy, Randolph B. *Prairie Traveller.* London: Trubner, 1863.

Marine, William Matthew. *The British Invasion of Maryland, 1812–1815.* Baltimore: Society of the War of 1812 in Maryland, 1913.

"Marine Borers in San Francisco Bay: Serious Losses to Railway and Other Waterfront Structures Threatened by Teredo." *Scientific American Monthly* 2 (May 1920), 446–450.

Marine Survey of India. London: HMG, 1886.

Marquand, John P. *Mr. Moto's Three Aces.* Boston: Little, Brown, 1938.

"Marshes, and Their Effects upon Human

Health." *American Journal of Agriculture and Science* 9 (November 1847), 283–285.

Martin, J. "Porn-Free." *Self* 15 (February 1993), 19.

Martyn, Payne. "Era of the Motor Boat." *Country Life* 12 (June 1907), 187–189, 230.

———. "Outdoor Swimming Pools— A Country Opportunity." *Country Life* 8 (July 1905), 316–318.

Masefield, John. *Jim Davis*. London: Wells, Gardner, n.d.

Mason, F. Van Wyck. *Golden Admiral*. Garden City, N.Y.: Doubleday, 1953.

Mason, Samuel K. "The Importance of Teaching School Children to Swim." *American City* 15 (September 1916), 314–316.

Massachusetts Bureau of Fisheries. *Report upon the Mollusk Fisheries*. Boston: Wright, 1912.

Masterson, William E. *Jurisdiction in Marginal Seas with Special Reference to Smuggling*. New York: Macmillan, 1929.

Mather, Cotton. *Magnalia Christi Americana*. 2 vols. [1724] Hartford: Andrus, 1853.

Maugham, W. Somerset. *The Moon and Sixpence*. [1919] New York: Random House, 1967.

Mayer, A. G. *Sea-Shore Life*. New York: New York Zoological Society, 1905.

Melville, Herman. *Moby-Dick: or, The Whale*. [1851] New York: Norton, 1967.

———. *Redburn: His First Voyage*. [1849] Chicago: Northwestern University Press, 1969.

Mendlowitz, Benjaimin. "Pete Culler's Workshop." *WoodenBoat* 27 (March– April 1979), 81–84.

Meyerowitz, Joel. *Cape Light*. Boston: Little, Brown, 1985.

Middleton, Arthur Pierce. "Yachting in Chesapeake Bay, 1676–1783." *American Neptune* 9 (July 1949), 180–184.

———. *Tobacco Coast: A Maritime History of Chesapeake Bay in the Colonial Era*. Baltimore: Johns Hopkins University Press, 1984.

Miller, Bob. "Cruising Maine's Non-Navigable Waters." *Messing About in Boats* 9 (May 15, 1991), 8–9.

Miller, Conrad. *Small Boat Engines: Inboard and Outboard*. New York: Sheridan, 1975.

Miller, Lewis B. "In the Quicksand." *Century* 66 (June 1903), 254–260.

Milton, John. *Poetical Works*. Edited by H. C. Beeching. Oxford: Oxford University Press, 1938.

Minnoch, J. E. *Aground: Coping with Emergency Groundings*. New York: DeGraff, 1985.

"Minnow." *Rudder* 19 (June 1908), 497– 498.

Mitchell, F. "Cape Cod." *Century* 26 (September 1883), 643–658.

Modern Mechanix. *How to Build Twenty Boats*. Minneapolis: Fawcett, 1933.

Morison, Samuel Eliot. [1921] *The Maritime History of Massachusetts, 1783– 1860*. Boston: Houghton, Mifflin, 1961.

Morris, Gerald E., and Llewellyn Howland III. *Yachting in America: A Bibliography*. Mystic, Conn.: Mystic Seaport, 1990.

Moulton, Ruth. "Now We Hunt Shells." *Leisure* 2 (June 1935), 10–12.

Mower, C. D. "How to Build an Eighteen-Foot Racing Cat." *Rudder* 13 (December 1902), 543–544.

Muller, Alfred. "A Little Talk on Catboats." *Rudder* 22 (August 1909), 91–94.

Muschamp, Herbert. "Don't Look Now." *Vogue* 78 (February 1991), 320–321, 357.

Must, Aviva, et al. "Long-term Mor-bidity and Mortality of Overweight Adolescents." *New England Journal of Medicine* 327 (November 5, 1992), 1350–1353.

"My First and Last Sea Fishing." *Knickerbocker* 19 (June 1842), 527–538.

Napier, Henry Edward. *New England Blockaded in 1814*. Edited by Walter Muir Whitehill. Salem, Mass.: Peabody Museum, 1939.

National Research Council. *Marine Structures: Their Deterioration and Preservation*. Washington, D.C.: National Research Council, 1924.

Naval Documents Related to the United States Wars with the Barbary Powers. 5 vols. Washington, D.C.: Government Printing Office, 1939–1944.

Neary, John S. "Sand-Dunes and the Camera." *Photo-Era* 45 (July 1920), 27–30.

New Standard Dictionary. New York: Funk & Wagnalls, 1896.

Nickerson, W. Sears. *Land Ho!—1620*. Boston: Houghton, 1931.

Niering, William A. "Vegetation Patterns and Processes in New England Salt Marshes." *Bioscience* 30 (May 1980), 301–307.

Noble, Louis Legrand. *After Icebergs with a Painter: A Summer Voyage to Labrador and Around Newfoundland*. New York: Appleton, 1862.

Nordhoff, Charles. *Cape Cod and All Alongshore*. New York: Harper, 1868.

Norris, Kathleen. "The Tide-Marsh." *Atlantic Monthly* 106 (December 1910), 745–753.

Northern Eye Institute. *Seeing Contour and Color*. Manchester: Northern Eye Institute, 1987.

"Novel Pleasures . . . And Other Great Escapes." *Savvy* 8 (July 1987), 39–43, 85.

Nutting, Wallace. *Connecticut Beautiful*.

Framingham, Mass.: Old America, 1921.

O'Brian, Patrick. *H.M.S. Surprise* [1973] New York: Norton, 1991.

———. *Master and Commander.* New York: Norton, 1970.

Ogburn, Charlton. *The Winter Beach.* New York: Morrow, 1966.

Olmsted, Frederick Law. *A Journey through Texas, or a Saddle Trip on the Southwestern Frontier.* [1857] Austin: University of Texas Press, 1978.

"On Boats." *Rudder* 22 (December 1909), 330–331.

"On the Beach." *Camera Notes* 6 (January 1910), 106.

"On the Nature of Soils." *Farmer's Cabinet* 1 (October 1, 1836), 81–82.

Ordway, Wallace B. "The Merrimac River Gundalow and Gundalow Men." *American Neptune* 10 (October 1950), 249–263.

Orton, Helen Fuller. *Mystery in the Pirate Oak.* New York: Scholastic, 1949.

Osborne, Duffield. "Surf and Surf-Bathing." *Scribner's Monthly* 8 (July 1890), 100–112.

"Our Sea-Coast Defense and Fortification System." *Putnam's Monthly* 7 (March 1856), 314–325.

Outboard Motor Boat Book. New York: Motor Boating, 1927.

"Outboard Motor the Angler's Hand." *Forest and Stream* 39 (May 1925), 302.

Outdoor Recreation Resources Review Commission. *Trends in American Living and Outdoor Recreation.* Washington, D.C.: Outdoor Recreation Resources Review Commission, 1962.

"Out-of-Doors Woman." *League of American Wheelmen Bulletin* 28 (October 14, 1898), 292.

Oxenhorn, Harvey. *Tuning the Rig: A Journey to the Arctic.* New York: Harper & Row, 1990.

Packard, John H. *Sea-Air and Sea-Bathing.* Philadelphia: Presley, 1880.

Page, Katherine Hall. *The Body in the Kelp.* New York: Avon, 1991.

Palmer, Raymond. *New England Tourist Camp and Cabin Directory.* Boston: National Tourist Camp Owner Association, 1930.

Pangborn, Frederic W. "Coast and Inland Yachting." *Century* 22 (May 1892), 14–26.

"Paraffin and Poison Protects Wood from Teredo." *Science*, n.s., 56 (November 24, 1922), supp. xii.

Paris, David M. *The Four Principal Battles of the Late War.* Harrisburg, Pa.: Baab, 1832.

Parker, Edwin S. "Yachts for All of Us." *Leisure* 1 (June 1934), 9–11, 43.

Parramon, José M. *Painting Seascapes in Oil.* New York: Watson-Guptill, 1988.

Pease, Howard. *Night Boat.* Garden City, N.Y.: Doubleday, 1944.

Peck, Nicholas. *Boat: A True Story.* Stockbridge, Mass.: Author, 1984[?].

Phinney, S. B. "Cultivation of the Pitch Pine on the Sea-Coast." *Report of the [Massachusetts] Board of Agriculture* 12 (1864), 208–213.

Pilkey, Orrin H., and William J. Neal. "Save Beaches, Not Buildings." *Issues in Science and Technology* 23 (Spring 1992), 36–41.

Pinkham, Paul. "A Chart of Nantucket Shoals." Boston: John Norman, 1791.

Plymouth Colony Records. 11 vols. Boston: William White, 1855–1861.

Poe, Edgar Allan. *Works.* 5 vols. New York: Collier, 1904.

"Popular Girl Must Learn Fancy Strokes." Boston *Sunday Post* 97 (April 16, 1911), 3.

"Port Jefferson." *Rudder* 23 (April 1910), 384.

"Power and the Editor." *Rudder* 13 (December 1902), 518.

"Protecting Piles from the Teredo." *Scientific American* 119 (August 17, 1918), 128.

Prout, Samuel. *Easy Lessons in Landscape Drawing.* London: Ackermann, 1820.

———. *Hints on Light and Shadow, Composition, etc., as Applicable to Landscape Painting.* London: Ackermann, 1838.

———. *Studies of Boats and Coast Scenery for Landscape and Marine Painters.* London: Ackermann, 1816.

Pyke, Rafford. "Summer Types of Men and Women." *Cosmopolitan* 35 (September 1903), 479–487.

Quamino. "A Handy Craft." *Rudder* 8 (September 1897), 273–274.

Quick, Herbert. *Virginia of the Air Lanes.* Indianapolis: Bobbs-Merrill, 1909.

Rabl, S. S. *Boatbuilding in Your Own Backyard.* New York: Cornell Maritime Press, 1947.

"Racing Catboat." *Rudder* 13 (July 1902), 331.

Railway Dry Docks. Boston: Crandall Engineering, 1929[?].

Ralph, Julian. "Old Monmouth." *Harper's Monthly* 89 (August 1894), 326–342.

Rattray, Everett T. *The South Fork: The Land and the People of Eastern Long Island.* New York: Random House, 1979.

Redfield, Alfred C. "Development of a New England Salt Marsh." *Ecological Monographs* 42 (Spring 1972), 201–237.

Reed, Douglas. "Rockport Names." *Historical Collections*, Essex Institute, 114 (January 1978), 24–31.

"Regulations for the Construction and Safe-Guarding of Piers and Wharves." *American City* 30 (June 1924), 621.

Reference Sheet E-73. Rochester, N.Y.: Eastman Kodak, 1985.

Reiger, George. *Profiles in Saltwater Angling: A History of the Sport, Its People and Places, Tackle and Techniques.* Englewood Cliffs, N.J.: Prentice-Hall, 1973.

"Remains of a Large Forest Not Far from Liverpool." *Gentleman's Magazine* 66 (July 1796), 549–551.

Report on a National Marina Survey of Motor Boat Harbors. New York: National Association of Engine and Boat Manufacturers, 1930.

"Return of the Wooden Ship." *Literary Digest* 54 (June 9, 1917), 1775–1776.

"Rheclair." *Rudder* 13 (April 1902), 128.

Riley, James Whitcomb. *Complete Works.* Indianapolis: Bobbs-Merrill, 1883.

Riley, P. M. "Masterpieces of Sea Photography." *Country Life* 20 (August 1, 1911), 41–42.

Robberson, Kay. "Where There's a Wind, There's a Way." *Women's Sports and Fitness* 11 (June 1989), 36–37.

Roberts, C. Harold. "The Old Steam Shed." *American Neptune* 7 (July 1947), 196–199.

Rogers, John G. *Origins of Sea Terms.* Mystic, Conn.: Mystic Seaport Museum, 1985.

Rohmer, Sax. *The Golden Scorpion.* New York: Burt, 1920.

———. *The Yellow Claw.* New York: McBride, 1915.

Romans, Bernard. *Chart of the Coast of East and West Florida.* New York: Romans, 1774.

Rood, Ogden N. *Modern Chromatics with Application to Art and Industry.* London: Kegan Paul, 1879.

Rosier, James. "A True Relation . . . of the Land of Virginia." [1605] *Collections,* Massachusetts Historical Society, 3d ser., 8 (1843), 111–146.

Rostnor, Louis. "Darwinism and Swimming." *Nineteenth Century* 34 (November 1893), 721–732.

"Rural Objects in England and America." *Putnam's Monthly Magazine* 7 (July 1855), 32–40.

Ruskin, John. *Elements of Perspective Arranged for the Use of Schools . . . with the First Three Books of Euclid.* London: Smith & Elder, 1859.

Sabatini, Rafael. *Captain Blood.* Boston: n.p., 1924.

"Salt Marshes." *Cultivator* 1 (October 1840), 303.

"Salt Marshes." *New England Farmer* 37 (April 1858), 185.

"Salting Hay—Marsh Hay." *New England Farmer* 37 (September 1858), 437.

Sangster, Margaret. "Open-Air for Women." *Collier's* 27 (April 27, 1901), 18.

———. "Open-Air Life for Women." *Collier's* 27 (April 13, 1901), 16.

Sauer, Carl O. *The Early Spanish Main.* Berkeley: University of California Press, 1966.

Schmitt, Peter J. *Back to Nature: The Arcadian Myth in Urban America.* [1969] Baltimore: Johns Hopkins University Press, 1990.

Schock, Edson B. *How to Build a Rowboat.* New York: Rudder, 1917.

Schult, Joachim. *Curious Boating Inventions.* New York: Taplinger, 1974.

Scott, Sir Walter. *The Bride of Lammermoor.* [1818] New York: Crowell, n.d.

"Sea." *Rural Repository* 6 (November 7, 1829), 95.

"Sea Views." *Household Words* 9 (July 15, 1854), 506–510.

"Sea-Gardens." *Household Words* 11 (September 27, 1856), 240–244.

"Seascapes." *Country Life* 20 (September 1, 1911), 47.

"Sea-Side Churchyard." *Household Words* 2 (July 1847), 257–262.

Sellers, Charles Coleman. *Charles Willson Peale.* Philadelphia: American Philosophical Society, 1947.

Shaheen, Jack G. *Nuclear War Films.* Carbondale: Southern Illinois University Press, 1978.

Shaler, Nathaniel Southgate. *Beaches and Tidal Marshes of the Atlantic Coast.* New York: American Book, 1895.

———. "Inundated Lands of Massachusetts." Massachusetts Board of Agriculture, *Report.* Boston: State Printers, 1892. Pp. 377–390.

———. *Sea-Coast Swamps of the Eastern United States.* Washington, D.C.: Government Printing Office, 1886.

Shannon, H. J. "Spring Awakening of the Sea." *Harper's Monthly* 116 (March 1908), 537–545.

Shaw, Thomas. *Travels or Observations Relating to Several Parts of Barbary and the Levant.* 2 vols. [1738] Edinburgh: Ritchie, 1808.

Shellenberger, William H. *Cruising the Chesapeake: A Gunkholer's Guide.* Blue Ridge Summit, Pa.: International Marine, 1990.

Sloane, Patricia. *Visual Nature of Color.* New York: Design, 1990.

Slotkin, Richard. *Regeneration Through Violence: The Mythology of the American Frontier.* Middletown, Conn.: Wesleyan University Press, 1973.

Smart, Charles Allen. *R. F. D.* New York: Norton, 1938.

Smith, E. Boyd. *The Seashore Book.* Fairfield, Conn.: Boy Scouts of America, 1912.

Smith, J. J. "Monokini Beaches." *Holiday* 57 (May 1976), 46.

Smith, John. *A Description of New England.* [1616] Boston: Veazie, 1865.

Smith, William. *A History of Chatham, Massachusetts.* 2 vols. Hyannis, Mass.: Gross, 1909–1913.

Smyth, William Henry. *The Sailor's Word-*

book: *An Alphabetical Digest of Nautical Terms*. London: Blackie, 1867.

Snow, Edward Rowe. *Mysterious Tales of the New England Coast*. New York: Dodd, Mead, 1952.

———. *True Tales of Buried Treasure*. Cornwall, N.Y.: Cornwall, 1960.

Sobol, Donald J. *Encyclopedia Brown Saves the Day*. New York: Dutton, 1970.

"Soliloquy on Bathing." *New England Farmer* 5 (September 22, 1826), 69.

Southack, Cyprian. *The New England Coasting Pilot*. London: n.p., 1730–1734.

Southwick, George N. "Our Defenceless Coasts." *North American Review* 162 (March 1896), 317–327.

"Sowing Rice: Reaping Ship-worms." *Literary Digest* 67 (October 2, 1920), 110–113.

Spears, John Randolph. *Our Navy in the War with Spain*. New York: Scribner's, 1898.

Spectre, Peter H. "The Lyman Legend: A Monument to Mass Production." *WoodenBoat* 82 (May–June 1988), 60–67.

———. "On the Waterfront." *Wooden-Boat* 92 (November–December 1992), 15.

Spenser, Edmund. *Poetical Works*. Edited by J. C. Smith and E. DeSelincourt. Oxford: Oxford University Press, 1924.

Spon, Ernest. *Dictionary of Engineering*. London: Spon, 1880.

Stebbins, L. "Photographer in Bermuda." *Photo-Era* 50 (March 1923), 144–147.

Stebbins, Theodore E. *Martin Johnson Heade*. Baltimore: University of Maryland Art Department, 1969.

Stein, Frank. "What's the Hurry?" *Rudder* 14 (July 1903), 411.

Stein, Roger B. *Seascape and the American Imagination*. New York: Whitney Museum, 1975.

Steinem, Gloria. *The Beach Book*. New York: Viking, 1963.

Stevens, Wallace. *Poems*. Edited by Samuel French Morse. New York: Random, 1959.

Stevenson, Robert Louis. *Treasure Island*. New York: Paglan [c. 1929].

———. *Treasure Island*. Boston: Roberts, 1895.

———. *Treasure Island*. Illustrated by Wal Paget. New York: Scribner's, 1902.

———. *Treasure Island*. New York: McLoughlin [c. 1915].

———. *Treasure Island*. Illustrated by Edward A. Wilson. New York: Heritage, 1941.

———. *Treasure Island*. Racine, Wis.: Whitman [c. 1956].

Stevenson, Robert Louis, and Lloyd Osbourne. *The Wrecker*. [1891] New York: Scribner's, 1910.

Stiles, Chester F. "Summer Work at the Seashore." *Photo-Era* 5 (July 1900), 1–4.

Stilgoe, John R. "Bikinis, Beaches, and Bombs." *Orion Nature Quarterly* 3 (Summer 1984), 4–17.

———. *Borderland: Origins of the American Suburb, 1820 to 1939*. New Haven and London: Yale University Press, 1988.

———. *Common Landscape of America, 1580 to 1845*. New Haven and London: Yale University Press, 1982.

———. "Popular Photography, Scenery Values, and Visual Assessment." *Landscape Journal* 3 (Autumn 1984), 111–122.

———. *Shallow-Water Dictionary: A Grounding in Estuary English*. Cambridge, Mass.: Exact Change, 1990.

Stoddard, Lothrop. *The Rising Tide of Color against White World-Supremacy*. New York: Scribner's, 1920.

Street, Donald M., Jr. "Improving the Classic Dinghy." *WoodenBoat* 50 (January–February 1983), 88–92.

Sturgis, Russell. *Dictionary of Architecture and Building*. New York: Macmillan, 1902.

Sullivan, John L. *A Description of the American Marine Rail-Way, Constructed at New York*. Philadelphia: Author[?], 1827.

"Summer on the Sands." *Munsey's Magazine* 17 (August 1897), 651.

Sumner, William Graham. *Social Darwinism: Selected Essays*. Edited by Stow Persons. Englewood Cliffs, N.J.: Prentice-Hall, 1963.

"Sun-Suit by Jantzen." *Saturday Evening Post* 201 (June 22, 1929), 99.

Swaringen, Richard G. "Fear." *Comfort* 34 (August 1921), 16.

Sweeney, John M. "Power-Boat Legislation." *Rudder* 19 (February 1908), 91.

"Swimming." *New England Magazine* 2 (June 1832), 506–514.

"Swimsuits through the Ages." *Life* 19 (July 9, 1945), 55–58.

Sylvester, Herbert Milton. *Maine Coast Romance*. 5 vols. Boston: Stanhope, 1904–1909.

Symonds, H. "Perils of Marine Photography." *Cosmopolitan* 42 (April 1907), 629–635.

Taber, Gladys. *My Own Cape Cod*. Philadelphia: Lippincott, 1959.

Taft, Henry W. "Rowing the Maine Coast." *WoodenBoat* 47 (July–August 1982), 52–57.

Taylor, David M. "Waterfront Development and Use Control." *Landscape Architect and Specifyer* 9 (February 1993), 20–26.

Taylor, E. G. R. *The Haven-Finding Art: A History of Navigation from Odysseus to Captain Cook*. New York: Abelard-Schuman 1957.

Taylor, Phoebe Atwood. *The Cape Cod Mystery*. [1931] Woodstock, Vt.: Foul Play, 1988.

———. *Going, Going, Gone*. [1943] Woodstock, Vt.: Foul Play, 1990.

Taylor, Richard Cowling. *Geology of East Norfolk*. London: Cochran, 1827.

Taylor, William H. "Motor Sailors." *Sportsman* 18 (November 1935), 42–43, 62.

Teal, John, and Mildred Teal. *Life and Death of the Salt Marsh*. New York: Atlantic, 1970.

Tetlock, Philip E., ed. *Behavior, Society, and Nuclear War*. New York: Oxford University Press, 1989.

Thackeray, William Makepeace. *A Shabby Genteel Story and Other Stories*. New York: Appleton, 1853.

Thomas, Barry. *Building the Crosby Catboat*. Mystic, Conn.: Mystic Seaport Museum, 1989.

Thomas, Stephen D. *The Last Navigator*. New York: Ballantine, 1987.

Thomas, Tony, et al. *The Films of Errol Flynn*. New York: Citadel, 1969.

Thompson, Winfield M. "A Catboat Sailor's Yarn." *Rudder* 17 (April 1906), 257–263; (May 1906), 319–323.

———. "Cats in Massachusetts Bay." *Rudder* 20 (July 1908), 12–17; (August 1908), 87–92.

———. "In the Isles of Shoals." *Rudder* 23 (January 1910), 1–12.

———. "Marblehead." *Rudder* 19 (March 1908), 112–128.

———. "Sea Wolf, Cruising Cat." *Rudder* 21 (January 1909), 124–130.

———. "Successful Small Tender." *Rudder* 17 (January 1906), 17–22.

Thoreau, Henry David. *Cape Cod*. [1865] New York: Penguin, 1987.

———. *Journal*. Edited by Bradford Torrey. 12 vols. Boston: Houghton Mifflin, 1906.

———. *A Yankee in Canada*. Boston: Houghton, Mifflin, 1887.

Thornton, Tamara Plakins. *Cultivating Gentlemen: The Meaning of Country Life among the Boston Elite, 1785–1860*. New Haven and London: Yale University Press, 1989.

Todd, Richard. "The Voyage of the Blue Blazers." *New England Monthly* 6 (August 1989), 45–53.

Tomlinson, H. M. *A Mingled Yarn: Autobiographical Sketches*. Indianapolis: Bobbs-Merrill, 1953.

Tooker, L. Frank. "A Boyhood Alongshore." *Century* 101 (February 1921), 500–508.

———. *The Middle Passage*. New York: Century, 1920.

Townsend, Charles Wendell. *Beach Grass*. Boston: Jones, 1923.

———. *Sand Dunes and Salt Marshes*. Boston: Page, 1913.

Treadwell, T. C. *Submarines with Wings: The Past, Present, and Future of Aircraft-Carrying Submarines*. London: Conway, 1985.

Trials of Eight Persons Indicted for Piracy. Boston: n.p., 1717.

Trott, Raymond H. "A Down East Merchant Fleet." *American Neptune* 4 (January 1944), 45–52.

"Trouble with Bikinis." *Life* 27 (September 12, 1949), 65–66.

Tudor, William. *Letters on the Eastern States*. New York: Kirk, 1820.

"Turn on the Sun." *Saturday Evening Post* 202 (September 14, 1929), n.p.

Turner, J. M. *Picturesque Views of the Southern Coast of England*. London: Arch, 1826.

"Twelve-foot Sailing Dingey." *Rudder* 18 (April 1907), 383–384.

Tyndall, John. *Forms of Water in Clouds and Rivers*. New York: Appleton, 1872.

United States Army. *Green Beret Fitness Program*. New York: Parallax, 1966.

United States Department of Commerce. *United States Coast Pilot*. Washington, D.C.: Government Printing Office, 1903.

"Unusual Use of Concrete in Repairing Wooden Ship." *Illustrated World* 31 (July 1919), 722–723.

Updegraff, Allan. "The Siren of the Air." *Century* 85 (December 1912), 283–288.

Varrell, William M. *Summer by the Sea: The Golden Era of Victorian Beach Resorts*. Portsmouth, N.H.: Strawbery Banke, 1972.

Varron, Benno Weiser. *Professions of a Lucky Jew*. New York: Cornwall, 1992.

Venk, Ernest. *The Complete Outboard Boating Manual*. Chicago: American Technical Society, 1958.

"Vigorous Physical Activity among High School Students." *Morbidity and Mortality Weekly Report* 41 (January 24, 1992), 33–35.

Villar, Roger. *Piracy Today: Robbery and Violence at Sea since 1980*. London: Conway Maritime Press, 1985.

Viscusi, W. Kip. *Risk by Choice*. Cambridge: Harvard University Press, 1983.

Visual Quality and the Coastal Zone: Conference Proceedings. Syracuse: SUNY School of Landscape Architecture, 1976.

Vocabulario queechi-español. Coban: A.V., 1930.

Von Scheliha, Viktor. *Treatise on Coast Defence*. London: Spon, 1868.

Wakeman, Frederic. *The Fabulous Train*. New York: Rinehart, 1955.

Walsh, Maurice. *Take Your Choice*. Philadelphia: Lippincott, 1954.

Wandelmaier, Roy. *Shipwrecked on Mystery Island*. Mahwah, N.J.: Troll, 1985.

Warner, Oliver. *A Secret of the Marsh*.

New York: Dutton, 1927.

Wasson, George S. *Cap'n Simeon's Store*. Boston: Houghton, Mifflin, 1903.

———. *Home from Sea*. Boston: Houghton, Mifflin, 1908.

———. *Sailing Days on the Penobscot: The Story of the River and the Bay in the Old Days*. New York: Norton, 1949.

Waters, Don. *Outboard Cruising*. New York: Furman, 1939.

Watts, Anthony. *Japanese Warships of World War II*. New York: Doubleday, 1967.

"We Must Have a Boat." *Rudder* 13 (July 1902), 15–16.

"We Nominate for the Hall of Fame." *Vanity Fair* 25 (September 1925), 63.

Webster, Noah. *An American Dictionary of the English Language*. [1824] New York: Allen, 1890.

———. *An American Dictionary of the English Language*. Springfield, Mass.: Merriam, 1852.

———. *An American Dictionary of the English Language*. Springfield, Mass.: Merriam, 1864.

Webster's New International Dictionary. Springfield, Mass.: Merriam-Webster, 1961.

Weightman, Charles. *The Art of Swimming*. New York: DeWitt, 1873.

Weiss, Howard F. "Wooden Ships and Ship Worms." *Scientific American* 116 (June 16, 1917), 592, 601.

Welty, Eudora. *Thirteen Stories*. New York: Harcourt, 1970.

Werth, Barry. "The Hidden Cape, the Unseen Cape." *New England Monthly* 7 (August 1990), 24–32.

Wharton, Edith. *A Backward Glance*. New York: Appleton, 1934.

"What Are the Young Men of the Cities Coming To?" *Rudder* 8 (May 1897), 134–135.

"What Female Complaints?" *Cosmopoli-*

tan 148 (April 1960), 4.

"What Shall She Wear?" *League of American Wheelmen Bulletin* 27 (March 18, 1898), 249–250.

Wheelwright, Jane Hollister, and Lynda Wheelwright Schmidt. *The Long Shore*. San Francisco: Sierra Club, 1991.

"Where the Cape Cats Breed." *Rudder* 13 (November 1902), 481–485.

Whitehouse, Arch. *Subs and Submarines*. Garden City, N.Y.: Doubleday, 1961.

Whittier, John Greenleaf. *Complete Poetical Works*. Boston: Houghton, Mifflin, 1892.

Wilcox, Stephen F. *The St. Lawrence Run*. New York: St. Martin's, 1990.

"Will There Be a Horsepower Race?" *Popular Science* 166 (March 1955), 160–162.

Williams, Charles. *The Diamond Bikini*. [1956] New York: Simon & Schuster, 1988.

Willis, Nathaniel Parker. *Hurry-Graphs; or, Sketches of Scenery, Celebrities, and Society*. Auburn, N.Y.: Alden, 1853.

Wilson, Harold. *The Story of the Jersey Shore*. Princeton, N.J.: Van Nostrand, 1964.

"Wind . . . Sun . . . and Complexions." *Vanity Fair* 18 (August 1922), n.p.

Winfield, Arthur M. [pseud. Edward Stratemeyer] *The Rover Boys on the Ocean*. New York: Grossett, 1899.

Wingert, H. Shindle. "Swimming on Dry Land." *Suburban Life* 10 (May 1910), 281.

Winsor, Justin. *A History of the Town of Duxbury*. Boston: Crosby and Nichols, 1849.

Winthrop, John. *The History of New England*. Edited by James Savage. Boston: Little, Brown, 1853.

———. *Journal*. Edited by James Savage. Boston: Little, Brown, 1853.

Winthrop, Theodore. *Life in the Open Air*.

Boston: Ticknor, 1863.

Wisby, Hrolf. "Learning to Swim." *Outing* 58 (July 1911), 457–463.

Wisser, John P. *The Tactics of Coast Defense*. Kansas City, Mo.: Hudson-Kimberly, 1902.

Wohlrabe, R. A. "Catching the Breath of the Sea." *Photo-Era* 60 (February 1928), 65–68.

Wood, J. G. *Common Objects at the Sea-Shore, Including Hints for an Aquarium*. London: Routledge, 1904.

Wood, William. *New-England's Prospect*. London: Bellamie, 1634.

Woodbury, G. *Great Days of Piracy in the West Indies*. New York: Norton, 1951.

"Wooden Ships Not Back Numbers." *Literary Digest* 69 (April 30, 1921), 53–54.

Woodworth, Samuel. *Poetical Works*. 2 vols. New York: Scribner, 1860.

———. *The War: Being a Faithful Record of the Transactions of the War Between the United States of America . . . and the United Kingdom*. 2 vols. New York: Woodworth, 1813–1814.

Worcester, Joseph. *Dictionary of the English Language*. Boston: Brewer and Tileston, 1874.

Works Progress Administration. *Hands That Built New Hampshire*. Brattleboro, Vt.: Stephen Daye, 1940.

Wright, Thomas. "A Curious and Exact Relation of a Sand-Cloud." *Philosophical Transactions*, Royal Society, 3 (1668), 722–725.

"Yes, Amy!" *Life* 94 (August 16, 1929), 14.

York, Charles. *Charlie York: Maine Coast Fisherman*. Edited by Harold B. Clifford. Camden, Maine: International Marine, 1974.

Young, Francis Brett. *The Ship's Surgeon's Yarn and Other Stories*. New York: Reynal & Hitchcock, 1940.

Young, James Rankin. *History of Our War with Spain*. Washington, D.C.: n.p., 1898.

Youth Physical Fitness in 1985. Washington, D.C.: President's Council on Physical Fitness and Sports, 1990.

Zeckhauser, Richard J., and W. Kip Viscusi. "Risk within Reason." *Science* 248 (May 4, 1990), 559–563.

Zurcher, Frederic. *Meteors, Aerolites, Storms, and Atmospheric Phenomena*. Translated by William Lackland. New York: Amber, 1870.

INDEX

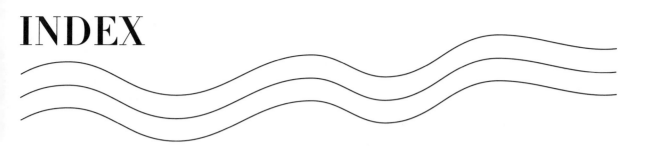